WIRELESS QUALITY OF SERVICE

WIRELESS NETWORKS AND MOBILE COMMUNICATIONS

Dr. Yan Zhang, Series Editor
Simula Research Laboratory, Norway
E-mail: yanzhang@ieee.org

Unlicensed Mobile Access Technology: Protocols, Architectures, Security, Standards and Applications
Yan Zhang, Laurence T. Yang and Jianhua Ma
ISBN: 1-4200-5537-2

Wireless Quality-of-Service: Techniques, Standards and Applications
Maode Ma, Mieso K. Denko and Yan Zhang
ISBN: 1-4200-5130-X

Broadband Mobile Multimedia: Techniques and Applications
Yan Zhang, Shiwen Mao, Laurence T. Yang and Thomas M Chen
ISBN: 1-4200-5184-9

The Internet of Things: From RFID to the Next-Generation Pervasive Networked Systems
Lu Yan, Yan Zhang, Laurence T. Yang and Huansheng Ning
ISBN: 1-4200-5281-0

Millimeter Wave Technology in Wireless PAN, LAN, and MAN
Shao-Qiu Xiao, Ming-Tuo Zhou and Yan Zhang
ISBN: 0-8493-8227-0

Security in Wireless Mesh Networks
Yan Zhang, Jun Zheng and Honglin Hu
ISBN: 0-8493-8250-5

Resource, Mobility and Security Management in Wireless Networks and Mobile Communications
Yan Zhang, Honglin Hu, and Masayuki Fujise
ISBN: 0-8493-8036-7

Wireless Mesh Networking: Architectures, Protocols and Standards
Yan Zhang, Jijun Luo and Honglin Hu
ISBN: 0-8493-7399-9

Mobile WIMAX: Toward Broadband Wireless Metropolitan Area Networks
Yan Zhang and Hsiao-Hwa Chen
ISBN: 0-8493-2624-9

Distributed Antenna Systems: Open Architecture for Future Wireless Communications
Honglin Hu, Yan Zhang and Jijun Luo
ISBN: 1-4200-4288-2

AUERBACH PUBLICATIONS

www.auerbach-publications.com
To Order Call: 1-800-272-7737 • Fax: 1-800-374-3401
E-mail: orders@crcpress.com

WIRELESS QUALITY OF SERVICE

Techniques, Standards, and Applications

Edited by
Maode Ma
Mieso K. Denko
Yan Zhang

CRC Press
Taylor & Francis Group
Boca Raton London New York

CRC Press is an imprint of the
Taylor & Francis Group, an **informa** business
AN AUERBACH BOOK

CRC Press
Taylor & Francis Group
6000 Broken Sound Parkway NW, Suite 300
Boca Raton, FL 33487-2742

First issued in paperback 2019

ISBN-13: 978-1-4200-5130-8 (hbk)
ISBN-13: 978-0-367-38692-4 (pbk)

Library of Congress Cataloging-in-Publication Data

Ma, Maode.
 Wireless quality of service : techniques, standards, and applications / Maode Ma, Mieso K. Denko, and Yan Zhang.
  p. cm. -- (Wireless networks and mobile communications)
 Includes bibliographical references and index.
 ISBN 978-1-4200-5130-8 (alk. paper)
 1. Wireless communication systems--Quality control. I. Denko, Mieso K. II. Zhang, Yan, 1977- III. Title. IV. Title: Wireless QoS, techniques, standards and applications. V. Series.

TK5103.2.M315 2008
621.384--dc22   2008020725

Visit the Taylor & Francis Web site at
http://www.taylorandfrancis.com

and the Auerbach Web site at
http://www.auerbach-publications.com

Contents

v

Editors

Maode Ma, chief editor of this book, received his BE degree in computer engineering from Tsinghua University in 1982, ME degree in computer engineering from Tianjin University in 1991, and PhD degree in computer science from Hong Kong University of Science and Technology in 1999. Dr. Ma is an associate professor at the School of Electrical and Electronic Engineering at Nanyang Technological University in Singapore. He has extensive research interests, including wireless networking, optical networking, grid computing, and bioinformatics. He has been a member of the technical program committee for more than 70 international conferences. He has been a technical track chair, tutorial chair, publication chair, and session chair for more than 30 international conferences. Dr. Ma has published more than 100 international academic research papers on wireless networks and optical networks. He currently serves as an associate editor for *IEEE Communications Letters*, an editor for *IEEE Communications Surveys and Tutorials*, and an associate editor for *International Journal of Wireless Communications and Mobile Computing, International Journal of Security and Communication Networks,* and *International Journal of Vehicular Technology.*

Mieso Denko is an associate professor in the Department of Computing and Information Science, University of Guelph, Ontario, Canada. He received his MSc degree form the University of Wales, United Kingdom and PhD degree from the University of Natal, South Africa, both in Computer Science. His current research interests include wireless mesh networks, mobile ad hoc networks, mobile and pervasive computing, and network security. He has published numerous referred articles in international journals and conferences in these areas.

Dr. Denko has served as program chair, program vice-chair, and technical program committee member of several international conferences, symposia, and workshops. Most recently he has been general cochair of IEEE PCAC'07, program vice-chair of IEEE AINA'08 workshop, cochair of MHWMN'08 at MASS, and publicity chair of IEEE PWN'08 at PerCom. He has served as technical program committee member of several international conferences including ICC'08-09, Globecom'08, ICC'08-09, and AINA'09. Dr. Denko is an associate editor of the *International Journal of Ubiquitous Multimedia Engineering* (IJMUE) and is on the editorial board of four other international journals. Since 2006, he has served as guest coeditor of six special issues in international journals including *Mobile Networking and Applications* (ACM/Springer) and the *International Journal of Communications Systems* (Wiley). Dr. Denko is a senior member of the ACM and IEEE and vice-chair of the IFIP.

Yan Zhang received a PhD degree from the School of Electrical and Electronics Engineering, Nanyang Technological University, Singapore. Since August 2006, he has worked at Simula Research Laboratory, Norway (http://www.simula.no/). He is associate editor of *Security and Communication Networks* (Wiley); and on the editorial boards of *International Journal of Network Security, International Journal of Ubiquitous Computing, Transactions on Internet and Information Systems (TIIS), International Journal of Autonomous and Adaptive Communications Systems (IJAACS),* and *International Journal of Smart Home (IJSH).*

Zhang currently serves as the book series editor for "Wireless Networks and Mobile Communications" (Auerbach Publications, CRC Press, Taylor & Francis Group). He serves as guest coeditor for *Wiley Security and Communication Networks* special issue on "Secure Multimedia Communication"; guest coeditor for *Springer Wireless Personal Communications* special issue on selected papers from ISWCS 2007; guest coeditor for *Elsevier Computer Communications* special issue on "Adaptive Multicarrier Communications and Networks"; guest coeditor for *Inderscience International Journal of Autonomous and Adaptive Communications Systems (IJAACS)* special issue on "Cognitive Radio Systems"; guest coeditor for *The Journal of Universal Computer Science (JUCS),* special issue on "Multimedia Security in Communication"; guest coeditor for *Springer Journal of Cluster Computing,* special Issue on "Algorithm and Distributed Computing in Wireless Sensor Networks"; and guest coeditor for *EURASIP Journal on Wireless Communications and Networking (JWCN),* special issue on "OFDMA Architectures, Protocols, and Applications."

Zhang serves as coeditor for several books: *Resource, Mobility and Security Management in Wireless Networks and Mobile Communications, Wireless Mesh Networking: Architectures, Protocols and Standards, Millimeter-Wave Technology in Wireless PAN, LAN and MAN, Distributed Antenna Systems: Open Architecture for Future Wireless Communications, Security in Wireless Mesh Networks, Mobile WiMAX: Toward Broadband Wireless Metropolitan Area Networks, Wireless Quality-of-Service: Techniques, Standards and Applications, Broadband Mobile Multimedia: Techniques and Applications, Internet of Things: From RFID to the Next-Generation Pervasive Networked Systems, Unlicensed Mobile Access Technology: Protocols, Architectures, Security, Standards and Applications, Cooperative Wireless Communications, WiMAX Network Planning and Optimization, RFID Security: Techniques, Protocols and System-On-Chip Design, Autonomic Computing and Networking, Security in RFID and*

Sensor Networks, Handbook of Research on Wireless Security, Handbook of Research on Secure Multimedia Distribution, RFID and Sensor Networks, Cognitive Radio Networks, Wireless Technologies for Intelligent Transportation Systems, Vehicular Networks: Techniques, Standards and Applications, and *Orthogonal Frequency Division Multiple Access (OFDMA).*

He serves as workshop general cochair for COGCOM 2008, workshop cochair for IEEE APSCC 2008, workshop general cochair for WITS-08, program cochair for PCAC 2008, workshop general cochair for CONET 2008, workshop chair for SecTech 2008, workshop chair for SEA 2008, workshop co-organizer for MUSIC'08, workshop co-organizer for 4G-WiMAX 2008, publicity cochair for SMPE-08, International Journals coordinating cochair for FGCN-08, publicity cochair for ICCCAS 2008, workshop chair for ISA 2008, symposium cochair for ChinaCom 2008, industrial cochair for MobiHoc 2008, program cochair for UIC-08, general cochair for CoNET 2007, general cochair for WAMSNet 2007, workshop cochair FGCN 2007, program vice cochair for IEEE ISM 2007, publicity cochair for UIC-07, publication chair for IEEE ISWCS 2007, program cochair for IEEE PCAC'07, special track cochair for "Mobility and Resource Management in Wireless/Mobile Networks" in ITNG 2007, special session co-organizer for "Wireless Mesh Networks" in PDCS 2006, and a member of Technical Program Committee for numerous international conference including ICC, PIMRC, CCNC, AINA, GLOBECOM, and ISWCS. Zhang received the Best Paper Award and Outstanding Service Award in the IEEE 21st International Conference on Advanced Information Networking and Applications (AINA-07). His research interests include resource, mobility, spectrum, energy and security management in wireless networks, and mobile computing. He is a member of IEEE and IEEE ComSoc. E-mail: yanzhang@ieee.org

Contributors

Dharma P. Agrawal
Center for Distributed and Mobile
  Computing
Department of Computer Science
University of Cincinnati
Cincinnati, Ohio

Antonios Alexiou
Research Academic Computer Tech-
  nology Institute and Computer
  Engineering and Informatics
  Department
University of Patras
Patras, Greece

Nilufar Baghaei
Department of Computer Science and
  Software
College of Engineering
University of Canterbury
New Zealand

Torsha Banerjee
OBR Center for Distributed and Mobile
  Computing
Department of Computer Science
University of Cincinnati
Cincinnati, Ohio

Jaume Barceló
Universitat Pompeu Fabra
Barcelona, Spain

Boris Bellalta
Universitat Pompeu Fabra
Barcelona, Spain

Christos Bouras
Research Academic Computer
  Technology Institute and
  Computer Engineering and
  Informatics Department
University of Patras
Patras, Greece

Carlos T. Calafate
Polytechnic University of Valencia
Camino de Vera
Valencia, Spain

Cristina Cano
Universitat Pompeu Fabra
Barcelona, Spain

Juan Carlos Cano
Polytechnic University of Valencia
Valencia, Spain

Iván Corredor
EUIT Telecomunicación—DIATEL
Universidad Politécnica de Madrid
Madrid, Spain

Nicola Cranley
Communications Network Research
    Institute
Focas Institute
Dublin, Ireland

Mark Davis
Communications Network Research
    Institute
Focas Institute
Dublin, Ireland

Ana B. García
EUIT Telecomunicación—DIATEL
Universidad Politécnica de Madrid
Madrid, Spain

Wei Guo
Mobile and Satellite Communications
    Research Centre
University of Bradford
Bradford, United Kingdom

Vicente Hernández
EUIT Telecomunicación
Universidad Politécnica de Madrid
Madrid, Spain

Jia Hu
Department of Computing
School of Informatics
University of Bradford
Bradford, United Kingdom

Ray Hunt
Department of Computer Science and
    Software
College of Engineering
University of Canterbury
Christchurch, New Zealand

Peng-Yong Kong
Institute for Infocomm Research
Singapore

Thomas Kunz
Department of Systems and Computer
    Engineering
Carleton University
Ottawa, Ontario

Dan Li
Institute for Infocomm Research
Singapore

Lourdes López
EUIT Telecomunicación
Universidad Politécnica de Madrid
Madrid, Spain

Manuel Pérez Malumbres
Universidad Miguel Hernández
Elche, Spain

Pietro Manzoni
Polytechnic University of Valencia
Valencia, Spain

Geyong Min
Department of Computing
University of Bradford
Bradford, United Kingdom

Miquel Oliver
Universitat Pompeu Fabra
Barcelona, Spain

José Fernan Martínez Ortega
EUIT Telecomunicación
Universidad Politécnica de Madrid
Madrid, Spain

Andreas Papazois
Computer Engineering and
    Informatics Department
University of Patras
Patras, Greece

Carlos H. Rentel
Department of Systems and Computer
    Engineering
Carleton University
Ottawa, Ontario

Anna Sfairopoulou
Universitat Pompeu Fabra
Barcelona, Spain

Antonio da Silva
EUIT Telecomunicación
Universidad Politécnica de Madrid
Madrid, Spain

Haitang Wang
Center for Distributed and Mobile
    Computing
Department of Computer Science
University of Cincinnati
Cincinnati, Ohio

Mike E. Woodward
Department of Computing
School of Informatics
University of Bradford
Bradford, United Kingdom

Bin Xie
OBR Center for Center for Distributed
    and Mobile Computing
Department of Computer Science
University of Cincinnati
Cincinnati, Ohio

Yan Zhang
Simula Research Laboratory
Fornebu, Norway

Chapter 1

Quality of Service Support in Mobile Multimedia Networks

Nilufar Baghaei and Ray Hunt

Contents

1.1   Introduction

Most current network architectures treat all packets in the same way—as a single level of service. Applications, however, have diverse requirements and may be sensitive to latency and packet losses. Examples include interactive and real-time applications such as Internet Protocol (IP) telephony; streaming services such as audio, video, and bulk data streaming; and interactive services such as voice, Web, and transaction service processing. When the latency or the loss rate exceeds certain levels, these applications become unusable. In contrast, best-effort services such as file transfer can tolerate a reasonable amount of delay and loss without much degradation of perceived performance.

The capability to provide resource assurance and service differentiation in a network is often referred to as quality of service (QoS). Resource assurance is critical for many new IP-based applications to succeed. The Internet will become a truly multiservice network only when service differentiation can be supported. Implementing these QoS capabilities has become one of the most difficult challenges in its evolution, particularly as this requires changes to its basic architecture.

The requirements for each type of traffic flow can be characterized by four primary parameters: reliability, delay, jitter, and bandwidth. Most IP-based networks rely on the Transmission Control Protocol (TCP) in the hosts to detect congestion in the network and reduce the transmission rates accordingly. TCP-based resource allocation requires all applications to use the same congestion control scheme. Although such cooperation is achievable within a small group, in a network as large as the Internet, it can be easily abused. Furthermore, many User Datagram Protocol (UDP)–based applications do not support TCP-like congestion control, and real-time mobile multimedia applications typically cannot cope with large fluctuations in the transmission rate.

The service currently provided by default is often referred to as best effort. When a link is congested, packets are simply discarded as the queue overflows. Because the network treats all packets equally, any flows could be hit by congestion, and this particularly impinges on wireless and mobile connections. Although best-effort service is sufficient for some applications that can tolerate large delay variation and packet losses, it does not satisfy the needs of many new applications and their users.

Resource assurance is critical for many new wireless applications. Although the Integrated Services (IntServ) and Differentiated Services (DiffServ) paradigms figure predominantly as QoS solutions, they focus on the IP layer, and it is necessary for the underlying layers to be able to respond to and configure such IP-based service requirements. The following sections address the specification and provisioning of these underlying QoS-based requirements for wireless local area networks (LANs) (Section 1.2), wireless personal area networks (PANs) (Section 1.3), wireless metropolitan area networks (MANs) (Section 1.4), and wireless wide area network (WAN) (3G) architectures (Section 1.5). Conclusions are discussed in the last section.

1.2 QoS in IEEE 802.11 Wireless LANs

In its current form, the IEEE 802.11 wireless LAN standard [IEEE, 1999] cannot provide QoS support for the increasing number of applications that demand QoS parameters—typical of many multimedia applications. A number of IEEE 802.11 QoS enhancement schemes have been proposed, each focusing on a particular mode of operation. This section first analyzes the QoS limitations of the IEEE 802.11 Medium Access Control (MAC) layer and then summarizes the QoS enhancement schemes necessary in wireless local area multimedia networks. Finally, it briefly covers the new IEEE 802.11e QoS enhancements.

1.2.1 An Overview of IEEE 802.11 MAC Operation

In general, the IEEE 802.11 WLAN standard covers the MAC sublayer and the physical (PHY) layer of the Open Systems Interconnection (OSI) network reference model. The Logical Link Control (LLC) sublayer is specified in the IEEE 802.2 standard. This architecture provides a transparent interface to higher-layer users: stations may move, roam through an IEEE 802.11 WLAN, and still appear as stationary to the IEEE 802.2 LLC sublayer and above. This allows existing network protocols (such as TCP/IP) to transparently operate over IEEE 802.11 WLAN without any special considerations.

At the PHY layer, the IEEE 802.11 standard provides three operational modes in the 2.4 GHz band: (1) infrared (IR) baseband PHY, (2) Frequency Hopping Spread Spectrum (FHSS) radio, and (3) Direct Sequence Spread Spectrum (DSSS) radio. All three PHY layers support both 1 and 2 Mbps operations. In 1999, the

IEEE defined an 11 Mbps 802.11b standard designed to operate in the 2.4 GHz free Industrial, Science, and Medical (ISM) band, and subsequently a 54 Mbps 802.11a orthogonal frequency-division multiplexing (OFDM) standard for the 5 GHz frequency band.

The IEEE 802.11 MAC sublayer defines two relative medium access coordination functions: the Distribution Coordination Function (DCF) and the optional Point Coordination Function (PCF) (Figure 1.1).

The IEEE 802.11 MAC protocol supports two types of transmission: asynchronous and synchronous [IEEE, 1999]. Asynchronous transmission is provided by the DCF, which implements the basic access method for the IEEE 802.11 MAC protocol. DCF is based on the Carrier Sense Multiple Access with Collision Avoidance (CSMA/CA) protocol and is the default implementation. The synchronous service (also called contention free service) is provided by PCF and implements a polling-based access method. The PCF uses a centralized polling approach that requires an access point (AP) to act as a point coordinator (PC). The AP cyclically polls stations to give them the opportunity to transmit packets. Unlike the DCF, the implementation of the PCF is not mandatory. In addition, the PCF itself relies on the underlying asynchronous service provided by the DCF. Although providing different service functions, neither DCF nor PCF+DCF has the ability to offer true QoS to wireless LAN multimedia applications.

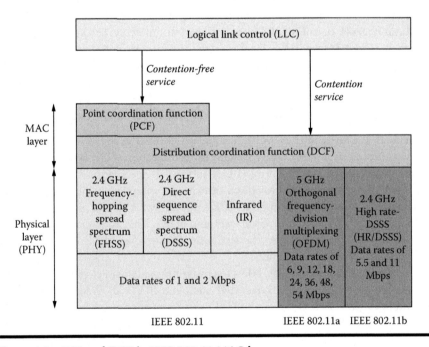

Figure 1.1   PCF and DCF in IEEE 802.11 MAC layer.

1.2.2 QoS Limitations of IEEE 802.11 MAC

In addition to providing channel access (via DCF or PCF+DCF), the wireless LAN MAC layer needs to provide facilities for:

- Maintaining QoS
- Providing security

Wireless links have specific characteristics such as high loss rate, packet reordering, large packet delay, and jitter. Furthermore, the wireless link characteristics are not constant and may vary over time and place. Mobility of users may cause the end-to-end path to change when users roam, and further, users will expect to receive the same QoS as they change from one AP to another. This implies that the new path should also support the existing QoS by service reservation, and problems may arise when the new path cannot support such requirements.

There are two ways to characterize QoS in WLANs: parameterized or prioritized QoS [Ni 2002; Ho 2002]. *Parameterized* QoS is a strict QoS requirement that is expressed in terms of quantitative values, such as data rate, delay bound, and jitter bound. In a traffic specification (TSpec), these values are expected to be met by the MAC data service in support of the transfer of data frames between peer stations. In a *prioritized* QoS scheme, the values of QoS parameters such as data rate, delay bound, and jitter bound may vary during the transfer of data frames, and without the need to reserve the required resources by negotiating the TSpec between the station and the AP.

1.2.2.1 QoS Limitations of DCF

DCF can only support best-effort services and does not provide any QoS guarantees for multimedia applications. Typically, time-bounded services such as Voice-over-IP, audio, and videoconferencing require specified bandwidths, delay, and jitter, but can also tolerate some loss. However, in DCF mode, all the stations in one basic service set or all the flows in one station compete for the resources and channel with the same priority. There is no differentiation mechanism to guarantee bandwidth, packet delay, and jitter for high-priority stations or multimedia flows [Aad 2001].

1.2.2.2 QoS Limitations of PCF

Although PCF has been designed by the IEEE working group to support time-bounded multimedia applications, this mode has some major problems, which leads to poor QoS performance. In particular, the central polling scheme is inefficient and complex and causes deterioration of the performance of PCF high-priority traffic under load. Additionally, all communications have to pass through the AP, which degrades the bandwidth performance [Lindgren, 2001].

1.2.3 QoS Enhancement Schemes for IEEE 802.11 MAC

QoS issues in wired Ethernet have been neglected due to the relative ease with which the PHY layer bandwidth has improved. Normally, the IP layer assumes that a LAN rarely drops or delays packets. However, in WLANs, the challenges of the wireless channel make PHY layer data rate improvements more difficult to achieve, particularly as the IEEE 802.11 WLAN standard was originally designed for best-effort services. The PHY layer's error rate can be more than three orders of magnitude larger than that of a wired LAN. Further, high collision rate and frequent retransmissions cause unpredictable delay and jitter, which further degrade the quality of real-time voice and video transmission. To address these issues, a number of proposals have been made and are detailed in the following sections.

1.2.3.1 Service Differentiation–Based Enhancement Schemes

QoS enhancement can be supported by adding service differentiation into the MAC layer. This can be achieved by modifying the parameters that define how a *station* or a *flow* should access the wireless medium. Current service differentiation–based schemes can be classified with respect to a multitude of characteristics. For example, a possible classification criterion can be based upon whether the schemes base the differentiation on per-station or per-queue (per-priority) parameters. Another classification depends on whether they are DCF (distributed control) or DCF+PCF (centralized control) enhancements. Figure 1.2 illustrates this classification. Previous research work has mainly focused on the station-based DCF enhancement schemes [Aad 2001; Deng 1999; Veres 2001], while other recent work has focused on queue-based hybrid coordination (combined PCF and DCF) enhancement schemes [Aad 2002; Mangold 2002; Romdhani 2002], because queue-based schemes perform more efficiently.

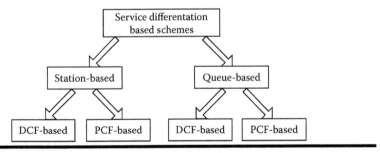

Figure 1.2 Classification of service differentiation–based schemes.

1.2.3.2 Error Control–Based Enhancement Schemes

In parallel, QoS enhancement can also be obtained by error control mechanisms. Because the network may occasionally drop, corrupt, duplicate, or reorder packets, the transport protocol (e.g., TCP) or the application itself (e.g., if UDP is being used) must recover from these errors on an end-to-end basis. Error recovery in the subnetwork is justified only to the extent that it can enhance overall performance. However, some subnetworks, such as wireless links, require link layer error recovery mechanisms to enhance performance, but these enhancements need to be lightweight [Ni, 2002]. For example, wireless links normally require link layer error recovery (such as IEEE 802.2 LLC) and MAC-level error recovery in the subnetwork.

1.2.3.3 IEEE 802.11e QoS Enhancement Standards

The focus of IEEE 802.11 TGe is to enhance the IEEE 802.11 MAC (DCF, PCF) to support QoS, providing classes of service, enhanced security, and authentication mechanisms in support of multimedia applications. It aims to enhance the ability of all the PHY layers (IEEE 802.11b, 802.11a, 802.11g) to deliver time-critical multimedia data, in addition to a best-effort data service. There are many new features in the IEEE 802.11e draft 3.0 [IEEE 2002] that enhance the existing DCF and PCF+DCF functionality to support new QoS applications [Ni 2002]. For more details, refer to [Baghaei 2004]. These include:

- HCF (Hybrid Coordination Function)
    - EDCA (Enhanced DCF Channel Access—prioritized QoS)
    - HCCA (HCF Controlled Channel Access—prioritized QoS plus a contention free period)
- Direct communication in infrastructure mode
- AP mobility
- MAC-level FEC (forward error correction)

1.3   QoS in IEEE 802.15 Wireless PANs

IEEE 802.15 is a communications specification that was approved in early 2002 by the IEEE Standards Association (IEEE-SA) for wireless personal area networks (WPANs). This group has currently defined three classes of WPANs that are differentiated by data rate, battery drain, and QoS. IEEE 802.15* is responsible for creating a variety of WPAN standards and is divided into four major task groups, which are described in Figure 1.3 [Ergen 2004].

---

* http://grouper.ieee.org/groups/802/15/.

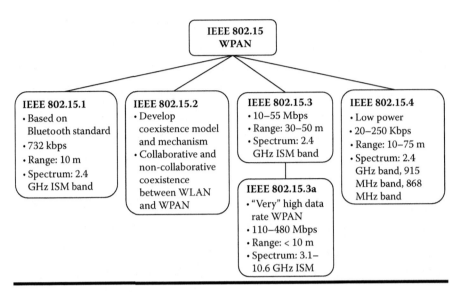

Figure 1.3   Organization of IEEE 802.15.

Whereas IEEE 802.11 was concerned with features such as Ethernet matching speed, long range (100 m), complexity to handle seamless roaming, message forwarding, and data throughput of 2–54 Mbps, WPANs are focused on a space around a person or object that typically extends up to 10 m in all directions. The focus of WPANs is low cost, low power, short range, and very small size [Ergen, 2004].

The initial version, 802.15.1, was adapted from the Bluetooth specification and is fully compatible with Bluetooth 1.1. Bluetooth is a well-known and widely used specification that defines parameters for wireless communications among portable digital devices, including notebook computers, cellular telephones, beepers, and consumer electronic devices. In addition, the specification allows for connection to the Internet.

The IEEE 802.15 working group proposes two general categories of 802.15, called TG4 (low rate) and TG3 (high rate). The low-rate WPANs (IEEE 802.15.4/ LR-WPAN/Zigbee) are intended to provide a set of industrial, residential, and medical applications with very low power consumption and cost requirements and with relaxed needs for data rate and QoS. The low data rate enables the LR-WPAN to consume very little power. The TG4 version provides data speeds of 20 or 250 Kbps [IEEE 802, 2006].

The high-data-rate WPAN (IEEE 802.15.3) supports data speeds ranging from 11 to 55 Mbps and is suitable for applications with very high QoS. The second standard in this usage segment, IEEE 802.15.3a (also called ultrawideband [UWB]), is designed for delivering multimedia services. UWB supports high data speeds of up to 480 Mbps, allowing for digital video disc (DVD) quality to be shared throughout the home. In this case, the PAN becomes a high-speed personnel area network.

1.3.1 IEEE 802.15.3 QoS Standard

While the IEEE 802.11 standard for WLANs is being extended to support QoS for multimedia applications, the high power consumption makes it less suitable for portable devices with limited battery power. On the other hand, although Bluetooth devices offer low power and low cost, they only support relatively low data rates. The increasing demand of low-power and low-cost devices supporting high data rates and QoS motivated the development of the IEEE 802.15.3 standard [IEEE 802.15.3 2003]. Interest in 802.15.3 has been rapidly growing in recent years because UWB is being considered the alternative PHY layer standard by task group 802.15.3a [IEEE 802.15.3a 2005]. The combination of ultrawide spectrum and very low power allows UWB transmissions to accomplish very high data rates over short distances in indoor wireless environments while keeping the level of interference very low. Thus, UWB offers a very promising solution for high-rate WPAN such as 802.15.3 [Porcino 2003], which can support mobile multimedia applications.

IEEE 802.15.3 WPANs are mainly organized as piconets. In each piconet devices exchange data in a peer-to-peer manner under the control of a piconet coordinator (PNC). QoS is supported by allocating guaranteed channel time for each traffic stream. Depending on the piconet size, some devices within the same piconet may be out of radio range with each other. Therefore, network layer routing may be necessary to ensure full piconet connectivity [Yin 2006].

1.3.2 Overview of IEEE 802.15.3 MAC

IEEE 802.15.3 supports various traffic types with different QoS requirements for multimedia applications. Designed for high-rate WPAN, the 802.15.3 MAC supports peer-to-peer communications under centralized control. A piconet is formed when a device, acting as the PNC, begins transmitting beacons. The PNC provides basic network timing for synchronization between devices, performs admission control, allocates network and channel time (CT) resources, manages power save requests, etc. Timing and data transmissions in the piconet are based on the superframe, which consists of three parts: the beacon, the optional contention access period (CAP), and the channel time allocation period (CTAP). Beacons are sent by the PNC to synchronize the piconet, set the timing allocations, and communicate management information. During a CAP, devices employ Carrier Sense Multiple Access with Collision Avoidance (CSMA/CA) to communicate command or a small amount of asynchronous data if the PNC allows data in the CAP. The CTAP consists of channel time allocations (CTAs), including management CTAs (MCTAs) assigned by means of time division multiple access (TDMA). For an isochronous stream or a large amount of data, a CTA should be allocated before transmission. The length of the allocation in the channel time request (CTR) is calculated by the originating device based on the traffic parameters. The PNC then allocates time in a CTA for the device if the resources are available. The guaranteed start time and duration for each CTA enable both power saving and good QoS characteristics [Yin 2006].

1.4 QoS in IEEE 802.16 Wireless MANs

IEEE 802.16 is a group of broadband wireless communications standards for metropolitan area networks (MANs) developed by an IEEE working group, as an alternative to traditional wired networks, such as Digital Subscriber Line (DSL) and cable modems. The original 802.16 standard, published in December 2001, included MAC and PHY layer specifications and specified fixed point-to-multipoint broadband wireless systems operating in the 10–66 GHz licensed spectrum. An amendment, 802.16a, approved in January 2003, specified non-line-of-sight extensions in the 2–11 GHz spectrum, delivering up to 70 Mbps at distances up to 50 km. Officially called the wireless MAN specification, 802.16 standards are expected to enable multimedia applications with wireless connection and, with a range of up to 50 km, provide a viable last-kilometer technology [IEEE 802, 2006].

IEEE 802.16 standards are expected to complement 802.11 specifications by enabling a wireless alternative to expensive 2 Mbps (T1/E1) links connecting offices to each other and the Internet. Even though the first amendments to the standard are only for fixed wireless connections, a further amendment, 802.16e, will enable connections for mobile devices [Summit Technical Media 2005].

A coalition of wireless industry companies, including Intel, Proxim, and Nokia, banded together in April 2001 to form Worldwide Interoperability for Microwave Access (WiMAX), an 802.16 advocacy group. The organization's aim is to actively promote and certify compatibility and interoperability of devices based on the 802.16 specification and to develop such devices for the marketplace [IEEE 802, 2006]. WiMAX and wireless MAN are generating great interest in two areas: as lower-cost alternatives to DSL or cable modem access and as an urban wireless access network operating in a city's main business district and other business centers. The latter application is usually intended to work in conjunction with 802.11 Wi-Fi hot spots and with 3G cellular high-speed data capabilities [Summit Technical Media 2005].

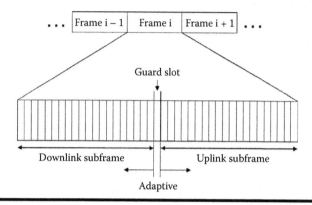

Figure 1.4 IEEE 802.16 TDD frame structure.

1.4.1 IEEE 802.16 QoS Mechanisms

Like ATM, the 802.16 standard was designed with a variety of traffic types in mind. The 802.16 standard has to handle the requirements of very-high-data-rate multimedia applications, such as Voice-over-IP (VoIP) and video or audio streaming, as well as low-data-rate applications, such as Web surfing, and handle very bursty traffic over the Internet. In addition, it may need to handle all of these at the same time. The 802.16 standard includes several QoS mechanisms at the PHY layer, such as time division duplex (TDD), frequency division duplex (FDD), and orthogonal frequency-division multiplexing (OFDM) [Wood 2006]. Each can help in providing the QoS necessary to support mobile multimedia applications. TDD can dynamically allocate uplink and downlink bandwidth, depending on their requirements. This is illustrated in Figure 1.4 [Maheshwari 2005]. Each 802.16 TDD frame is one downlink subframe and one uplink subframe, which are separated by a guard slot. The 802.16 standard adaptively allocates the number of slots for each, depending on their bandwidth needs. In FDD, base stations transmit on different subbands and therefore do not interfere with each other. This allows for even more bandwidth allocation flexibility.

Another QoS mechanism provided in the PHY level is adaptive burst profiles. Both TDD and FDD configurations support adaptive burst profiling, in which the modulation and coding schemes are specified in a burst profile, where they can be adjusted individually to each subscriber station (SS) on a frame-by-frame basis. The burst file allows the modulation and coding schemes to be adaptively adjusted according to link conditions [IEEE 802.16, 2004].

The 802.16 standard incorporates a number of other mechanisms to provide QoS; refer to [Wood 2006] for more details:

■ Adaptive modulation ([Quadrature Phase-Shift Keying] QPSK to [Quadrature Amptitude Modulation] QAM 16 to QAM 64)
■ Fast fourier transform (FFT)
■ Forward error correction (FEC)

These mechanisms are already well established in the wireless technology industry, and they have been proven to reduce latency, jitter, and packet loss, which are all goals of QoS. Every 802.16 implementation will utilize some combination of these mechanisms to achieve QoS. They are all implemented in the PHY layer, and their parameters are based on the QoS requirements handed down by the higher layers, and implemented through QoS provisioning.

1.4.2 IEEE 802.16 QoS Provisioning

The 802.16 standard has three main methods for QoS provisioning in support of multimedia applications, which were approved in 2003 [Wood 2006]:

■ Service flow classification
■ Dynamic service establishment
■ Two-phase activation model

1.4.2.1 Service Flow Classification

The main feature of 802.16 QoS provisioning, and what distinguishes it from its competitors (i.e., 802.11 and 3G), is that it associates each packet with a service flow. The 802.16 standard is contention oriented at the MAC layer, where each connection is assigned a unique connection ID (CID) and a service flow ID (SFID) with an associated service class. The upper part of the MAC maps data into a QoS service class. In addition, external applications can request service flows with desired QoS parameters using a named service class.

The 802.16 standard provides four scheduling services, each with an associated service class: UGS, rtPS, nrtPS, and BE. These are described in Table 1.1 [Wood 2006]. Each network application has to register with the network, where it will be assigned one of these service flow classifications with an SFID.

QoS mapping in the form of classification of higher layer data is provided in the upper part of the MAC. When the application is required to send data packets, the service flow is mapped to a connection using a unique CID [Ganz 2004].

Table 1.1   IEEE 802.16 QoS Service Classes

QoS Service Class	Description
Unsolicited grant service (UGS)	Supports CBR services, such as T1/E1 emulation and VoIP without silence suppression
Real-time polling service (rtPS)	Supports real-time services with variable-size data on a periodic basis, such as Motion Picture Experts Group (MPEG) and VoIP with silence suppression
Non-real-time polling service (nrtPS)	Supports non-real-time services that require variable-size data grant bursts on a regular basis, such as FTP
Best effort (BE)	For applications that do not require QoS, such as Web surfing

1.4.2.2 Dynamic Service Establishment

The 802.16 standard provides a signaling function for dynamically establishing service flows and requesting QoS parameters. There are three types of control messages for service flows:

- Dynamic service activate (DSA)—Activate a service flow
- Dynamic service change (DSC)—Change an existing service flow
- Dynamic service delete (DSD)—Delete a service flow

New connections may be established when a customer's needs change. This may be initiated by the base station (BS). The BS sends a control message called a DSA-REQ, which can contain the SFID, CID, and a QoS parameter set. The subscriber station (SS) then sends a DSA-RSP message to accept or reject the service flow.

This mechanism allows an application to acquire more resources when required. Multiple service flows can be allocated to the same application, so additional service flows can be added if needed to provide good QoS for multimedia applications.

1.4.2.3 Two-Phase Activation Model

Activation of a service flow proceeds in two phases: admit first, then activate. This is facilitated via an authorization module in the BS, which approves or rejects a request regarding a service flow. The authorization module can activate a service flow immediately or defer activation to a later time [Ganz 2004].

Table 1.2   IEEE 802.16 QoS Parameter Sets

QoS Parameter Set	Description
ProvisionedQoSParamSet	A set of external QoS parameters provided to the MAC, for example, by the network management system
AdmittedQoSParamSet	A set of QoS parameters for which the BS and possibly the SS are reserving resources, because the associated service flows have been admitted by the BS
ActiveQoSParamSet	A set of QoS parameters that reflect the actual service provided to the associated service flow

Once the service flow has been admitted, both the BS and SS can reserve resources for it, which are not limited to bandwidth and can include other resources, such as memory. Dynamic changes to the QoS parameters of an existing service flow are also approved by the authorization module. QoS parameter changes are requested with dynamic service flow messages sent between the BS and SS.

A QoS parameter set is associated with each service flow as shown in Table 1.2. The type of QoS parameter set distinguishes the status granted by the authorization module (admitted or active). The standard defines three types of QoS parameter sets as shown in Table 1.2.

The method for determining which QoS parameters will be allowed depends on the authorization model. The 802.16 standard recognizes two authorization models:

■ Provisioned authorization: QoS parameters are provided by the network management system upon setup and remain static.
■ Dynamic authorization: Changes to QoS parameters can be requested, and the authorization module issues its decisions.

Therefore, the 802.16 standard provides some flexibility in its QoS provisioning. The QoS requirements are determined by the higher-layer application. For instance, a VoIP application may require a real-time service flow with fixed-size bandwidth allocation, whereas an FTP application may use a non-real-time service flow with variable-size bandwidth allocation. If the application requires QoS, it can define the QoS parameter set, or it can imply a set of QoS parameters with a service class name. Depending on the available network resources, the network then decides if it can meet the QoS requirements of the application. If so, the QoS parameters are handed down through the MAC layer [Wood 2006].

1.5   QoS in 3G Wireless Networks

2G networks such as Global System for Mobile Communications (GSM)/code division multiple access (CDMA) have essentially only one QoS option, viz., speech at full-rate coding in GSM. Subsequently, a half-rate service was introduced, thus offering a new QoS. In reality, however, this was done to save network capacity, and therefore serving more users in congested hot spots, rather than offering a new grade of service to users. The user was not offered the choice of full or half rate, but more often, those with half-rate-capable mobile phones were put onto half rate, without the subscriber knowing that the speech quality was deliberately lowered by the network being used.

In 2.5G networks such as general packet radio service (GPRS), there has been a deliberate attempt to introduce mechanisms whereby the subscriber can request a different QoS (average/peak data throughput, packet delay, etc.). In principle, this QoS requirement can be established at the beginning of the data transfer session (at Packet Data Protocol [PDP] context setup). For example, a user intending to use an interactive service (such as Web surfing) may want to use a service with a faster reaction time/lower round-trip delay. He or she can then ask for a smaller packet delay at PDP context setup time, and the network can confirm whether this request is accepted or rejected.

3G is a wireless industry term for a collection of international standards and technologies aimed at improving the performance of mobile wireless networks. 3G wireless services offer packet data enhancements to applications, and these include higher speeds, increased capacity for voice and data services, as well as QoS facilities in support of multimedia service applications. The two main 3G technologies for which QoS is being standardized are

- Universal Mobile Telecommunications System (UMTS)
- Code Division Multiple Access 2000 (cdma2000)*

Several common applications in wireless WAN (3G) are listed in Table 1.3 along with the stringency of their requirements.

Table 1.4 shows QoS-based application requirements in terms of bandwidth, delay, and losses for different categories such as data, real-time traffic, non-real-time traffic, games, and network services in 3G networks (Fitzek 2002).

---

* 3GPP is responsible for the UMTS standards specification, while 3GPP2 is responsible for the cdma2000 standards specification. This has resulted in two 3G standards being released: UMTS and cdma2000.

**Table 1.3   Common Wireless WAN (3G) Applications and Their
Requirements**

Application	Loss Rate	Delay	Throughput	Jitter	Disconnect Probability
Control	Null	Low	—	—	Null
Real-time multimedia	Low-medium	Low	Medium-high	Low	Low
Remote login	Low	Low	Low	—	Null-low
FTP	Low-medium	High	High	—	Medium-high
Web	Low-medium	Medium	Medium	—	Low
E-mail, news	Low	High	Low	—	Medium-high

Table 1.4   Typical QoS Application Requirements in 3G

Type of Application and Example			(Kbps)	Losses (%)	Delay (ms)
Data		FTP	High rate desirable	0	Contolled by TCP Timer
Real-time multimedia	Audio	Voice	-64	10^{-4}	-300
		Voice-over-IP	10–64	5.10^{-2}	-300
	Video	MPEG-4	-2,000	10^{-2}	-40
		H.320	-64	10^{-4}	-40
Non-real-time	Audio	CD	150	10^{-4}	Buffer size
	Video	MPEG-4	High rate desirable	10^{-2}	Buffer size

1.5.1   UMTS/3GPP-Defined QoS

Third Generation Partnership Project (3GPP)* has standardized a common QoS
framework for IP-based data services. They have defined a comprehensive framework for end-to-end QoS covering all subsystems in a UMTS network, including
core network, wireless and universal terrestrial radio access networks, etc. UMTS
is the first wireless data service that offers a comprehensive QoS specification across
a wireless wide area network infrastructure. This is a fundamental requirement
for the provisioning of multimedia application support. In addition, the specification provides for control signaling, user plane transport, and QoS management
functionality.

---

* http://www.3GPP.org.

QoS enables a network to deliver classes of service (CoS), i.e., different prioritized treatments to different services or different groups of users. QoS allocates network capacity according to the type of traffic required for a certain type of service, while CoS provides preferred allocation of the network resources in a manner similar to that of DiffServ for IP-based services. CoS is implied in a QoS policy associated with a subscriber. It is used by the network to provide differential QoS treatments to different services subscribed by different users.

UMTS defines QoS classes [ETSI 2001]. Users of these services may communicate with both fixed networks and other mobiles; therefore, end-to-end performance is also influenced by the features of these networks on which other parties may be situated.

The 3GPP end-to-end QoS specification, which includes the definition of UMTS QoS architecture, bearer services, and recommendations for supporting QoS mechanisms, also establishes four overriding UMTS QoS classes or traffic classes for mobile/wireless data, taking into account the restrictions and limitations of the air interface. The characteristics of these four QoS classes are described in the following sections.

1.5.1.1 UMTS QoS Basic Classes

The basic classes defined by UMTS/3GPP are [ETSI 2002; Baghaei 2004]:

1. Conversational
2. Streaming
3. Interactive
4. Background

The main distinguishing factor between these traffic classes lies with sensitivity to delay.

1.5.1.1.1 Conversational Class

This class applies to any application that involves real-time person-to-person communication such as audio voice, videophone, etc. The basic qualities required for speech are low delay, low jitter, reasonable clarity (common codecs and quality), and absence of echo. In the case of multimedia applications, such as videoconferencing, it is also necessary to maintain synchronization of the different media streams. Failure to provide low enough transfer delay will result in unacceptable lack of quality. This class is tolerant of some errors, e.g., voice packet corruption lasting for up to 20 ms. However, the degree of error protection required varies with applications.

1.5.1.1.2 Streaming Class

The streaming class consists of real-time applications that exchange information between viewer and listener, without any human response. Examples of this include video on demand, live MPEG4 listening, Web radio, news streams, and multicasts. Because of the absence of interaction, there is no longer a need for low delay, but the requirements for low jitter and media synchronization remain. Error tolerance is a function of the audio application. The removal of the low-delay criterion makes it possible to use buffering techniques in the end-user equipment, so the acceptable level of network jitter is higher than that for the conversational class.

1.5.1.1.3 Interactive Class

This class covers both humans and machines that interact with another device. Examples of this include some games, network management systems polling for statistics, and Web browsing or database retrieval. Applications in this class are characterized by the request-response pattern of the end user. Round-trip delay and tolerance to packet loss are key QoS characteristics.

1.5.1.1.4 Background Class

The background class covers all applications that either receive data passively or actively request it, but without any immediate need to handle this data. Examples of this include e-mails, short message service, and file transfers. The only requirement is for data integrity, although large file transfers will also require an adequate throughput.

Table 1.5 summarizes the characteristics of each of the above four classes.

1.5.1.2 UMTS QoS Parameters and Attributes

There are many QoS parameters and attributes defined for UMTS, which are necessary for the support of multimedia services:

- Maximum bit rate (Kbps)
- Guaranteed bit rate (Kbps)
- Delivery order (yes/no)
- Maximum service data size (octets)
- service data unit size format information (bits)
- Service data unit size error ratio
- Residual bit error ratio
- Delivery of erroneous service data units (yes/no)
- Transfer delay (ms)

Table 1.5   UMTS QoS Traffic Classes

	Real-Time		Best Effort	
Traffic Class	*Conversational*	*Streaming*	*Interactive*	*Background*
Fundamental characteristics	Preserve timing of stream Conversational pattern—stringent, low delay	Preserve time relation (variation) between information entities of the stream	Request response pattern Preserve payload content	Destination does not care about arrival time Preserve payload content
Application example	Voice	Streaming video	Web browsing	Background, e.g., e-mails

- Traffic handling priority
- Allocation/retention priority

For definitions of these parameters, refer to [Xiao, 2005].

In Table 1.6, the defined UMTS bearer attributes and their relevancy for each bearer traffic class are summarized. For definitions of these parameters, refer to Xiao (2005).

1.5.2 cdma2000 QoS

Third Generation Partnership Project 2 (3GPP2), the standards body in charge of cdma2000 standards, has issued a series of specifications that describe requirements necessary to support end-to-end QoS in the cdma2000 wireless IP network [3GPP2 2004a, 2004b]. The requirements are based on leveraging and extending where applicable the standard Internet Engineering Task Force (IETF) protocols for QoS, such as IntServ and DiffServ. The proposed functionalities include the use of IntServ, DiffServ, IntServ-to-DiffServ interworking, network policy and subscriber profile, network provisioning, and link layer to upper-layer QoS adaptation [Zhao 2005].

With respect to the other QoS attributes (bandwidth, delay, jitter, packet loss, and priority), 3GPP2 defines cdma2000 QoS classes of service similar to UMTS basic classes (described in Section 1.5.1.1), i.e., conversational class, streaming class, interactive class, and background class [3GPP2, 2004a]. The main difference between these QoS classes relates to the parameters, which affect delay sensitivity.

QoS signaling is used to enforce QoS parameters between endpoints and is conducted in the application layer, network layer, and link layer. The Session Initiation Protocol (SIP) [Rosenberg 2002] is used as the application-level signaling

Table 1.6 UMTS QoS Attributes Defined for Each Class

Traffic Class	Conversational Class	Streaming Class	Interactive Class	Backround Class
Maximum bitrate	X	X.	X	X
Delivery order	X	X	X	X
Maximum service data unit size	X	X	X	X
Service data unit format information	X	X		
Service data unit error ratio	X	X	X	X
Residual bit error ratio	X	X	X	X
Delivery of erroneous service data units	X	X	X	X
Transfer delay	X	X	X	X
Guaranteed bitrate	X	X		
Traffic handling priority			X	X
Allocation/ retention priority	X	X	X	X

Note: CS, conversational class; SC, streaming class; IC, interactive class; BC, background class.

protocol to create, modify, and terminate multimedia sessions with one or more participants. SIP runs on top of different transport protocols, e.g., TCP or UDP.

QoS parameters are negotiated between endpoints running SIP user agents through the SIP proxy and Authentication, Authorization, Accounting (AAA) server. The policy decision point (PDP) is co-located with the SIP proxy to determine the allowed QoS parameters based on SIP negotiation and local policy of the network. Session-specific QoS parameters are exchanged via the Session Description Protocol (SDP) or SIP header fields, and QoS parameters are enforced using the policy enforcement point (PEP) as part of packet data serving node (PDSN) in cdma2000 [Siddiqui 2004].

The end-to-end QoS support in the cdma2000 network tries to reserve the necessary resources to ensure that the requested QoS requirements for a user's application are satisfied. If the necessary resources are not available, an attempt should be made to negotiate a lower QoS. However, service differentiation based on a set of traffic classes requires a simple and reliable translation mechanism between the different domains involved, and the network must be well monitored and managed to ensure the implementation of the users' agreements [Zhao 2005]. For details on cdma2000 end-to-end QoS reference model and end-to-end QoS architecture, refer to [3GPP2 2004a, 2004b].

The QoS targets [3GPP2 2002] for audio and video streaming are:

■ **Synchronization:** For transmission of combined audio and video streams, the intermedia skew should be kept below 20 ms.
■ **Bandwidth:** The service shall be able to provide bandwidth allocation of up to 2 Mbps for streams with video and audio contents.
■ **Play-out delay:** The video streaming service shall be able to provide service of reasonable end-to-end delay to accommodate data transfer from the source to the mobile terminal and shall support buffering at the terminal to accommodate transmission path degradations to a specific level. The recommended maximum play-out delay is 30 s.
■ **Delay jitter:** The system shall be able to operate under delay jitter of three times the Radio Link Protocol (RLP) retransmission time in the network with retransmission activated.
■ **Error rate:** The service shall operate over channels with end-to-end bit error rate in the order of 10^{-3} (for circuit-switched network services) and frame error rate in the order of 10^{-2} (for packet-switched network services).

For multimedia applications such as videoconferencing the targets are similar, except the play-out delay has to be much less so that end-to-end delay does not exceed 400 ms. The degree of jitter that must be compensated is up to 200 ms. Throughput must range from 32 Kbps upwards, including the specific rates of 384 and 128 Kbps for packet- and circuit–switching, respectively.

1.6 Conclusions

Providing QoS for modern audio- and video-based multimedia applications is a key challenge for today's wireless mobile networks. Limited bandwidth, varying channel conditions, mobility, as well as QoS interface requirements between a variety of wireless and wired network infrastructures are very complex problems to solve.

Table 1.7 QoS Advantages and Disadvantages of Competing Technologies

Technology	QoS Advantages/Disadvantages
IEEE 802.11	Contention-based MAC, requires acknowledgments, which causes overhead, latency, timeouts; uses time slots, no preemption; fixed channel size
3G	Still not an "all IP" solution; IP QoS must be mapped onto circuit-switching layer, leads to corruption; mapping point may be far away, causing queuing and scheduling inefficiencies; most parameters are fixed, not adaptive
Bluetooth	Interference on ISM band, limited range, maximum of eight devices/network and master; high setup latency; ISM band; simple ad hoc networking
IEEE 802.16	Connection-oriented protocol, provides service flows; grant-based MAC allows centralized control and eliminates overhead and delay of acknowledgments; reacts to QoS needs in real-time; OFDM, FEC, and adaptive modulation for flexible and efficient QoS

Table 1.8 Wireless Access Technologies: Standards and Bandwidth

Network	Standard	Bandwidth
Cellular	GSM	6, 9 Kbps
	GPRS	128 Kbps
	EDGE	384 Kbps
	UMTS/cdma2000	2 Mbps
WLAN	IEEE 802.11b	1, 2, 5.5, 11 Mbps
	IEEE 802.11a/g	54 Mbps
Bluetooth	IEEE 802.15.1	2 Mbps
Zigbee	IEEE 802.15.4	250 Kbps
WiMAX	IEEE 802.16e	70 Mbps

This chapter has addressed the fundamental concepts of QoS provisioning in wireless LANs, PANs, and MANs, and wireless 3G networks. Much work has yet to be carried out to offer this same service across a concatenation of fixed/mobile and wired/wireless networks. Table 1.7 summarizes the advantages and disadvantages of 802.11, 802.16, and 3G technologies. Only QoS aspects are listed.

Table 1.8 summarizes some common and emerging technologies and their bandwidth.

Although advances have been made in the bandwidth capacity of 3G networks, the problems become much more complex as issues of diverse and multiple networks are added. Network operators rarely have end-to-end control over a data path, and the problems of guaranteeing IP-based QoS across multiple networks remains.

While mechanisms such as Resource Reservation Protocol (RSVP) offer QoS guarantees, this still relies upon these mechanisms being implemented by the service providers across multiple wired/wireless networks and the expectation that the underlying lower-layer infrastructure can respond to these stringent requirements.

References

3GPP2. 2002. *Video conferencing services—Stage 1.* Technical Report S.R0021/R0022. http://www.3gpp2.org.

3GPP2. 2004a. *Support for end-to-end QoS stage 1 requirements.* Technical Report S.R0079, v1.0.

3GPP2. 2004b. *cdma2000 wireless IP networks standard.* Technical Report P.S0001-B, v2.0.

Aad, I., and Castelluccia, C. 2001. Differentiation mechanisms for IEEE 802.11. Paper presented at IEEE Infocom 2001, Anchorage, AL.

Aad, I., and Castelluccia, C. 2002. Remarks on per-flow differentiation in IEEE 802.11. In *Proceedings of European Wireless (EW2002),* Florence, Italy, 1–6.

Baghaei, N., and Hunt, R. 2004. Review of quality of service performance in wireless LANs and 3G multimedia application services. *Computer Communications* 27:1684–92.

Deng, J., and Chang, R. S. 1999. A priority scheme for IEEE 802.11 DCF access method. *IEICE Transactions in Communications* 82:B(1).

Ergen, S. C. 2004. *ZigBee/IEEE 802.15.4 summary.* University of Berkeley, California. http://www.eecs.berkeley.edu/~csinem/academic/publications/zigbee.pdf.

ETSI. TS 123 107. *QoS concepts and architecture.* ETSI TS 123 107, v4.3.0 (2001). http://www.etsi.org (accessed December, 2006).

ETSI. TS 122 105. *Services and service capabilities.* ETSI TS 122 105, v5.2.0 (2002). http://www.etsi.org (accessed December, 2006).

Fitzek, F., Krishnam, M., and Reisslein, M. 2002. Providing application-level QoS in 3G/4G wireless systems: A comprehensive framework based on multi-rate CDMA. *IEEE Wireless Communications* 9:42–47.

Ganz, A., Phonphoem, A., and Ganz, Z. 2001. *Robust super-poll with chaining protocol for IEEE 802.11 wireless LANs in support of multimedia applications,* 65–73. Wireless Networks version 7: 65–73.

Ganz, A., Ganz, Z., and Wongthavarawat, K. 2004. *Multimedia wireless networks: Technologies, standards, and QoS,* 119–39. Englewood Cliffs, NJ: Prentice-Hall.

Held, G. 2001. The ABCs of IEEE 802.11. *IT Professional* 3:49–52.

Ho, J. M. 2002. *Some comments on 802.11e draft 2.0.* IEEE 802.11e Working Document 80211-02/005r0.

IEEE 802.11 WG. 1999. *Reference number ISO/IEC 8802-11:1999(E).* IEEE Standard 802.11.

IEEE 802.11 WG. 2002. *Draft supplement to standard for telecommunications and information exchange between systems—LAN/MAN specific requirements: Wireless medium access control (MAC) and physical layer (PHY) specifications: Medium access control (MAC) enhancements for quality of service (QoS).* IEEE 802.11e/Draft 3.0, Part 11.

IEEE. 2003. *Wireless medium access control (MAC) and physical layer (PHY) specifications for high rate wireless personal area networks (WPANs).* IEEE Standard 802.15.3.

IEEE. 2004. *IEEE standard for local and metropolitan area networks: Air interface for fixed broadband wireless access systems.* IEEE 802.16-2004 (802.16REVd) Specification, Part 16, pp. 31–207.

IEEE 802.15 WPAN High Rate Alternative PHY Task Group 3a (TG3a). 2005. http://www.ieee802.org/15/pub/TG3a.html.

IEEE 802.15.3, IEEE Standard 802.15.3, /Wireless medium access control (MAC) and phycical layer (PHY) specifications for high rate wireless personal area networks (WPANs)/, Sept. 2003.

IEEE 802.16, 2004. < . . / . . / Application%20 Data/Application %20 Data/Microsoft/Word/Qos_over_802_16.html#IEEE802.16-04> IEEE 802.16-2004 (802.16REVd) Specification, /802.16-2004 IEEE Standard for Local and Metropolitan Area Networks/, Part 16, Air Interface for Fixed Broadband Wireless Access Systems, 31–207, June 2004.

IEEE 802 Working Group. 2006. http://www.ieee.org/portal/pages/about/802std/index.html (accessed December, 2006).

Lindgren, A., Almquist, A., and Schelen, O. 2001. Evaluation of quality of service schemes for IEEE 802.11 wireless LANs. In *Proceedings of the 26th Annual IEEE Conference on Local Computer Networks*, Tampa, FL, 15–16.

Maheshwari, S. 2005. An efficient QoS scheduling architecture for IEEE 802.16 wireless MANs. Masters thesis, Indian Institute of Technology, Bombay.

Mangold, S., Choi, S., May, P., Klein, O., Hietz, G., and Stibor, L. 2002. IEEE 802.11e wireless LAN for quality of service. Paper presented at Proceedings of European Wireless (EW2002), Florence, Italy.

Ni, Q., Romdhani, L., Turletti, T., and Aad, I. 2002. *QoS issues and enhancements for IEEE 802.11 wireless LAN.* Sophis-AntipolisFrance: INRIA (Institut National de researche en Informatique Automatique).

Porcino, D., and Hirt, W. 2003. Ultra-wideband radio technology: Potential and challenges ahead. *IEEE Communications Magazine* 41:66–74.

Romdhani, L., Ni, Q., and Turletti, T. 2002. *AEDCF: Enhanced service differentiation for IEEE 802.11 wireless ad-hoc networks.* INRIA Research Report 4544.

Rosenberg, J., Schulzrinne, H., Camarillo, G., Johnston, A., Peterson, J., Sparks, R., Handley, M., and Schooler, E. 2002. *SIP: Session initiation protocol.* RFC-3261.

Siddiqui, M. A., Guo, K., Rangarajan, S., and Paul, S. 2004. End-to-end QoS support for SIP sessions in CDMA2000 networks. *Bell Labs Technical Journal* 9(3).

Summit Technical Media. 2005. *802.16 wireless MAN, 802.15 WPAN and ZigBee.* Summit Technology report. http://www.highfrequencyelectronics.com/Archives/Feb05/HFE0205_TechReport.pdf (Accessed december, 2006).

Veres, A., Campbell, A. T., Barry, M., and Sun, L. H. 2001. Supporting service differentiation in wireless packet networks using distributed control. *IEEE Journal of Selected Areas in Communications (JSAC)* 19:2094–104.

Wilson, A., Lenaghan, A., and Maylan, R. 2005. Optimising wireless access network selection to maintain QoS in heterogeneous wireless environments. In *Proceedings of Wireless Personal Multimedia Communications (WPMC'05)*, Aalborg, Denmark.

Xiao, Y., Leung, K., Pan, Y., and Du, X. 2005. Architecture, mobility management, and quality of service for integrated 3G and WLAN networks. *Wiley Journal of WCMC.* http://www.commsp.ee.ic.ac.uk/~kkleung/papers/yang_integrate_3G_WLAN.pdf (accessed December, 2006).

Yin, Z., and Leung, V. 2006. IEEE 802.15.3: Intra-piconet route optimization with application awareness and multi-rate carriers. Paper presented at IWCMC'06, Vancouver, British Columbia.

Zhao, H., Luo, X., and Tang, X. 2005. Providing end-to-end quality of service in CDMA2000 networks. *International Conference on Mobile Technology, Applications and Systems*, 1–4.

Chapter 2

Policy-Based QoS Provision in WLAN Hotspots

Boris Bellalta, Cristina Cano, Jaume Barceló,
Anna Sfairopoulou, and Miquel Oliver

Contents

2.1 WLANs: A Broadband Access to Internet

Wireless local area networks (WLANs) are becoming a new worldwide technology to access the Internet in different scenarios, ranging from home users or organizational communication infrastructure to public hotspots. The latter, in which users can obtain broadband access to the Internet, has been growing in the last few years, ranging from 14,752 in 2002 [1] to approximately 142,332 nowadays [2] (more than a 900 percent increase), and it is expected to continue growing.

Another important question is how Internet usage is evolving from traditional Web browsing and e-mail transfer to a more detailed content, including multimedia traffic. Several studies that analyze user behavior [3, 4] report that more than 90 percent of the total traffic is Transmission Control Protocol (TCP) based (basically due to Hypertext Transfer Protocol (HTTP) transactions), although the presence of peer-to-peer (P2P) traffic is also remarkable, reaching values higher than those obtained by e-mail or File Transfer Protocol (FTP) services. The multimedia content is also becoming an important part of the Internet traffic. A clear example is YouTube, which serves more than 70 million videos daily [5]. An important place is also occupied by Voice-over-IP (VoIP) telephony traffic; Skype, for instance, has multiplied its number of users concurrently connected from 1 million to 9 million in only three years [6].

The heterogeneous characteristics of the traffic and the growing number of WLAN hotspots call for an efficient management of this mixture of traditional, P2P, and multimedia services in wireless networks. The challenge lies in trying to provide the quality of service (QoS) that the sensitive or real-time services need, maximizing at the same time the throughput achieved for best-effort traffic. The main problems in WLANs to support this kind of traffic are the variable characteristic of the wireless channel (depends on several factors, like transmission power, noise, and interferences) and the fair access nature of the Medium Access Control (MAC) protocol used in current WLAN hotspots, the Distributed Coordination Function (DCF) [7]. Basically, the DCF is a Carrier Sense Multiple Access with Collisim Avoidance (CSMA/CA) protocol that provides fair channel sharing among mobile terminals with data ready to be transmitted, without any kind of flow prioritization mechanism to guarantee the QoS requirements of the multimedia traffic flows.

To solve these problems, next-generation WLAN access points (APs) and mobile terminals (STAs) will use the Wireless Multimedia (WMM) specification, which implements a subset of the Enhanced Distributed Channel Access (EDCA) standard [8], which provides static traffic differentiation, allowing the changing of the channel access probability defining different MAC parameters for each traffic profile. Apart from that, the system performance can be improved by using a proper set of MAC parameters for each situation. For instance, consider a group of mobile stations transmitting best-effort traffic. In this case, the aim is to maximize the system throughput or to minimize the entire flow transfer delay, without considering the transmission delay of a single packet. Therefore, it is essential to reduce to the minimum the access overhead by increasing the number of packets transmitted each time a STA gets access to the channel, reducing also the time spent in collisions.

This chapter intends to provide a clear view of the EDCA capabilities to enhance current WLAN hotspots, providing an open set of possibilities to improve their performance. The main goals are:

1. Analyze how the hotspots are evolving (in terms of traffic and user behavior) and their current performance limitations.
2. Study and identify potential improvements to enhance WLAN utilization in the presence of heterogeneous traffic flows.
3. Provide basic knowledge about EDCA, including a brief description of how to model mathematically the new QoS enhancements.
4. Introduce the benefits of properly tuning the EDCA parameters.
5. Present an effective EDCA parameter tuning algorithm and demonstrate how it is able to satisfy the desired hotspot QoS requirements.

2.2 Providing QoS in a WiFi Hotspot Using EDCA

The Distributed Coordination Function (DCF) is a distributed MAC random access protocol designed to share efficiently and in a distributed way a wireless channel among multiple nodes. However, due to the dynamics of the CSMA/CA protocol, especially when used in an infrastructure configuration, several considerations about the hotspot performance are required, such as the uplink/downlink throughput long-term unfairness [9], the link adaptation performance anomaly [10], and the inability to differentiate traffic flows with different QoS requirements [11, 12].

The equal probability to access the channel for all the nodes provided by DCF has a big impact when downlink traffic is considered. The AP typically transmits the aggregated traffic to all the nodes in the network, so in most of the cases the amount of traffic this node has to transmit is higher than the traffic offered by the mobile nodes. As all nodes have the same probability to access the channel, the AP may not be able to achieve its throughput requirements in the presence of uplink flows (it will depend on the number of STAs transmitting and the traffic characteristics of the traffic flows of each STA). EDCA also tries to minimize this uplink/downlink unfairness by giving to the AP a higher probability to access the channel with regard to the STA and therefore reducing this problem [13].

In a multirate environment where nodes transmit at different data rates due to the capacity-varying characteristic of wireless channels or incompatibility problems (due to the use of different IEEE 802.11 physical (PHY) standards), we also face some problems related to the DCF scheme. In this kind of scenario, nodes that are transmitting at low data rates impact the other nodes in the network, which could see their available bandwidth reduced due to the high amount of time spent in slow transmissions [10]. This problem could also find a possible solution with the improvements introduced by EDCA by designing priority access schemes to nodes with different data rates [14].

Finally, one of the most important problems of DCF is the failure to provide traffic differentiation, a key issue to supply QoS guarantees, which makes difficult the coexistence between sensitive/real-time and best-effort flows. The new IEEE 802.11e EDCA solves this problem by using different access categories (ACs) with different MAC parameters for each, which are used to govern the transmission attempts. Then, an incoming traffic flow is assigned to use an AC considering its QoS requirements. An overview of this DCF limitation and how EDCA can solve it is shown in [11], including several results to show the EDCA capabilities to provide traffic prioritization and differentiation. EDCA also provides call admission control signaling that could be used to guarantee the system stability and the QoS of the flows that are already active in the system [12, 13, 15].

2.2.1 An Example: A Hotspot with VoIP Calls and Elastic Traffic

A very simple classification for the traffic flows can be done by considering the transport protocol used: User Datagram Protocol (UDP) (*rigid*) and TCP (*elastic*) flows. The main difference between them is that for rigid flows the bandwidth/delay requirements remain constant during the time that the flow is alive, while on the other hand, the main characteristic of the elastic flows is their ability to adapt their rate (bandwidth demand) to the network state without having to consider the transfer delay as a stringent requirement.

Consider the scenario depicted in Figure 2.1, where STAs carrying rigid flows (with the required bandwidth of each indicated) share the WLAN resources with STAs transmitting at maximum available bandwidth (elastic flows). The AP transmits both type of flows in the downlink (from fixed network to STAs). The basic traffic characteristics for the VoIP calls and the elastic flows are shown in Table 2.1. Note that a VoIP call implies a bidirectional flow. Furthermore, the scenario is evaluated using the analytical model introduced in Section 2.3.

If the AP and STAs use the DCF with the default parameters for the IEEE 802.11b specification [7] and a single data rate equal to 2 Mbps (the basic rate is set to 1 Mbps), the maximum number of VoIP calls, without elastic traffic, using the G.711 VoIP codec is equal to 5 [15]. However, a key issue considering the deployment of VoIP services in WLANs is that the same wireless channel will be shared simultaneously between the VoIP traffic and the data (elastic) traffic. The coexistence between them is difficult, with VoIP flows being starved by the elastic ones [15]. For example, see Figure 2.2, where the performance of rigid (VoIP) flows

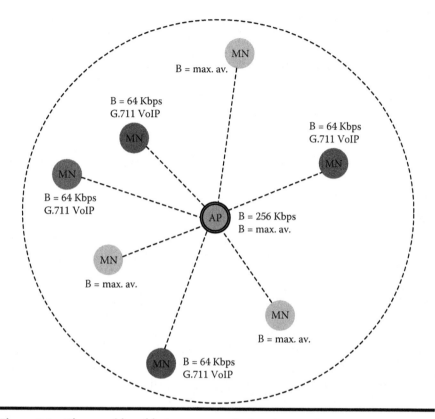

Figure 2.1   The considered heterogeneous hotspot scenario.

Table 2.1   Traffic Profiles

Traffic type	Application /codec	AC (only if EDCA is used)	Packet length	Bandwidth
VoIP	G.711	AC_VO	160 bytes	64 Kbps
Elastic	E.g., P2P	AC_BE	1500 bytes	Max. available

in the presence of elastic flows is shown, both for uplink (a single flow, $n_{e,u} = 1$) and downlink (four flows, $n_{e,d} = 4$) directions. With just a single uplink elastic flow, only one or two VoIP calls will perform correctly as the AP reaches saturation. Moreover, the four downlink elastic flows also suffer from the AP starvation

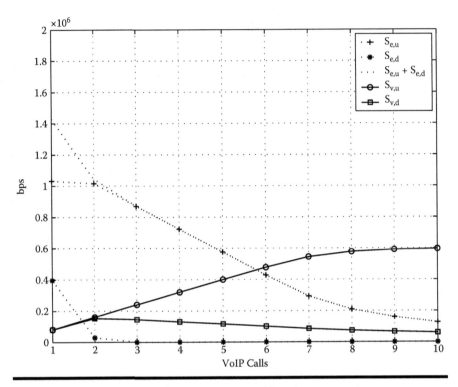

Figure 2.2 VoIP performance with elastic traffic using DCF.

caused by the uplink elastic flow, and the fact that the AP is saturated by the VoIP packets, which results in a very low downlink elastic throughput compared with the uplink elastic flows. With $S_{e,d}$ ($S_{e,u}$), it is referred to as the downlink (uplink) elastic throughput, and with $S_{v,d}$ ($S_{v,u}$), the downlink (uplink) VoIP throughput. Furthermore, the throughput values are calculated at the MAC layer, so they include the IP, UDP/TCP, and Real-Time Transport Protocol (RTP) headers when required. In this chapter, the AP is referred by the subindex d (downlink) and the STAs by the subindex u (uplink), and the AC_BE queue by the letter e (elastic) and the AC_VO queue by the letter v (voice). The other variables follow the same nomenclature.

In Figure 2.2 it is clear that the AP is the bottleneck for the VoIP traffic as it gets saturated before the rest of the mobile nodes (the STAs). This is because the AP has to carry alone the same traffic as all the STAs together, although it has the same probability to access the channel. For elastic throughput this is also true, with the AP being affected by the increment of the number of uplink elastic flows, as well as the saturation of voice packets in the AP queue. Therefore, two main issues arise:

(1) providing differentiation between the rigid and elastic flows, and (2) providing throughput fairness between the uplink and downlink.

2.2.2 Enhanced Distributed Channel Access (EDCA)

The EDCA mode of operation of IEEE 802.11e is an extension of the DCF with the goal to provide priorities and traffic differentiation in wireless access. To achieve this traffic differentiation, the Medium Access Control protocol classifies each traffic flow in an access category (AC). Four ACs are defined, each one associated to one MAC transmission queue. Each AC has its own MAC parameters and behaves independently of others. Letting $AC_{i,j}$ be the access category j of i-*STA*, the basic MAC parameters of each access category are labeled as: *arbitration interframe space* $AIFS_{i,j}$, the binary exponential backoff (BEB) parameters (*minimum contention window* $CW_{min,i,j}$ and *maximum contention window* $CW_{max,i,j}$), and *transmission opportunity* $TXOP_{i,j}$.

In Figure 2.3 a synthetic representation of the different ACs and the virtual collision handler implemented by the EDCA are depicted. In Table 2.2, the recommended static parameters are shown for the different queues.

When node i receives a packet from the network layer, it sends the packet to the corresponding $AC_{i,j}$ queue. For each AC, when there is at least one packet to be transmitted and according to the basic access (BA) mechanism, the node starts to sense the channel to determine its state, which can be either *busy* or *free*. If the channel is detected busy, the node waits until the channel is released. When the channel is detected free for a period of time larger than the $AIFS_{i,j}$ duration, a new backoff instance is generated, which consists of a counter set to a random value. The random value is chosen from a uniform distribution in the range

$$CW_{i,j}(k) = [0, min(2^k CW_{min,i,j} - 1, CW_{max,i,j} - 1)] \, ,$$

where k is the current packet transmission attempt. For each packet to be transmitted, k is initially set to 0 and is increased by one at each failed transmission until a maximum number of retransmissions, called retry limit, is reached and the packet is dropped.

The backoff counter is decreased by one each time-slot that the channel is sensed free, until the countdown reaches zero, at which the node starts the packet transmission on the channel. If, during the backoff countdown the channel is sensed busy, the backoff is suspended until the channel is detected free again. The $AIFS_{i,j}$ value is computed using a nonnegative integer $AIFSN_{i,j}$ specific for each $AC_{i,j}$ $AIFS_{i,j} = SIFS + AIFSN_{i,j} \, \sigma$ (where σ is an empty SLOT duration). Once a node gets the channel, it can transmit up to $B_{i,j}$ MAC Protocol Data Unit (*MPDU*) packets ($TXOP_{i,j}$ limit). This limit is expressed in time units (ms) and corresponds to the consecutive

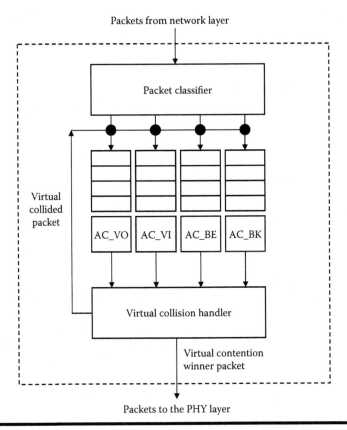

Packets from network layer

Packet classifier

Virtual
collided
packet

AC_VO | AC_VI | AC_BE | AC_BK

Virtual collision handler

Virtual contention
winner packet

Packets to the PHY layer

Figure 2.3   Access categories and virtual collision handler.

Table 2.2   Default EDCA Parameter Set Element Values for the 802.11b Specification

AC	$AIFSN_j$	$TXOP_{limit}$ (ms)	$CW_{min,j}$	$CW_{max,j}$
0 (background: BK)	7	0	CW_{min}	CW_{max}
1 (best effort: BE)	3	0	CW_{min}	CW_{max}
2 (video: VI)	2	6.016	$CW_{min}/2$	CW_{min}
3 (voice: VO)	2	3.264	$CW_{min}/4$	$CW_{min}/2$

time that a node can transmit few (large) or several (small) packets. In Figure 2.4 $B_{i,j}$ is computed by considering the average duration of the packets transmitted by node i for the normal and block Acknowledge (ACK) configurations [8] and two data rates (2 and 11 Mbps).

A channel collision occurs if two nodes transmit at the same time, i.e., a backoff instance from two nodes reaches 0 at the same time. After the data packet is transmitted to the channel by the sender, the receiver waits for a short interframe space (SIFS) time and sends a MAC layer ACK to acknowledge the correct reception of the data packet. In case the sender does not receive the ACK frame, it starts the retransmission procedure. After discarding or successfully transmitting a packet, if more packets are ready to be transmitted, the node starts the transmission procedure again. Otherwise, it waits for a new packet from the network layer. Another EDCA feature is the use of different ACK policies (no ACK transmission or ACKs aggregation), which can also be used to improve the system performance, but will not be analyzed here. As an alternative to the BA mechanism, nodes can employ a Request to Send (RTS)/Clear to Send (CTS) protocol to access the channel, so as to reduce the hidden terminal effect.

Therefore, using EDCA it is expected that the low VoIP capacity of the previous example could be improved. Results are shown in Figure 2.5. Notice that for a single uplink elastic flow, the system is able to carry up to $n_v = 5$ VoIP calls. Moreover, the uplink/downlink unfairness has been mitigated as the AP is not starved, at the cost of a lower aggregated elastic throughput.

When a hotspot is configured to support VoIP traffic, it is expected that the wireless service provider will give the maximum priority to VoIP traffic with the goal to maximize the number of simultaneous calls. Therefore, comparing Figures 2.2 and 2.5, it is clear how EDCA is able to provide a better performance than the DCF, allowing more simultaneous calls in all situations considered. However, the cost of this greater number of VoIP calls is that the best-effort or data throughput is reduced.

At that point, a question arises: Is it possible to improve the performance of EDCA? The answer is that it is possible as EDCA, in low loaded conditions, assigns more than the required resources to each VoIP call. This situation can be observed in Table 2.3, where the utilization of the AC_VO queue at the AP is shown. For queue utilization values lower than a certain threshold ρ_{th} (the value of this threshold can be set to guarantee a maximum queuing delay), the VoIP flows receive higher resources than required. Therefore, if the unnecessary resources assigned to VoIP calls are reduced and given to best-effort flows, the quality of the VoIP calls can be guaranteed while increasing the best-effort traffic, improving the overall hotspot performance.

How can it be done? The answer is by tuning the EDCA parameters to values other than the ones defined in the standard. However, to do it in the correct way, the impact of changing the MAC parameters of each access category (AC) must be known, for the same AC nodes and also for the other nodes that are using other

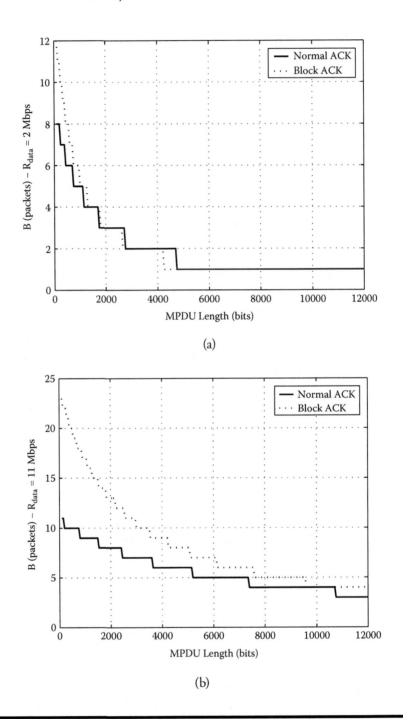

(a)

(b)

Figure 2.4 Maximum number of packets transmitted ($B_{i,j}$) at each TXOP for data rates equal to (a) 2 Mbps and (b) 11 Mbps.

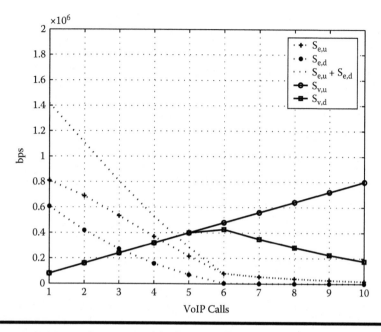

Figure 2.5 VoIP performance with elastic traffic using EDCA.

Table 2.3 AC_VO Queue Utilization $\rho_{v,d}$ for Different $n_{e,u}$, and with $n_{e,d} = 4$

n_v	0	$N_{e,u}$ 4	8
1	0.084	0.331	0.460
2	0.181	0.548	0.690
4	0.441	0.851	0.976
6	0.880	0.999	1.000

ACs. With this knowledge it is possible to design algorithms that optimize the WLAN operation, mitigating at the same time the different set of impairments.

2.3 Modeling the EDCA in Nonsaturated Conditions

In [15] an analytical model of the DCF was presented and validated for heterogeneous traffic scenarios. In [12] that model is extended to introduce the EDCA enhancements, such as different backoff values, multiple AIFS values, and the burst transmission mechanism. The effect of the new EDCA enhancements is illustrated in Figure 2.6, where three STAs compete to access the channel. First, STA uses the AC_VO queue, which, using lower backoff parameters, lower AIFS, and a higher TXOP duration than the ones used by the AC_BE queue, gets a higher priority to transmit its packets.

The stochastic model of the backoff presented by Bianchi [16], allows the steady-state probability that a node transmits in a given slot to be obtained. From a different point of view, but numerically equivalent, Cali et al. [17] and Tay and Chua [18] also derive expressions for that parameter. In equation 2.1 [18] the average number of slots that a node i using the $AC_{i,j}$ waits before transmitting the packet is shown, which depends on both $CW_{min,i,j}$ and $m_{i,j} = log_2 \frac{CW_{max,i,j}}{CW_{min,i,j}}$ (number of backoff stages):

$$EB_{i,j} = \frac{1 - p_{i,j} - p_{i,j}(2p_{i,j})^{m_{i,j}}}{1 - 2p_{i,j}} \frac{CW_{min,i,j}}{2} - \frac{1}{2} \qquad (2.1)$$

However, these expressions are obtained under the assumption of independent and steady-state behavior of the nodes contending for the channel at any slot (a node/AC collides with fixed and constant probability $p_{i,j}$ independently of the backoff stage). This assumption proves to be very accurate for the CW_{min} and CW_{min} values considered in the DCF [16]. However, as the EDCA reduces these values, its accuracy is also reduced, as it is justified in [19], but it is enough to capture the joint dynamics of the EDCA parameters. Following the renewal theory, a node/

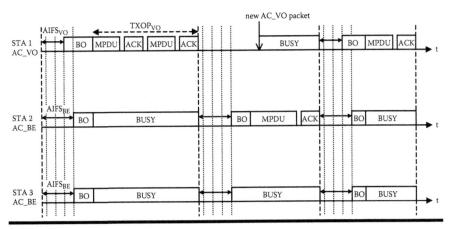

Figure 2.6   Sketch of EDCA temporal evolution.

AC will transmit after waiting an average of $EB_{i,j}$ slots; therefore, the transmission probability is

$$\tau_{i,j} \approx \frac{\rho_{i,j}}{EB_{i,j}+1} \qquad (2.2)$$

where $\rho_{i,j}$ is the probability that a node has at least a packet ready to be transmitted (queue utilization).

Then, with probability $\tau_{i,j}$, a node/AC transmits to the channel. It can collide with probability $p_{i,j}$ if at least one more node (or another AC of the same node) decides to transmit in the same slot. Otherwise, it will be able to transmit $n_{i,j}$ consecutive packets until the $TXOP_{limit,i,j}$ is reached. To model this behavior, each mobile node is approximated by a finite-length queue with bulk- and network-dependent service time ($M/M^{[1,B_{i,j}]}/1/K_{i,j}$), as we are considering that packets with average length $L_{i,j}$ arrive to node i and at $AC_{i,j}$ with average rate $\alpha_{i,j}$. Both the time between packet arrivals and the bulk service time are assumed to be exponentially distributed, with different mean values depending on the bulk size. The state space of the queue and transitions among them are depicted in Figure 2.7.

From Figure 2.7, notice that in low loaded conditions, the number of packets transmitted at each TXOP is always almost 1. Thus, the use of higher TXOPs is only noticeable when the system is near saturation or when it accommodates bulk packet arrivals. Another possibility is to block the packet/s transmission until there are at least $B_{min,i,j}$ (in Figure 2.7, $B_{min,i,j}$ = 1) packets stored in the queue, which reduces the number of transmission attempts but introduces a higher queuing delay [20].

In Figure 2.6, the two STAs using the AC_BE queue have a higher AIFS (one extra SLOT time, 20 µs higher) value than the STA that uses the AC_VO queue. In [21, 22] the AIFS impact is modeled by setting different contention zones (based on the set of AIFS values). These models are very accurate, but they lack flexibility (the ability to set arbitrarily different AIFSN values), which is crucial for a first evalu-

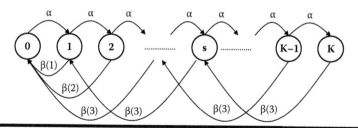

Figure 2.7 $M/M^{[1,B]}/1/K$, for values of B = 3; $\beta(.)$ is the packet departure rate.

ation of the impact of using different AIFSN values. The approximation used in equation (2.3) is similar to the one in [23]. The number of transmissions observed by a node during its backoff is $p_{tr,i,j} \cdot EB_{i,j}$, and the backoff counter will be maintained on hold for $p_{tr,i,j} \cdot EB_{i,j} \cdot AIFSN_{i,j}$ slots. Then the average total number of slots (including the backoff slots) to wait is

$$\zeta_{i,j} \approx \left(EB_{i,j} + AIFSN_{i,j} \right) + p_{tr,i,j} \cdot EB_{i,j} \cdot AIFSN_{i,j} \tag{2.3}$$

where $p_{tr,i,j}$ is the probability that at least one more node/AC transmits in a given slot. Note that higher AIFS values are translated to higher average backoff values.

Notice that the impact of the AIFS parameter is load dependent (depends on the behavior of the other nodes/ACs), as it is affected by the value of $p_{tr,i,j}$. Nodes/ACs with higher AIFS values will suffer lower transmission opportunities when they compete for the channel with nodes with a lower AIFS value, but also with lower CW_{min} values. Consider the extreme case in which two nodes/ACs A and B compete for the channel and $AIFS_A + CW_{min,A} \cdot T_{slot} < AIFS_B$. In this case, A will always obtain the channel.

The simultaneous use of different $TXOP_{limit}$ values, AIFSN and BEB parameters, allows the management of the number of transmission opportunities and the number of packets transmitted at each attempt, resulting in different performances that satisfy the existence of different traffic classes. Thus, high-priority traffic (e.g., VoIP) will use lower CW_{min} and CW_{max} values, lower AIFSN, and higher TXOP than low-priority classes (e.g., best effort).

In [24] the effect of the MAC parameters is shown in different scenarios with saturated and unsaturated sources using different ACs and EDCA parameters for each one. Moreover, the impact of different combinations of parameters is shown in [25].

2.4 Tuning the EDCA Parameters

A lot of research has been done in adjusting the MAC parameters, focusing on fairness between the downlink and uplink (e.g., [26–29]), maximizing the throughput/minimizing the transfer delay (e.g., [30–32]), or providing traffic differentiation capabilities (e.g., [33, 34]). Other works also focus on dynamically adjusting the MAC parameters to mitigate the multirate effect (e.g., [14, 35]).

Major parts of these proposals (e.g., [12, 27, 32, 35, 36]) focus on tuning the BEB parameters,* especially CW_{min}, assuming that the binary exponential increase is disabled (i.e., $CW_{max} = CW_{min}$), or at least has a minor impact on the overall performance. Medepalli and Tobagi [37] show how this assumption improves the short-term fairness and system stability. Thus, the use of a single backoff stage could be beneficial, especially in nonsaturated situations, so as to provide the desired short-term fairness. About modifying the AIFS, in [26] the authors propose that the AP (DCF based) will use PIFS time instead of DIFS. This solution is similar to that suggested in EDCA allowing the AP to use the defined AIFS values minus 1, so $AIFS_{AP} = AIFS_{STA} - 1$, for each AC. Finally, the TXOP parameter is also considered in [29, 38], where it is adjusted according to the load of each STA (at the MAC queue). When using a combination of these parameters, a heuristic algorithm to adjust them properly is presented in [28].

Moreover, several authors link the parameter tuning algorithm with the admission control (e.g., [12, 13, 36, 39]), with these solutions a complementary part of improving system performance.

Making a first classification of these papers based on when the MAC parameters are updated, there are two basic categories:

1. **Continuous parameter updating (measurement based):** A major part of the previously cited works can be classified here, as they measure several parameters, such as the instantaneous load at each AC, the collision rate, or the number of contending stations [40], to adjust the MAC parameters properly.

2. **Parameter updating at fixed periods:** The algorithm is only activated when a flow arrives or departs the system, or when one of the active flows observes a change on the channel state or on its traffic profile.

A second classification could be done based on where the new parameters are computed. Here two possibilities exist:

1. **Distributed:** The parameters are adjusted by each STA itself. The algorithms of this group are measurement based.

2. **Centralized:** The new parameters are computed at the AP. Both types, measurement and parameter updating at fixed periods, are considered.

A third classification, based on EDCA, refers to the parameters that are updated each time the process is activated. The classification is as follows:

---

* This is motivated by the fact that it was considered that only the BEB parameters could be modified in the DCF.

1. **Static parameters:** This solution is considered by current EDCA implementations, as it is the simplest: the default parameters are always used. The main drawback of this solution is that it is unable to react to different load or channel situations. However, the standard defined parameters are already a good trade-off to protect rigid flows, while they also provide good performance for elastic ones.

2. **Tuning the AC parameters:** A more efficient way is to adjust the parameters of each AC, so that all flows using the same AC in different nodes can view the same changes on the parameters. If the flows that use the same AC have different traffic profiles, the computation of the new parameters has to be done to satisfy the flow with the highest requirements, which might result in low performance.

3. **Tuning each AC/STA parameter:** The MAC parameters for any AC of any STA are modified independently from the others. However, this solution is the most flexible—and the most complex—as any change on the network state affects all flows, which implies the readjustment of all parameters again.

There are very few works that focus on centralized MAC parameter tuning algorithms. Pong and Moors [39] presented a MAC parameter tuning algorithm based on reducing CW_{min} and increasing the packet length. Freitag et al. [28] proposed another parameter tuning algorithm that tunes the parameters of each AC according to the load of each flow and the number of STAs contending for the channel (similar ideas are used in Ksentini et al. [38] and Ma et al. [32]).

2.5 Designing a MAC Parameter Tuning Algorithm

The goal of the proposed algorithm is to provide the values for the EDCA parameters (which are different for the AP and the STAs) to satisfy both the flow (user) requirements and the system constraints. It is a partially heuristic algorithm, which refers to a *rule of thumb* for adjusting EDCA parameters, based on the common sense behind the rationale of EDCA dynamics. Once the EDCA parameters are selected, they are set in the AP and sent to the STAs through beacon frames [8]. All STAs are configured with the same EDCA parameters.

Let θ_o be the initial set of EDCA parameter values, θ_i the value of the parameters at iteration i, and θ^* the chosen EDCA parameters combination, which better fits the required policy (set of requirements and constraints). To evaluate the EDCA performance (i.e., to check the different parameter combinations), the analytical model introduced in Section 2.3 is used. It reports the achieved performance (throughput, queue utilization, delay, and losses) by each flow (Ω_j). This information is used as feedback for the next iteration. The block diagram to select the EDCA parameters is depicted in Figure 2.8.

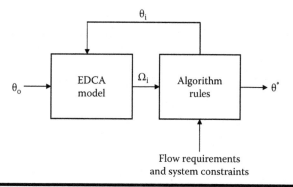

Figure 2.8   Algorithm block diagram.

2.5.1   The Hotspot Policy

The first step in the design of a tuning algorithm is to define the policy that will govern the hotspot. A policy is described by a set of constraints and objectives that the hotspot has to meet. Two types of requirements are considered for each possible state: hard and soft. A *hard* requirement means that it must be satisfied by any system state to accept that state as feasible. A *soft* requirement means that it is a desirable requirement.

In this chapter, a policy for VoIP and best-effort traffic is defined. The policy is designed to guarantee the maximum achievable quality for VoIP calls while trying to also maximize best-effort (data) traffic, giving priority (soft) to the downlink (AP to STAs) direction. The basic points of the considered policy are

1. (**soft**) A queue utilization of rigid flows lower than ρ_{th}. For example, a value of $\rho_{th} = 0.8$ will ensure that the losses and average delays are limited. Notice that modeling the AC_VO queue by a **M**/*M*/1/*K* queue with space for $K = 20$ packets, the probability to lose a packet is $P_b = 2.3273 \cdot 10^{-3}$.

2. (**hard**) A percentage of losses lower than 1 percent (a VoIP call supports up to 3 percent in losses).

3. (**hard**) An average queuing delay (including service time) lower than 50 ms, as it has to be lower than 150 ms for good voice quality [41], but the wireless hop is only a part of the total end-to-end delay. Moreover, a jitter lower than 75 ms is expected.

4. (**soft**) Maximum throughput for the elastic flows, while previous constraints are achieved.

5. (**soft**) Try to ensure that the downlink best-effort throughput will be greater than the uplink best-effort throughput.

Notice that the objective of guaranteeing the bandwidth required by VoIP flows is already included in objectives 1 and 2. Objective 1 is soft, as values higher than ρ_{th} are possible if they satisfy objectives 2 and 3. Objective 5 is also soft, as it is not always possible to satisfy.

To implement this policy, this set of constraints and requirements must be mapped to a set of values for the EDCA parameters, which, adjusted for each state, are able to satisfy them. This will be done by the EDCA parameters tuning algorithm.

2.5.2 Set of EDCA Parameters

Considering only two ACs, AC_BE and AC_VO, there are 16 possible parameters (4 + 4 for the downlink and 4 + 4 for the uplink). The four parameters are CW_{max}, CW_{min}, *AIFSN*, and $TXOP_{limit}$. The range of values that are considered for each parameter is shown in Table 2.4.

The 16 different parameters and the suggested range of values give a huge number of combinations for any given state (number of uplink/downlink flows). Then, a brute-force method, based on checking all combinations, is completely infeasible. Thus, the introduction of some heuristics about the joint behavior of the parameters under some assumptions will reduce dramatically the computational cost of finding a feasible parameter combination that provides near-optimal values at a lower computational cost and delay.

2.5.3 Building the Algorithm

There are several assumptions or considerations that are necessary to understand how the proposed algorithm works. The algorithm could be signaled by a SIP proxy each time a new VoIP call arrives or departs [42] and periodically at intervals of δ (for example, every minute) to adjust the AC_BE parameters based on the number of active downlink/uplink TCP connections. Moreover, the algorithm can be used

Table 2.4   Range of Values of EDCA Parameters

	Downlink		Uplink	
Parameter	*AC_VO*	*AC_BE*	*AC_VO*	*AC_BE*
CW_{min}	8	[16–1,024]	[8–1,024]	[16–1,024]
CW_{max}	32	1,024	[32–1,024]	1,024
AIFSN	2	[2–8]	2	[3–9]
TXOPlimit (packets)	4	[8–1]	4	[1–8]

as a part of a call admission control, which allows the system to accept and reject calls, guaranteeing the system stability.

Thanks to the TXOP option, an EDCA node might be allowed to transmit various consecutive packets. In the proposed algorithm, the AC_BE queue at AP uses a $TXOP_{limit}$ equal to $B_{e,d} = 8$ packets. However, it is assumed that the AP has the ability to cancel an ongoing downlink AC_BE transmission that includes more than a single packet. This mechanism is implemented to avoid an arrived downlink AC_VO packet suffering higher delays. This mechanism allows the AP to use higher TXOP values for the AC_BE, which increases the downlink throughput. Moreover, it can be useful to mitigate the uplink/downlink throughput unfairness.

The AC_VO parameters are assumed to be static (see Table 2.4). However, as pointed out in [15], the VoIP capacity could be slightly improved if the existing uplink/downlink unfairness is solved by adjusting the BEB parameters of the uplink STAs. This means to adjust the AC_VO CW_{min} parameter of the STAs and the AP to ensure that the transmission probability of the AP is equal to the aggregate transmission probability of all STAs, so $\tau_{v,d} \approx n_v \tau_{v,u}$, as $\tau_{v,u} \ll 1$. However, assuming that the AC_VO queue is unsaturated, the transmission probability of both the uplink and downlink depends on their queue utilization, $\rho_{v,d}$ and $\rho_{v,u}$, respectively. Therefore, $CW_{min,v,u}$ is computed from

$$EB_{v,u} = \frac{n_v \cdot \rho_{v,u} \left(EB_{v,d} + 1 \right)}{\rho_{v,d}} - 1 \qquad (2.4)$$

under the assumption of a conditional collision probability equal to 0 to simplify its computation. Note that in nonsaturated conditions it is expected that the conditional collision probability is very low. Thus, $EB_{v_r} = \frac{CW_{min,v_r} - 1}{2}$. Moreover, when the AP and STAs queue are near saturation, the $CW_{min,v,u}$ is approximately equal to $n_v \cdot CW_{min,v,d}$.

The AC_VO parameters are assumed to be independent of the best-effort load. Thus, as the best-effort load increases, the system has to reduce the impact of the data traffic over the VoIP calls. There are two parameters that reduce the impact of the AC_BE traffic over the AC_VO performance: (1) increase the AIFS and (2) increase $CW_{min,e}$. Then, the proposed algorithm starts to check iteratively combinations of $AIFSN_e$ and $CW_{min,e}$ until a feasible combination is found. Notice that if a joint combination is chosen and the system load increases, the next combination has to be equal to or greater than the one selected, thus reducing the number of required iterations. At each iteration i, the $AIFSN_e$ and $CW_{min,e}$ values are selected simultaneously (see Table 2.5). For example, for $i = 2$, $AIFSN_e = 4$ and $CW_{min,e} = 32$.

To provide uplink/downlink fairness, with priority for downlink, the algorithm controls the transmission probability of the AP and STAs to be proportional to the number of active flows at each direction. Let $n_{e,d}$ and $n_{e,u}$ be the downlink and uplink elastic active flows, respectively (remember that each STA carries a single flow). Following the same reasoning used for adjusting the CW_{min} of AC_VO, the $CW_{min,e}$ parameter is updated. Now, the relation between $\tau_{e,d}$ (transmission probability of the AP) and $\tau_{e,u}$ (transmission probability of a STA) must be

$$\frac{\tau_{e,d}}{\tau_{e,u}} \approx n_{e,d} \qquad (2.5)$$

As both the AC_BE queues of the AP and STAs are saturated, $CW_{min,e}$ is increased proportionally to the number of downlink elastic flows ($n_{e,d}$). However, to avoid very large $CW_{min,e,u}$ parameters, the decrease of $CW_{min,e,d}$ is also considered, as shown in the pseudo-code algorithm (algorithm 2.1). Moreover, for any AIFSN value, $AIFSN_{e,d} = AIFSN(i) - 1$ and $AIFSN_{e,u} = AIFSN(i)$.

To guarantee the uplink/downlink fairness soft condition, the following criteria should be enforced: $S_{e,d}/n_{e,d} \geq S_{e,u}/n_{e,u}$, where $S_{e,d}$ ($S_{e,u}$) is the aggregate best-effort downlink (uplink) throughput. Then, once the algorithm has selected the combination of $AIFSN_e$ and $CW_{min,e}$ that guarantees that $\rho_{v,d} < \rho_{th}$, the AC_BE $TXOP_{limit}$ at both the AP and STAs can be adjusted to provide the desired uplink/downlink throughput fairness. Remember that the AC_BE $TXOP_{limit}$ at the AP is initially set at eight packets, and only one packet at the STAs. Then, the algorithm starts to increase the $TXOP_{limit,e}$ at STAs and decrease the $TXOP_{limit,e}$ at the AP. The increase of the $TXOP_{limit,e}$ at STAs could have a negative impact on the jitter properties of the AC_VO queue. As an example, if AC_BE_u has assigned a $TXOP_{limit}$ of 54 ms, and fully utilizes its right to transmit eight consecutive packets of 12,000 bits each, a VoIP packet arriving at its queue at the beginning of the TXOP will suffer a delay of at least this value. This effect occurs only in uplink flows because, as mentioned above, the AP can intercept its AC_BE queue even if it has been granted

Table 2.5 Set of Possible $AIFSN_e$ and $CW_{min,e}$ Values

	Iteration					
Parameter	1	2	3	4	5	6
$AIFSN_e$	3	4	5	6	7	8
$CW_{min,e}$	16	32	64	128	256	512

a long TXOP period. However, the above described delays are not expected to be very important because the condition $\rho_{v,d} < \rho_{th}$ implicitly limits the $TXOP_{limit}$ of AC_BE_u to acceptable values for the delay and jitter of the VoIP packets. Moreover, the probability that best-effort packets win a contention against VoIP packets is very low.

Finally, the pseudo-code of the proposed algorithm is shown in algorithm 2.1.

Algorithm 2.1: EDCA* Iterative Algorithm

```
 1. Set the default AC_VO parameters for the AP and STAs.
 2. i=1
 3. while EMBED Equation.DSMT4 and i<7
 4. Pick the CWmin(i) and AIFSN(i) values from Table 2.5
 5. if  EMBED Equation.DSMT4 then
 6. EMBED Equation.DSMT4
 7. EMBED Equation.DSMT4
 8. else
 9. EMBED Equation.DSMT4
10. EMBED Equation.DSMT4
11. end if
12. ComputeModel()
13. i=i+1
14. end while
15. while EMBED Equation.DSMT4 and  EMBED Equation.DSMT4 do
16. EMBED Equation.DSMT4
17. EMBED Equation.DSMT4
18. ComputeModel()
19. end while
20. Update the selected parameters.
```

2.6 Performance Results

The proposed algorithm (EDCA*) is evaluated in a heterogeneous scenario with simultaneous VoIP and best-effort flows. Each VoIP call consists of one uplink and one downlink rigid flow. These calls coexist with uplink and downlink elastic flows. The performance results are compared with the ones obtained by the standard EDCA. The goal of this section is to show the benefits of tuning/adjusting the EDCA parameters based on policy directives to solve the current hotspot impairments for both VoIP calls and TCP flows.

Two scenarios are considered: (1) increase the number of VoIP (n_v, from 1 to 10) calls in the presence of uplink elastic flows, and (2) increase the number of uplink elastic flows ($n_{e,u}$, from 1 to 10) in the presence of VoIP calls. In both cases the number of downlink elastic flows is set to $n_{e,d} = 4$. The traffic parameters are shown in Table 2.1. Through this work the data rate of 2 Mbps has been considered, which is among those defined in the IEEE 802.11b [7] standard.

2.6.1 Capacity for VoIP Calls

In Figure 2.9 the aggregated VoIP throughput in the downlink direction (AP to STAs) is shown. Becauase the AP saturates before the STAs, effectively becoming the bottleneck, only the downlink traffic is depicted. Using EDCA, the system admits six calls before showing symptoms of saturation in the absence of uplink elastic flows. As the number of elastic flows increases, the maximum number of VoIP calls is further decreased. The introduction of the described algorithm EDCA* allows the increasing of the number of calls to seven, thanks to the $CW_{min,v,u}$ adaptation. Moreover, the proposed algorithm is able to guarantee the VoIP capacity independently of the number of active uplink elastic flows. For example, with eight uplink elastic flows, $n_{e,u} = 8$, the VoIP capacity using EDCA reduces to three calls,

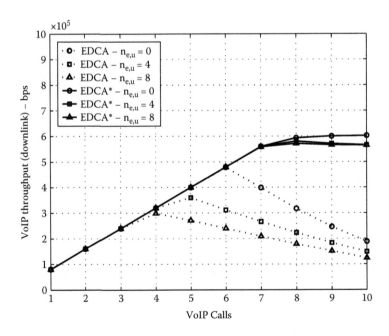

Figure 2.9 VoIP throughput for different $n_{e,u}$ flows.

while using EDCA* it remains equal to seven calls. For both EDCA and EDCA*, it is assumed that the VoIP packets benefit from strict priority in the downlink.

Concerning the best-effort throughput, EDCA* is designed to satisfy two requirements: (1) provide uplink/downlink throughput fairness with higher bandwidth for the downlink when possible, and (2) maximize the aggregate best-effort traffic. As shown in Figure 2.10, the aggregate traffic ($S_E = S_{e,d} + S_{e,u}$) provided by EDCA* is higher than that provided by EDCA, until the system is near saturation. At that point, EDCA* reduces the resources assigned to the AC_BE queue to increase the protection of the VoIP calls. Moreover, in Table 2.6, the throughput fairness index $S = S_{e,d}/S_{e,u}$ for different n_v and $n_{e,u}$ combinations is shown. Note how the use of EDCA results in very low values of the fairness index S as the aggregate throughput is basically the elastic uplink throughput due to the starvation of the AP, caused by both the saturated uplink elastic flows and the prioritization of the VoIP traffic. However, EDCA* is able to guarantee S values near 1 until the AP is saturated by the VoIP traffic. Thus, EDCA* is able to provide equal downlink and uplink elastic throughput, which will mitigate some of the impairments that the TCP downlink/uplink flows suffer [43, 44].

Finally, the AP AC_VO queue utilization is plotted in Figure 2.11. It satisfies the requirement of resulting in lower values than $\rho_{th} = 0.8$ in all cases where it is possible. Without the requirement of uplink/downlink throughput fairness, a better fit to ρ_{th} could be achieved, which would also result in a higher elastic

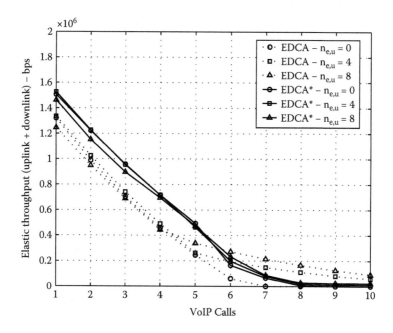

Figure 2.10   Elastic throughput for different $n_{e,u}$ flows.

Table 2.6 Throughput Fairness Index $S = S_{e,d}/S_{e,u}$ Index

	EDCA		EDCA*	
nVoIP	$N_{e,u} = 4$	$N_{e,u} = 8$	$N_{e,u} = 4$	$N_{e,u} = 8$
1	$1.6249 \cdot 10^{-1}$	$6.8284 \cdot 10^{-2}$	$1.0259 \cdot 10^{0}$	$9.8819 \cdot 10^{-1}$
3	$8.4964 \cdot 10^{-2}$	$2.5192 \cdot 10^{-2}$	$1.2709 \cdot 10^{0}$	$7.9418 \cdot 10^{-1}$
5	$7.2297 \cdot 10^{-3}$	$2.9932 \cdot 10^{-5}$	$1.3675 \cdot 10^{0}$	$6.2044 \cdot 10^{-1}$
7	$1.0851 \cdot 10^{-7}$	$3.9602 \cdot 10^{-10}$	$3.0205 \cdot 10^{-1}$	$1.1139 \cdot 10^{-1}$

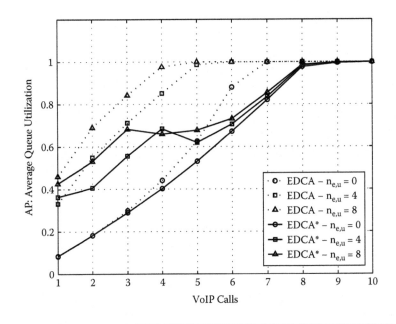

Figure 2.11   AP AC_VO queue utilization for different ne,u $n_{e,u}$ flows.

throughput. However, the considered policy gives priority to fairness over through-put maximization.

2.6.2 Providing Protection to VoIP Calls

Similar conclusions are obtained in the second scenario: EDCA* offers a higher protection for VoIP calls in the presence of data traffic, as well as better uplink/downlink fairness. See the evolution of both uplink/downlink throughput for the

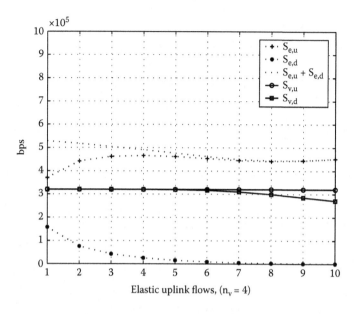

Figure 2.12   Elastic and VoIP throughput for different $n_{e,u}$ (EDCA).

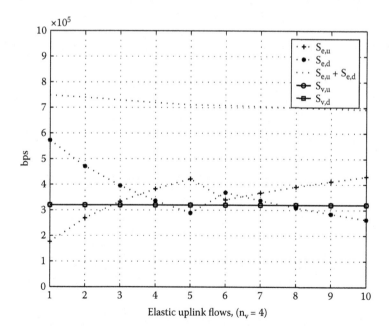

Figure 2.13   Elastic and VoIP throughput for different $n_{e,u}$ (EDCA*).

VoIP calls and elastic flows when EDCA (Figure 2.12) or EDCA* (Figure 2.13) is used. In both cases, the number of VoIP calls is fixed to $n_v = 4$, and the downlink elastic flows to $n_{e,d} = 4$. As expected, using EDCA*, the n_v VoIP calls are not affected by the uplink elastic flows, the throughput fairness is near 1 for all $n_{e,u}$, and the aggregated elastic traffic is increased.

2.7 Conclusions

Future hotspots will have to handle traffic heterogeneity. Traditional data transfers such as HTTP or e-mail will be combined with P2P and VoIP as emerging services. However, the distributed random access protocol of WLAN introduces several performance impairments that reduce the efficient usage of the wireless bandwidth. The basic performance impairment is the starvation of the downlink due to the activity of the STAs, which has major consequences in both TCP and VoIP traffic.

IEEE 802.11e was introduced to deal with those impairments. However, the parameters suggested by EDCA are designed to provide a strict traffic differentiation, which could reduce unnecessarily the performance of best-effort traffic, without solving the existing uplink/downlink unfairness. To solve this situation, a tuning algorithm has been presented that, compared with EDCA: (1) increases the maximum number of feasible VoIP calls, (2) provides uplink/downlink fairness, and (3) increases the throughput for the best-effort active flows.

Moreover, the presented tuning parameters algorithm is based on a set of assumptions/heuristics that make its implementation feasible as the possible state space of the parameters is reduced. This algorithm could be combined with a call admission control to guarantee system stability and optimize the network resources simultaneously.

References

[1] Gartner Dataquest. 2003. Gartner says simplistic focus on hot spot profits misguided, rationales for growth are more complex. http://www.gartner.com/press_releases/pr30june2003a.html (accessed April 2007).

[2] Jiware. 2007. WiFi hotspot finder. http://www.jiwire.com/search-hotspot-locations.htm (accessed April 2007).

[3] C. Na, J. K. Chen, and T. S. Rappaport. 2004. Hotspot traffic statistics and throughput models for several applications. In *IEEE Globecom* 5: 3257–3263, Dallas, TX.

[4] S. Choi, K. Park, and C. K. Kim. 2005. On the performance characteristics of WLANs: Revisited. In *ACM Sigmetrics*, 33: 97–106. Banff, Alberta.

[5] YouTube. 2007. Broadcast yourself. http://www.youtube.com/t/about (accessed April 2007).

[6] Skype. 2007. Skype. http://www.skype.com/share/ (accessed April 2007).

[7] IEEE. 1999. *Wireless LAN medium access control (MAC) and physical layer (PHY) specifications.* ANSI/IEEE Standard 802.11, 1999 edition (revised 2003).

[8] IEEE. 2005. *Wireless LAN medium access control (MAC) and physical layer (PHY) specifications: Medium access control (MAC) quality of service enhancements.* IEEE Standard 802.11e, Amendment.

[9] J. Ha and C.-H. Choi. 2006. TCP fairness for uplink and downlink flows in WLANs. In *IEEE Globecom 2006*, San Francisco.

[10] M. Heusse, F. Rousseau, G. Berger-Sabbatel, and A. Duda. 2003. Performance anomaly of 802.11b. In *IEEE INFOCOM 2003* 2: 836–843, San Francisco.

[11] Q. Ni. 2005. Performance analysis and enhancements for IEEE 802.11e wireless networks. *IEEE Network* 19(4): 21–27.

[12] B. Bellalta, M. Meo, and M. Oliver. 2007. VoIP call admission control in WLANs in presence of elastic traffic. *Journal of Communications Software and Systems* 2(4).

[13] C. Cano, B. Bellalta, and M. Oliver. 2007. Adaptive admission control mechanism for IEEE 802.11e WLANs. In *18th Annual IEEE International Symposium on Personal, Indoor and Mobile Radio Communications (PIMRC'07)*, Athens.

[14] V. A. Siris and G. Stamatakis. 2006. Optimal CWmin selection for achieving proportional fairness in multi-rate 802.11e WLANs: Test-bed implementation and evaluation. In *ACM WinTECH 2006*, Los Angeles.

[15] B. Bellalta and M. Meo. 2006. Call admission control in WLANs. In *Resource, mobility and security management in wireless networks and mobile communications* 3–4. Boca Raton, FL: Auerbach Publications/CRC Press.

[16] G. Bianchi. 2000. Performance analysis of the IEEE 802.11 distributed coordination function. *Journal on Selected Areas in Communications* 18(3).

[17] F. Cali, M. Conti, and E. Gregori. 2000. Dynamic tuning of the IEEE 802.11 protocol to achieve a theoretical throughput limit. *IEEE/ACM Transactions on Networking* 18:785–99.

[18] Y. C. Tay and K. C. Chua. 2001. A capacity analysis for the IEEE 802.11 MAC protocol. *Wireless Networks* 7:159–71.

[19] I. Tinnirello and G. Bianchi. 2005. On the accuracy of some common modeling assumptions for EDCA analysis. In *CITSA*, Orlando, FL.

[20] S. Kuppa and G. R. Dattatreya. 2006. Modeling and analysis of frame aggregation in unsaturated WLANs with finite buffer stations. In *IEEE International Communications Conference (ICC 2006)*, 3: 967–972. Istanbul.

[21] J. W. Robinson and T. S. Randhawa. 2004. Saturation throughput analysis of IEEE 802.11e enhanced distributed coordination function. *Journal on Selected Areas in Communications* 22(5).

[22] D. Malone, K. Duffy, and D. J. Leith. 2005. Modeling the 802.11 distributed coordination function with heterogenous finite load. In *Workshop on Resource Allocation in Wireless Networks*, 15(1): 159–172. Trento, Italy.

[23] P. E. Engelstad and O. N. Østerbø. 2005. Non-saturation and saturation analysis of IEEE 802.11e EDCF with starvation prediction. In *ACM/IEEE MSWIM 2005*, Montreal.

[24] F. Roijers, H. van den Berg, X. Fan, and M. Fleuren. 2006. A performance study on service integration in IEEE 802.11E wireless LANs. *Computer Communications* 29:2621–33.

[25] L. Scalia and I. Tinnirello. 2004. Differentiation mechanisms for heterogeneous traffic integration in IEEE 802.11 networks. In *IEEE BroadNets, Workshop on Broadband Wireless Multimedia*, San José, CA.

[26] S. W. Kim, B.-S. Kim, and Y. Fang. 2005. Downlink and uplink resource allocation in IEEE 802.11 wireless LAN. *IEEE Transactions on Vehicular Technology* 54(1).

[27] J. Jeong, S. Choi, and C. K. Kim. 2005. Achieving weighted fairness between uplink and downlink in IEEE 802.11 DCF-based WLANs. In *IEEE QoS in Heterogeneous Wired/Wireless Networks (QShine'05)* 10 pp., Orlando, FL.

[28] J. Freitag, N. L. S. da Fonseca, and J. F. de Rezende. 2006. Tuning of 802.11e network parameters. *IEEE Communications Letters* 10(8).

[29] G. Hanley, S. Murphy, and L. Murphy. 2005. Adapting WLAN MAC parameters to enhance VoIP call capacity. In *Proceedings of the 8th ACM International Symposium on Modeling, Analysis and Simulation of Wireless and Mobile Systems* 250–2554, Montreal.

[30] F. Cali, M. Conti, and E. Gregori. 2000. Dynamic tuning of the IEEE 802.11 protocol to achieve a theoretical throughput limit. *IEEE/ACM Transactions on Networking* 8(6).

[31] L. Bononi, M. Conti, and E. Gregori. 2004. Runtime optimization of IEEE 802.11 wireless LANs performance. *IEEE Transactions on Parallel and Distributed Systems* 15(1).

[32] H. Ma, X. Li, H. Li, P. Zhang, S. Luo, and C. Yuan. 2004. Dynamic optimization of IEEE 802.11 CSMA/CA based on the number of competing stations. In *IEEE International Conference on Communications (ICC)* 1: 191–195, Paris.

[33] A. Banchs, X. Pérez, M. Radimirsch, and H. J. Stüttgen. 2001. Service differentiation extensions for elastic and real-time traffic in 802.11 wireless LAN. In *IEEE Workshop on High Performance Switching and Routing (HPSR 2001 245–249)*, Dallas, TX.

[34] L. Romdhani, Q. Ni, and T. Turletti. 2003. Adaptive EDCF: Enhanced service differentiation for IEEE 802.11 wireless ad-hoc networks, New Orleans.

[35] G. R. Cantieni, Q. Ni, C. Barakat, and T. Turletti. 2005. Performance analysis under finite load and improvements for multirate 802.11. *Elsevier Computer Communications Journal* 28:1095–109.

[36] A. Banchs, X. Pérez-Costa, and D. Qiao. 2003. Providing throughput guarantees in IEEE 802.11e wireless LANs. In *ITC Specialist on Providing QoS in Heterogeneous Environments Seminar*, Berlin.

[37] K. Medepalli and F. A. Tobagi. 2006. On optimization of CSMA/CA based wireless LANs: Part I: Impact of exponential backoff. In *IEEE International Conference on Communications (ICC)* 5: 2089–2094, Istanbul.

[38] A. Ksentini, A. Guéroui, and M. Naimi. 2005. Adaptive transmission opportunity with admission control for IEEE 802.11e networks. In *ACM/IEEE MSWIM 2005* 239–241, Montreal.

[39] D. Pong and T. Moors. 2003. Call admission control for IEEE 802.11 contention access mechanism. In *IEEE Globecom 2003* 1: 174–178, San Francisco.

[40] J. Galtier. 2004. Optimizing the IEEE 802.11b performance using slow congestion window decrease. In *Proceedings of the 16th ITC Specialist Seminar on Performance Evaluation of Wireless and Mobile Systems* 165–176, Antwerpen, Belgium.

[41] C. Casetti and C.-F. Chiasserini. 2004. Improving fairness and throughput for voice traffic in 802.11e EDCA. *IEEE PIMRC 2004*, Barcelona.

[42] P.-Y. Wu, Y.-C. Tseng, and H. Lee. 2005. Design of QoS and admission control for VoIP services over IEEE 802.11e WLANs. In *National Computer Symposium*, Taiwan.

[43] S. Pilosof, R. Ramjee, D. Raz, Y. Shavitt, and P. Sinha. 2003. Understanding TCP fairness over wireless LAN. In *IEEE INFOCOM 2003* 2: 863–872, San Francisco.

[44] D. J. Leith and P. Clifford. 2005. Using the 802.11e EDCF to achieve TCP upload fairness over WLAN links. In *Modeling and Optimization in Mobile, Ad Hoc, and Wireless Networks (WiOpt'05)* 109–118, Riva del Garda, Italy.

Chapter 3

QoS for Multimedia Streaming Applications over IEEE 802.11b and 802.11e WLANs

---

Nicola Cranley and Mark Davis

Contents

3.1 Introduction

Streaming multimedia over wireless networks is becoming an increasingly important service [1]. This trend includes the deployment of wireless local area networks (WLANs) that enable users to access various services, including those that distribute rich media content anywhere, anytime, and from any device, e.g., in-home wireless entertainment systems. There are many performance-related issues associated with the delivery of time-sensitive multimedia content using current IEEE 802.11 WLAN standards. Among the most significant are low delivery rates, high error rates, contention between stations for access to the medium, backoff mechanisms, collisions, signal attenuation with distance, signal interference, etc. Multimedia applications, in particular, impose onerous resource requirements on bandwidth-constrained WLANs. Moreover, it is difficult to provide quality of service (QoS) in WLANs as the capacity of the network also varies with the offered load [2].

Providing QoS is difficult because different users, service providers, network administrators, and applications have diverse and sometimes conflicting QoS requirements [3]. For real-time multimedia applications, packet loss and packets dropped due to excessive delay are the primary factors affecting user-perceived quality. Real-time multimedia is particularly sensitive to delay, as it has a strict bounded end-to-end delay constraint. Every multimedia packet must arrive at the client before its playout time with enough time to decode and display the contents of the packet. For video streams, the delay incurred transmitting the entire video frame from the sender to the client is of particular importance. The loss rates incurred due to packets being delayed past their playout time are heavily dependent on the delay constraint imposed on the video stream. Video streaming applications typically impose an upper limit on the tolerable packet loss. Specifically, the packet loss ratio is required to be kept below a threshold to achieve acceptable visual quality. For example, a large packet loss ratio can result from network congestion, causing severe degradation of multimedia quality. Although WLANs allow for packet retransmissions in the event of an unsuccessful transmission attempt, the retransmitted packet must arrive before its playout time or within a specified delay constraint. If the packet arrives too late for its playout time, the packet is useless and effectively lost.

In this chapter we shall demonstrate the challenges of providing QoS for video streaming applications over IEEE 802.11b and how such challenges can be met through the QoS-enabling features of IEEE 802.11e. The remainder of this chapter is structured as follows. In Section 3.2 we shall briefly describe the operation of IEEE 802.11b and IEEE 802.11e. Section 3.3 provides an overview of multimedia streaming applications. In Section 3.4 we show that video streaming applications exhibit a sawtooth delay characteristic over WLAN. In Section 3.5 we experimentally demonstrate the two primary sources of congestion in WLANs. The first is where the AP becomes saturated due to a heavy downlink load, which results in packets being dropped from its transmission buffer and manifests itself as bursty losses and increased delays. The second case is where there are a large number of wireless stations contending for access to the medium, and this results in an increased number of deferrals, retransmissions, and collisions on the WLAN medium. In Section 3.6 we show how the delivery of video streaming applications can be improved by appropriately tuning the TXOP limit parameter in IEEE 802.11e. Finally, we present some conclusions and directions for future work.

3.2 Overview of IEEE 802.11b and 802.11e

3.2.1 IEEE 802.11b

The IEEE 802.11b standard is currently the most popular and widely deployed wireless LAN (WLAN) technology. The IEEE 802.11b operates in the unlicensed industrial, scientific, and medical (ISM) band at 2.4 GHz and supports a mandatory bit rate of 1 Mbps and an optional higher rate of 2 Mbps. In September 1999 the Institute of Electrical and Electronics Engineers (IEEE) approved the HR, or "high rate" extension, to the standard, known as IEEE 802.11b, which supports data rates up to 11 Mbps. The WLAN standard uses the 802 Logical Link Control (LLC) protocol but provides an independent physical layer (PHY) and Medium Access Control (MAC) sublayer specification. There are two modes of operation in WLAN, the Distributed Coordination Function (DCF) and the Point Coordination Function (PCF). Neither DCF nor PCF provides service differentiation mechanisms that can be used to ensure QoS guarantees such as bounded delays or loss or throughput constraints.

3.2.1.1 DCF

The basic access scheme used in 802.11 WLANs is the Distributed Coordination Function (DCF). Stations (STAs) can access the medium without the need for a centralized controller using an access mechanism known as Carrier Sense Multiple Access with Collision Avoidance (CSMA/CA). This allows for asynchronous data transfer on a best-effort basis where all STAs must contend with each other to access the medium to transmit their data. CSMA/CA is a "listen before talk" access protocol whereby any STA wishing to transmit must first use the carrier sense mechanism to determine whether the medium is busy or idle. If the medium is busy, the STA defers its transmission until the medium has been idle for a period of time equal to Distributed Inter-frame Soace (DIFS) (or Extended Inter-frame Space (EIFS) in the case of an incorrectly received frame). The deferral process uses a collision avoidance mechanism where the STA randomly selects a backoff counter (BC) value in units of time slots (TSs) (i.e., BC*TS, where each TS is 20 μs) for the contention window (CW) that is between [0, CW], where CW is initially set to a CWmin value that is doubled when transmission fails up to the maximum value defined by CWmax. In IEEE 802.11b WLAN the CWmin is 31 and CWmax is 1,023. The BC is decremented when the medium is idle, paused when the medium is sensed as busy, and restarted when the medium is sensed idle again for a period of time that is at least DIFS (or EIFS as appropriate). When the BC reaches zero, the STA can initiate the transmission of its frame. In DCF all STAs have equal probability of gaining access to the medium and share it according to equal data frame rate and not according to equal throughput. When multiple STAs are deferring and go into random backoff, the STA selecting the smallest BC value will win the right to transmit. If two or more STAs choose the same BC value, this will lead to a collision whereby the STAs involved will transmit their frames at the same time. To resolve collisions between STAs, an exponential backoff scheme is adopted whereby the size of the CW is doubled after each unsuccessful transmission.

Packet priorities are implemented by defining three different-length interframe spaces (IFSs) between the frame transmissions as shown in Figure 3.1. The IFS intervals are mandatory periods of idle time on the medium. The 802.11 standard defines four different IFS intervals as follows:

- Short interframe space (SIFS): Used for the highest-priority transmissions (i.e., control frames), such as ACK and RTS/CTS frames. In 802.11b, SIFS = 10 μs.

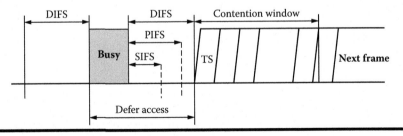

Figure 3.1   IEEE 802.11b interframe spaces.

- PCF interframe space (PIFS): Used by the Point Coordination Function (PCF) during contention free operation. STAs with data to transmit in the contention free period can transmit after PIFS has elapsed and pre-empt any contention-based traffic. In 802.11b, PIFS = 30 μs.
- DCF interframe space (DIFS): The minimum idle time for contention-based (i.e., DCF) services. After this interval has expired, any DCF mode frames can be transmitted asynchronously according to the CSMA backoff mechanism. DIFS is determined as SIFS + 2*TS = 50 μs.
- Extended interframe space (EIFS): Used to recover from a failed transmission attempt. It is derived from the SIFS, DIFS, and time required to transmit an ACK frame at the basic rate of 1 Mbps.

3.2.1.2 PCF

The Point Coordination Function (PCF) supports prioritized access by employing a contention free service. The point coordinator (PC) periodically sends a beacon frame to broadcast network identification and management parameters specific to the wireless network. PCF splits the time into a contention free period (CFP) and a contention period (CP). Only STAs polled by the PC may transmit during the CFP. The CFP ends after the time announced by the beacon frame or by a CF end frame. Although PCF can offer some priority access, it cannot differentiate between traffic sources with time-sensitive data. Furthermore, the start time and duration of the CFP vary because the PC must contend with other STAs to gain control of the medium.

3.2.2 IEEE 802.11e

A significant limitation of IEEE 802.11b is its inability to enable QoS or take into consideration the characteristics and performance requirements of the traffic. DCF provides channel access with equal probabilities to all stations contending for the channel access in a distributed manner regardless of the requirements of the traffic. The IEEE 802.11e QoS MAC enhancement standard enables traffic differentiation by allowing for up to four different transmit queues with different access priorities [4], allowing the AP to provide differentiated service to different applications and enabling them to meet their target QoS requirements.

3.2.2.1 EDCA

Enhanced Distributed Channel Access (EDCA) is designed to provide differentiated, distributed channel accesses. EDCA can be used to provide eight different levels of priority (from 0 to 7) by enhancing the DCF. EDCA is not a separate coordination function. Rather, it is a part of a single coordination function, called the hybrid controller (HC) of the 802.11e MAC. The HC combines the aspects of both DCF and PCF. The 802.11e standard defines four AC queues into which different traffic streams can be directed: voice (VO), video (VI), best effort (BE), and background (BK), as shown in Figure 3.2. Each frame arriving at the MAC with a priority is mapped into a particular AC.

Each Access Categories (AC) is configured with the EDCA parameters: AIFS[AC], CWmin[AC], CWmax[AC], and TXOP[AC]. The duration of AIFS[AC] is determined by the AIFSN[AC]. In 802.11b the duration of DIFS had an AIFSN[AC] of at least 2. In 802.11e the duration of AIFS[AC] is determined by SIFS + AIFSN[AC]*TS. The smaller the AIFSN[AC], the higher the medium access priority. The backoff period of each AC is chosen according to a uniform distribution over the interval [0, CW[AC]]. The CW size is initially assigned a CWmin value that is doubled when transmission fails up to the maximum value defined by CWmax.

Each AC behaves as a single enhanced DCF contending entity where each AC has its own EDCA parameters and maintains its own backoff counter (BC). When two or more competing ACs finish the backoff process at the same time, the collision is handled in a virtual manner. The frame from the highest-priority AC is chosen and transmitted while the lower-priority ACs perform a backoff with increased CW values. The EDCA parameters can be used to differentiate the channel access among different priority traffic. Smaller AIFSN and CWmin values reduce the channel access delay and provide a greater capacity share for the AC. However, using smaller values of CWmin increases the probability of collisions. The EDCA parameters are announced by the AP via beacon frames and can be dynamically adapted to meet the traffic requirements and network load conditions.

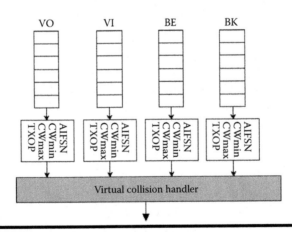

Figure 3.2   IEEE 802.11e access categories.

3.2.2.2 TXOP

The IEEE 802.11e standard also defines a transmission opportunity (TXOP) as the interval of time during which an AC has the right to initiate transmissions without having to recontend for access. During an EDCA TXOP, an AC is allowed to transmit multiple MAC Protocol Data Units (MPDUs) from the same AC with a SIFS time gap between an ACK and the subsequent frame transmission [5]. The duration of the TXOP is determined by the value of the TXOP limit parameter. Figure 3.3 shows the transmission of two data frames during an EDCA TXOP, where the whole transmission time for two data and ACK frames is less than the EDCA TXOP limit announced by the AP. The TXOP limit parameter is an integer value in the range (0, 255) and gives the duration of the TXOP interval in units of 32 µs. If the calculated TXOP duration requested is not a factor of 32 µs, that value is rounded up to the next higher integer that is a factor of 32 µs. The maximum allowable TXOP limit is 8,160 µs, with a default value of 3,008 µs [5].

$$TXOP_N = \lceil N_P * T_P \rceil \tag{3.1}$$

When there are no more packets to be sent during the TXOP interval and the channel becomes idle again, the 802.11 hybrid controller (HC) may sense the channel and reclaim the channel after a duration of PIFS after the TXOP.

3.3   Introduction to Wireless Multimedia Streaming

3.3.1 Multimedia Streaming Networks

Video streaming is a server/client technology that allows multimedia data to be transmitted and consumed. Streaming applications include e-learning, video conferencing, video on demand, etc. The main goal of streaming applications is that the stream should arrive and play out continuously without interruption; however, this is constrained by fluctuations in network conditions. An adaptive streaming server keeps track of the network conditions and adapts the quality of the stream to minimize interruptions and stalling. Real-time streaming can be delivered by either

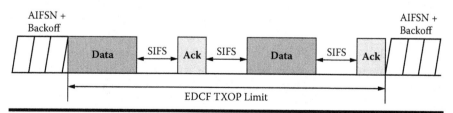

Figure 3.3   IEEE 802.11e TXOP facility.

peer-to-peer (unicast) or broadcast (multicast). There are two types of real-time streaming services [6, 7], on-demand or live streaming. In addition to the different types of streaming, there are a large and diverse number of variables that must be taken into consideration when evaluating the performance of such applications. Such variables include:

■ The actual content and complexity of the content streamed, which in turn affects the efficiency of the encoder to compress the stream. For example, if two different video clips were encoded using the exact same encoding configuration, they would have very different bit rate variations over time.
■ The compression scheme used, that is, different compression schemes have differing levels of efficiency. For example, a 512 Kbps Motion Picture Experts Group (MPEG)-2 stream will have very different characteristics from a 512 kbps MPEG-4 stream.
■ The encoding configuration [2]. There could be any number of possible encoding configurations possible, such as the error resilience, frame rate, I frame rate, quantization parameter, target bit rate (if any) supplied, and target stream type, i.e., variable bit rate (VBR), constant bit rate (CBR), or near CBR.
■ If the file to be streamed is MP4 or .3gp, then a hint track must be prepared that indicates to the server how the content should be streamed.
■ The streaming server used, the rate control adaptation algorithm used, and the methods of bit rate adaptation used by the server [8, 9].

3.3.2 MPEG-4

MPEG-4 dramatically advances audio and video compression, enabling the distribution of content and services from low bandwidths to high-definition quality across broadcast, broadband, wireless, and packaged media [10]. MPEG-4 decomposes a scene into media objects, each with its own audio and video track that will vary over time. The visual part of a media object is known as video object planes (VOPs). In this chapter we consider only rectangular-shaped VOPs that correspond to the entire video image and shall refer to them as video frames throughout the remainder of this chapter. In the MPEG-4 standard, there are a number of profiles that determine the required capabilities of the player to decode and play out the content. The purpose of these profiles is that a codec only needs to implement a subset of the MPEG-4 standard while maintaining interworking with other MPEG-4 devices built to the same profiles. The most widely used MPEG-4 visual profiles are the MPEG-4 Simple Profile (SP) and the MPEG-4 Advanced Simple Profile (ASP), and are part of the nonscalable subset of visual profiles. The main difference between MPEG-4 SP and ASP is that SP contains only I and P frames, whereas ASP contains I, P, and B frames.

MP4 files comprise a hierarchy of data structures called atoms, and each atom has a header, which includes its size and type [11, 12]. A parent atom is of type moov and contains the following child atoms: mvhd (the movie header), a series of trak atoms (the media tracks and hint tracks), and a movie user data atom udta. A trak represents a single independent data stream, and an MP4 file may contain any number of video, audio, hint, binary format for scenes (BIFS), or object descriptor (OD) tracks.

3.3.3 Hint Tracks for Streaming

Within an MP4 file, each video and audio track must have its own associated hint track. Hint tracks are used to support streaming by a server and indicate how the server should packetize the data. As with MP4 streaming, .3gp files use the hint track mechanism for streaming the content, although in .3gp files the BIFS and OD tracks are optional and can be ignored. Streaming media requires that the media be sent to the client as quickly as possible with strict delay requirements. Hint tracks allow a server to stream media files without requiring the server to understand media types, codecs, or packing. Each track in a media file is sent as a separate stream, and the instructions for packetizing each stream are contained in a corresponding hint track [13]. Each sample in a hint track tells the server how to optimally packetize a specific amount of media data. The hint track sample contains any data needed to build a packet header of the correct type, and also contains a pointer to the block of media data that belongs in the packet. For each media track to be streamed there must be at least one hint track. It is possible to create multiple hint tracks for any track, each optimized for streaming over different networks. Hint tracks have the same structure as media tracks and are atoms of type trak. Hint samples are protocol specific by specifying the protocol to be used and providing the necessary parameters for the server. The stsd child atom contains transport-related information about the hint track samples. It specifies the data format (currently only Real-Time Transport Protocol [RTP] data format is defined), the RTP timescale, and the maximum packet size in bytes (maximum transmission unit [MTU]). The hint track MTU setting means that the packet size will not exceed the MTU size.

Hint track settings are required for streaming MP4 and .3gp multimedia files and are particularly important for audio streaming because multiple audio samples can be packetized into one packet. In general, most video frames are quite large, and so at most one video frame can be packetized into a single 1024 B packet. If the video frame is larger than the packet, several packets are required to send the video frame, resulting in a group of packets the size of the hint track MTU setting and a smaller packet containing the remaining information.

3.3.4 Experimental Test Bed

3.3.4.1 Video Content Preparation

In the experiments described in this chapter, the video content was encoded using the commercially available X4Live MPEG-4 encoder from Dicas. DH is an extract from the film *Die Hard*, DS is an extract from the film *Don't Say a Word*, EL is an extract from the animation film *The Road to Eldorado*, FM is an extract from the film *Family Man*, and finally JR is an extract from the film *Jurassic Park*. It is necessary to repeat the experiments for a number of different video content types because the characteristics of the streamed video have a direct impact on its performance in the network. Each video clip has its own unique signature of scene changes and transitions that affect the time-varying bit rate of the video stream. Animated videos are particularly challenging for encoders because they generally consist of line art and, as such, have greater spatial detail.

The video content was encoded as MPEG-4 SP and ASP with a target bit rate of 1 Mbps using two-pass encoding, Common Interchange Format Resolution (CIF352 × 288), frame rate of 25 fps, and a refresh rate of one I frame every ten frames. The video clips were prepared for streaming by creating an associated hint track using MP4Creator from MPEG4IP. The hint track tells the server how to optimally packetize a specific amount of media data. The hint track MTU setting means that the packet size will not exceed the MTU size. The mean packet sizes for video with hint track settings of 1024B and 512B are 912B and 468B, respectively.

3.3.4.2 Delay Measurement

Because delay has a significant impact on the quality of the multimedia streaming application, in this chapter we observe performance-related issues by measuring the end-to-end delay for video streaming applications. To measure the end-to-end delay, both the video client and server were configured with the packet monitoring tool WinDump [14] to record the packets sent and received. The delay is measured as the difference between the time at which the packet was received at the link layer of the client and the time it was transmitted at the link layer of the sender. The clocks of both the client and server are synchronized before each test using Net-Time [15]. However, in spite of the initial clock synchronization, there was a noticeable clock skew observed in the delay measurements, and this was subsequently removed using Paxson's algorithm as described in [16].

3.3.4.3 Streaming Server

There are two open-source streaming servers available, Helix from Real [17, 18] and Darwin Streaming Server (DSS) from Apple [19, 20]. DSS is an open-source, standards-based streaming server that is compliant to MPEG-4 standard profiles,

Internet Streaming Media Alliance (ISMA) streaming standards, and all Internet Engineering Task Force (IETF) protocols. The DSS system is a client/server architecture where both client and server consist of the Real-Time Transport Control Protocol (RTP)/User Datagram Protocol (UDP)/Internet Protocol (IP) stack with RTCP/UDP/IP to relay feedback messages between the client and server. The server is configured with an RTSP timeout of 180 s and RTP timeout of 120 s. The client can be any QuickTime Player or any player that is capable of playing out ISMA-compliant MPEG-4 or .3pg content. The client connects to and interacts with the server via Real-Time Streaming Protocol (RTSP) to establish a unicast video streaming session. In addition, RTSP can be used by the client as a network remote control to fast forward, rewind, or skip to any location in a pre-encoded video clip with a 3 s prebuffering delay.

3.3.4.4 IEEE 802.11 WLAN Equipment

The Cisco Aironet 1200 access point (AP) with firmware version IOS 12.3(8)JA is used for all experimental work described in this chapter. The AP can be configured to operate in IEEE 80.11b or IEEE 802.11e/WMM mode [21]. The AP is configured with a QoS policy where the Differentiated Services Code Point (DSCP) values in the IP header are used to apply a particular class of service (CoS) to the incoming packets. For IEEE 802.11b mode all DSCP values are mapped to a single CoS, which is then subsequently mapped to a single transmission buffer with the standard settings for DIFS, CWmin, and CWmax. For IEEE 802.11e/WMM mode different DSCP values are mapped to a different CoS. Each CoS is then mapped to a particular AC where the CWmin, CWmax, AIFSN, and TXOP limit parameters can be configured.

3.4 Characteristics of Video Streaming over WLAN

There are many performance-related issues associated with the delivery of time-sensitive multimedia content using current IEEE 802.11 standards. Among the most significant are low delivery rates (e.g., theoretically up to 11 Mbps for IEEE 802.11b, but in practice only a maximum throughput of approximately 6 Mbps can be achieved due to the protocol overhead), high error rates due to media characteristics, contention between stations for access to the medium, backoff mechanisms, collisions, signal attenuation with distance, signal interference, etc. Multimedia applications, in particular, impose significant resource requirements on bandwidth-constrained WLANs [2, 3]. Under these conditions it is difficult to provide any QoS guarantees. Every multimedia packet must arrive at the client before its playout time with enough time to decode and display the packet. If the multimedia packet does not arrive on time, the packet is effectively lost. In a WLAN environment, lost or corrupted packets are repeatedly retransmitted until either the retransmitted packet is successfully ACKed by the receiving station or the retransmission limit has been reached. If a packet has expired, there is no need for it to continue along its path to the client because its contents will be worthless when it arrives. The end-to-end delay plays a crucial role in the performance of real-time and near-real-time streaming applications.

For video streaming applications, not only is the end-to-end delay important, but also the delay incurred transmitting the entire video frame from the sender to the client. Video streaming is often described as "bursty," and this can be attributed to the frame-based nature of video. Video frames are transmitted with a particular frame rate and are generally large, often exceeding the MTU of the network, which results in a number of packets being transmitted in a burst for each video frame. A video frame cannot be decoded or played out at the client until all or most of the constituent video packets for the frame are received correctly and on time. Although, error resilient–encoded video and systems that include error concealment techniques allow for a certain degree of loss tolerance [6], the ability of these schemes to conceal bursty and high loss rates is limited.

In a WLAN environment, the bursty behavior of video traffic has a sawtooth-like delay characteristic. Consider a burst of packets corresponding to a video frame arriving at the AP. The arrival rate of the burst of packets is high, and typically these packets are queued consecutively in the AP's transmission buffer. For each packet in the queue, the AP must gain access to the medium by deferring to a busy medium and decrementing its MAC backoff counter between packet transmissions. This process occurs for each packet in the queue at the AP, causing the end-to-end delay to transmit the entire video frame and to vary with a sawtooth characteristic.

To describe this sawtooth characteristic we have defined the Interpacket Delay (IPD) as the difference in the measured delay between consecutive packets within a burst for a video frame at the receiver. In our analysis, we focus on the video Frame Transmission Delay (FTD), i.e., the end-to-end delay incurred in transmitting the entire video frame. The video frame delay is related to the number of packets required to transmit the entire video frame. The FTD is measured as the sum of the IPD for each packet required to transmit the entire video frame where the frame consists of N packets.

$$FTD = \sum_{i=2}^{N} IPD_i \qquad (3.2)$$

The QFTD is the FTD plus the transmission delay (D_1) for the first packet of the video frame to reach the client:

$$QFTD = D_1 + FTD \qquad (3.3)$$

Figure 3.4 shows the relationship between the IPD, FTD, and QFTD for a single video frame. The sawtooth delay characteristic was measured experimentally

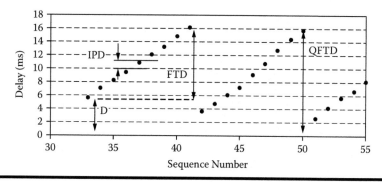

Figure 3.4 Relationship among IPD, FTD, and QFTD.

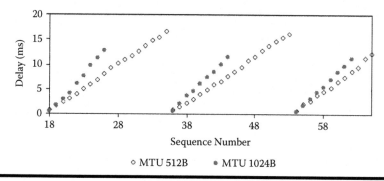

Figure 3.5 Sawtooth delay characteristic for three video frames.

as shown in Figure 3.5. This graph shows packet delay measured for equally sized video frames streamed using a hint track MTU of 1024 and 512 B. It can be clearly seen that when using a hint track MTU setting of 512 B, the delay is much greater to send one complete frame despite the fact that the IPD is less. However, when using a larger hint track MTU setting, it takes more time to send each individual packet, but because there are fewer packets in the video frame, overall it takes less time to send the complete video frame.

For 1024 B-sized packets, the mean IPD is 1.34 ms and varies in the range (1.0 ms, 1.66 ms). For 512B-sized packets, the mean IPD is 0.96 ms and varies in the range (0.64 ms, 1.28 ms). This IPD delay range includes the DIFS and SIFS intervals, data transmission time, including the MAC acknowledgment, as well as the randomly chosen backoff counter values of the 802.11b MAC mechanisms contention windows in the range 0–31 [22]. This can be seen in Figure 3.6, where there is an upper plateau with 32 spikes corresponding to each of the possible 32 backoff counter values with a secondary lower plateau that corresponds to the proportion of packets that were required to be retransmitted through a subsequent doubling of the contention window under the exponential binary backoff mechanism employed in the 802.11b MAC.

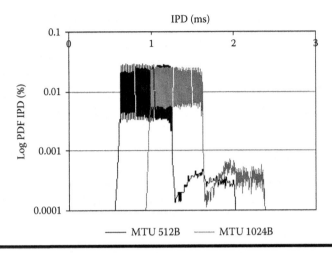

Figure 3.6 PDF of IPD for video packets.

3.5   Multimedia Streaming over IEEE 802.11b

In IEEE 802.11b WLANs, the AP is a critical component that determines the performance of the network as it carries all of the downlink transmissions to wireless clients and is usually where congestion is most likely to occur. There are two primary sources of congestion in WLANs. The first is where the AP becomes saturated due to a heavy downlink load, which results in packets being dropped from its transmission buffer and manifests itself as bursty losses and increased delays [23]. In contrast, the second case is where there are a large number of wireless stations contending for access to the medium, and this results in an increased number of deferrals, retransmissions, and collisions on the WLAN medium. The impact of this manifests itself as significantly increased packet delays and loss. For video streaming applications, this increased delay results in a greater number of packets arriving at the player too late for playout and being effectively lost.

3.5.1 AP Saturation

The end-to-end delay for video streaming applications over WLAN is affected by the video frame rate, frame size, packet rate, and packet size. In particular, the performance is affected when there is a downlink background traffic load. There is a critical threshold load value that is related to the packet size and offered load. Once this threshold background load value has been exceeded, the video streaming application experiences excessive delays.

Given the large number of encoding parameters that can be varied while preparing the video content for streaming over the network, only the frame rate of the video and the size of the video frames were varied. As a result, the mean transmitted bit rate varies in an additive increase proportional decrease (AIPD) manner and reaches a maximum bit rate of 2.1 Mbps after 1,700 s, as shown in Figure 3.7. The mean video frame sizes were varied from 3.1, 6.1, and 9.2 kb every 100 s, and the frame rate was increased from 10 to 30 fps in steps of 5 fps every 300 s. As the mean video frame size was increased, the number of packets required to transmit the video frame, n_{VID}, also increased. The video was streamed with a hint track setting, S_{VID}, of 1024B. When using S_{VID} of 1024B, n_{VID} varied from {3, 6, 9} packets required to transmit each video frame.

In this section we shall show the combined effect of the background traffic load and the packet size on the end-to-end delay of the video stream. The traffic generator MGEN was used to create a source of background traffic with loads of 1, 3, and 5, respectively, using a packet size, S_{BAK}, of 512B and 1024 B. This background traffic is streamed via the wired network to the AP and received by a sink station in the WLAN.

In the best-case scenario, once the AP has serviced all the packets relating to a video frame, there remains unused or idle times when the AP can transmit other traffic before the next video frame arrives. In an ideal situation, the background traffic is perfectly interleaved with the video stream, that is, after the AP has serviced all packets in the queue relating to the video frame, the unused time between sending the video frame and the arrival of the next video frame is given to the background traffic.

This load represents the ideal maximum background traffic load that can be serviced. For example, given a video stream encoded with a video frame rate of X_{FPS}, where the number of packets required to transmit the entire video frame is n_{VID}, and an IPD, IPD_{VID}, that is related to the packet size, S_{VID}, it takes the AP ($n_{VID} \times IPD_{VID}$) ms to send the video frame. The idle time between sending the video frame and the arrival of the next video frame can then be used to service the background traffic. The time interval for background traffic in this case is therefore:

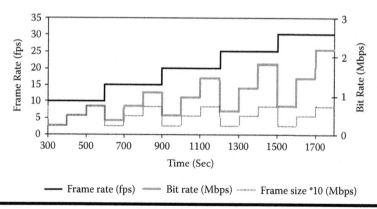

Figure 3.7 Offered video traffic characteristics.

$$BakInterval = \left(1\,sec\!\!\left/X_{FPS}\right.\right) - \left(n_{VID} \times IPD_{VID}\right) \tag{3.4}$$

During this interval a number of background traffic packets n_{BAK} can be sent. However, this varies with the packet size S_{BAK}, of the background traffic, which in turn affects the mean IPD per background packet, IPD_{BAK},

$$n_{BAK} = \left\lfloor BakInterval\!\!\left/IPD_{BAK}\right.\right\rfloor \tag{3.5}$$

where is the floor function. Given that for every video frame there is a corresponding interval during which the background packet can be sent, this results in a total ideal background traffic load:

$$TotalIdealLoad = n_{BAK} \times IPD_{BAK} \times 8 \times Xfps \tag{3.6}$$

The ideal background traffic load represents the maximum load that can be supported in such a way so as to not negatively impact on the video stream where the video is transmitted with a variety of frame rates and packet sizes.

Let us consider an interval ST of 1,000 ms at the AP. We can predict that excessive delays will occur when the total service time for the video stream ST_{VID} and the total service time for the background traffic load ST_{BAK} exceed the interval of interest. For example, the total service time at the AP ST_{VID} is 302 ms to send video with X_{FPS} of 25 fps, n_{VID} of 9 packets per video frame, and a mean IPD_{VID} of 1.34 ms. The total service time at the AP, ST_{BAK}, is 703 ms to send a background traffic load of 3 Mbps with S_{BAK} of 512B. Thus, the total service time for all offered traffic, ST_{TOTAL} (i.e., $ST_{VID} + ST_{BAK}$), exceeds ST; then it can be expected that excessive delays will be experienced by both the video stream and the background traffic. In this work, there are no other stations contending for access to the medium, which gives the AP full use of the service time during the interval of interest. However, when there are other stations contending for access to the WLAN medium, the service time at the AP is reduced.

Using this approach, we compare the predicted intervals of excessive delay with observed intervals of excessive delay. The predicted intervals of excessive delays are defined as those intervals where the total service time for all offered traffic, ST_{TOTAL}, exceeds ST. It can be seen that in all cases, when the total combined bit rate of the video stream and background traffic exceeds the threshold bit rate value, the video stream experiences excessive delays. However, once the bit rate of the video falls below this threshold value, the delay of the video stream returns to a low value.

It was observed that with a background traffic load of 1 Mbps, regardless of the packet size of both the video stream and the background traffic, the video stream was unaffected. However, as the background traffic load is increased to 3 Mbps, depending on the packet sizes of both the background traffic and the video traffic, the video stream experiences excessive delays when the total load reaches some threshold value.

The experimentally observed end-to-end delays are shown on the left-hand side of Figure 3.8, while the predicted intervals of excessive delay are shown on the right-hand side of Figure 3.8. The observed results show the offered bit rate of the video stream indicated by a thick gray line and the end-to-end delay indicated by a thin black line. It can be seen that the bit rate of the video increases due to an increasing video frame rate and video frames size; once the bit rate of the video exceeds a certain threshold bit rate value, the video stream experiences large delays. Once the video bit rate falls below this threshold value, the delay returns to a low value. This threshold bit rate value varies with the packet size of both the video and background traffic. It can be seen that when using a large packet size for the background traffic, the threshold bit rate of the video is higher. Figure 3.8[(a), (c)] show the case for S_{VID} of 1,024B with a background load of 3 Mbps; the bit rate threshold of the video stream is 1.84 and 2.21 Mbps when using S_{BAK} of 512B and 1024B, respectively. With a background load of 5 Mbps the video stream cannot be supported at all using S_{BAK} of 512B, as shown in Figure 3.8(c); however, when using S_{BAK} of 1,024B, the threshold value is 0.98 Mbps, as shown in Figure 3.8(d). In Figure 3.8(d) at interval 1,100–1,200 ms it can be seen that the delay is gradually increasing over time as the AP can service more packets than are arriving, allowing the AP time to clear the backlog of queued packets in the transmission buffer, resulting in the number of queued packets to slowly decrease over time. It can be seen that there is a good correlation between the predicted and observed intervals of excessive delay. This basic mechanism provides a simple and convenient means to determine the offered loads that will affect the video stream causing excessive delays and, as a consequence, resulting in poor QoS.

3.5.2 Contention

The IEEE 802.11b MAC mechanism is considered to be fair in the sense that all stations contending for access to the medium have an equal probability of winning a transmission opportunity. However, although stations enjoy the same probability of winning access opportunities, they do not share the bandwidth equally, as this depends on the size of the transmitted packet. When a station has gained access to the medium, all other stations must pause their backoff process until the medium becomes idle again. In this way, contention for access increases the end-to-end delay for each packet as stations are forced to wait longer for a transmission opportunity. As the level of contention increases, it takes longer to win a transmission opportunity, and consequently, the maximum achievable service rate is reduced, which increases the probability of buffer overflow. In this section we shall demonstrate the effect of contention on a single downlink video stream by varying the number of contending stations. In this way we can affect the service rate of the buffer and

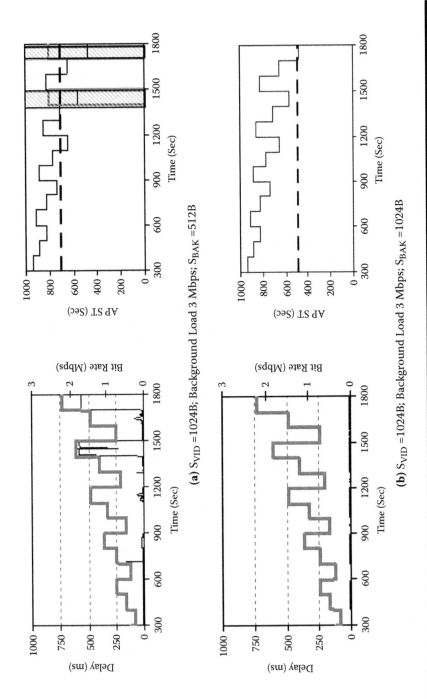

(a) S_{VID} =1024B; Background Load 3 Mbps; S_{BAK} =512B

(b) S_{VID} =1024B; Background Load 3 Mbps; S_{BAK} =1024B

Figure 3.8 Comparison of observed and predicted intervals of excessive delay.

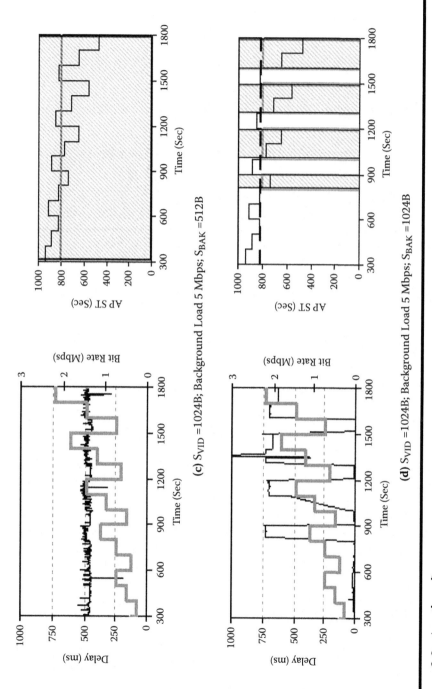

(c) $S_{VID} = 1024B$; Background Load 5 Mbps; $S_{BAK} = 512B$

(d) $S_{VID} = 1024B$; Background Load 5 Mbps; $S_{BAK} = 1024B$

Figure 3.8   (continued)

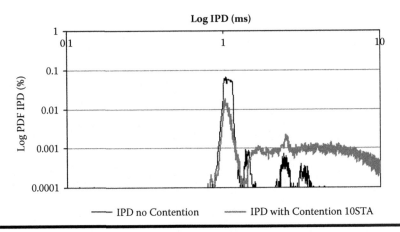

Figure 3.9   PDF of IPD with and without contention.

thereby its ability to manage the burstiness of the video stream. This can be seen in
Figure 3.9, where there is a long tail in the distribution of IPD values for the ten-
station case. In this case, ten wireless background traffic stations are transmitting
packets to the wired network via the AP's receiver. The aggregate load from these
stations is held constant as the number of background stations is increased.

To experimentally demonstrate the effects of contention on video streaming
applications, we focus on a single video clip DH being streamed from the wired
network via the AP to a wireless client. This particular clip was chosen because it is
representative of a typical nonsynthetic video stream. Table 3.1 presents the mean
performance values for the video clip DH over the test period with increased contention. It can be seen that the mean delay, loss rate, QFTD, and IPD increase with
increased contention. In this work the video content has a playout delay constraint
of 500 ms, which is a typical delay constraint for low-latency real-time interactive
video. The loss rate therefore corresponds to packets that have failed to be successfully received, as well as those packets that have been dropped as a result of
exceeding the playout delay constraint. If packets arrive too late, exceeding this
constraint, these packets are effectively dropped by the player because they have
arrived too late to be played out.

It can be seen that when there are no background contending stations, the
mean packet delay is approximately 10 ms. As the number of contending stations
increases from 3 to 7 to 10, the mean packet delay increases from 30 to 106 to 395
ms, respectively. This can be explained from the growing tail of the IPD distribution, as shown in Figure 3.9. In addition to the increased delays to gain access to
the medium, the mean loss rate also increases from 1 to 15 to 41 percent for 3, 7,
and 10 contending STAs, respectively. This is due to the fact that many packets do

Table 3.1 Mean Performance Values for Clip DH with Increased Contention

No. contending STA	Mean delay (ms)	Mean IPD (ms)	QFTD (ms)	Mean loss rate
0	10.43	1.24	11.5	0
3	29.62	3.73	36.62	0.01
4	30.97	3.75	37.96	0.01
5	37.91	3.97	45.39	0.03
6	63.63	4.34	71.76	0.08
7	105.75	4.82	115.61	0.15
8	174.91	5.27	186.05	0.23
9	311.71	5.66	325.01	0.34
10	395.27	5.95	406.83	0.41

not arrive within the given delay constraint and are effectively dropped at the client because they have arrived too late for playout.

3.6 Multimedia Streaming over IEEE 802.11e

In Section 3.5.1 we showed that the AP is a critical component that determines the performance of the network because it carries all of the downlink transmissions to wireless clients and is usually where congestion is most likely to occur. Congestion manifests itself as bursty losses and increased delays for multimedia traffic, which has a serious impact on multimedia streaming applications. This situation, however, need no longer apply following the approval of the IEEE 802.11e QoS MAC enhancement standard, which allows for up to four access categories (ACs), with different access priorities [4], allowing the QoS-enabled AP (QAP) to provide differentiated service to different applications. The Enhanced Distributed Channel Access (EDCA) mechanism of the IEEE 802.11e standard also defines a transmission opportunity (TXOP) as the interval of time during which a particular QoS-enabled station (QSTA) has the right to initiate transmissions without having to recontend for access. During an EDCA TXOP, a QSTA is allowed to transmit multiple MPDUs from the same AC with a SIFS time gap between an ACK and the subsequent frame transmission [5]. The duration of the TXOP is determined by the value of the TXOP limit parameter.

This TXOP mechanism is particularly suited to video streaming applications. Video streaming is often described as "bursty," and this can be attributed to the frame-based nature of video. Video frames are transmitted with a particular frame rate. In general, video frames are large, often exceeding the MTU of the network, and result in several packets being transmitted in a burst for each video frame where the frequency of these bursts corresponds to the frame rate of the video. The TXOP feature can be used to transmit a burst of video packets corresponding to a single video frame during the allocated TXOP interval.

3.6.1 TXOP for Video Streaming

The TXOP has been investigated in a number of previous works primarily through simulation. Suzuki et al. [24] have investigated the IEEE 802.11e QoS capabilities through simulation using the default values for the TXOP, but do not optimize its value. Kim and Suh [25] have used the TXOP limit parameter as a means to provide bandwidth fairness among contending stations. However, not all applications exhibit a bursty nature, and consequently, stations may not need to avail of the TXOP facility to transmit a burst of packets in a transmission opportunity. In [26] the authors describe a cross-layer adaptive video streaming system that adapts the TXOP limit parameter for layered encoded video streaming applications. Such a scheme is dependent on the adaptive capabilities of the end-to-end video streaming system. However, multicast video streaming applications have limited adaptive functionality.

The distribution of the frame size is used to correctly dimension the TXOP limit parameter as it statistically describes the encoding characteristics of the video stream and the time required to transmit the video frame. The time it takes to transmit a single video packet (T_p) during a TXOP interval is related to the packet size (PSz) and the physical line rate $(Rate)$, which for 802.11b has a maximum value of 11 Mbps [27].

$$T_P = \left(PSz\!\big/\!Rate\right) + (2 * SIFS) + Ack \tag{3.7}$$

N_p is the number of packets required to transmit the video frame of size FSz and is given by

$$N_P = \left(FSz\!\big/\!PSz\right) \tag{3.8}$$

The TXOP limit parameter is set to the number of packets required to transmit the video frame N_p multiplied by the time it takes to transmit each packet T_p during the TXOP interval. If the calculated TXOP duration requested is not a factor of 32 μs, that value is rounded up to the next-higher integer that is a factor of 32 μs. The maximum allowable TXOP limit is 8,160 μs, with a default value of 3,008 μs [5].

$$TXOP_N = \lceil N_P * T_P \rceil \tag{3.9}$$

Usage of the TXOP is not wasteful because when the AC_VI queue has won a TXOP and has no more packets to send during the TXOP interval, the hybrid controller (HC) may sense the channel and reclaim it after a duration of PIFS after the TXOP.

Figure 3.10 shows the CDF of the number of packets required to transmit video frames for the video stream EL encoded at 1 Mbps. It can be seen that the number of packets required to transmit I frames is significantly higher than that for B or P frames. However, because I frames have a lower frequency, they pull the CDF averaged over all frames only slightly to the right. In contrast, B frames have the highest frequency and pull the CDF of the frame sizes to the left. By dimensioning the TXOP limit parameter based upon the mean number of packets per video frame, 60 percent of video frames can be delivered in a single TXOP, which translates to 3, 26, and 74 percent of I, P, and B frames, respectively. However, if the mean plus one standard deviation of the frame size is used, 92 percent of video frames can be delivered in a single TXOP, which translates into 13, 81, and 98 percent of I, P, and B frames.

In all cases, the AC queues were configured with IEEE 802.11b settings for CWmin, CWmax, and AIFSN while the value for the TXOP limit parameter was varied. Before video streaming can be optimized using multiple IEEE 802.11e parameters, it is important that the behavior of a single parameter is known under a diverse range of test conditions. The purpose of this is so that the effects of varying the TXOP limit parameter can be observed in isolation. The 802.11e standard defines a number of AC queues into which different traffic streams can be directed: voice (AC_VO), video (AC_VI), best effort (AC_BE), and background (AC_BK). In this section we experimentally demonstrate a number of different scenarios and methods of setting the TXOP limit parameter as shown in Table 3.2.

For the purposes of comparison, cases A and B represent the best- and worst-case scenarios, respectively. Cases C, D, and E use just two AC queues: the AC_VI

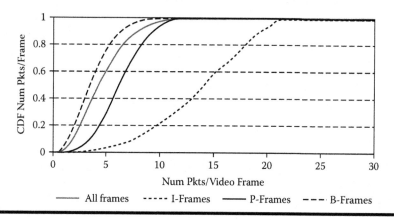

Figure 3.10   CDF of number of packets per video frame.

and AC_BK queues. Cases F and G utilize the full availability of the four AC queues: AC_VO, AC_VI, AC_BE, and AC_BK under the 802.11e standard.

Figure 3.11 shows the mean QFTD for each video-encoded bit rate for each of the different test cases averaged over all frames. As expected, cases A and B provide the best- and worst-case values for QFTD. The QFTD was found to be much larger for I frames because typically more packets are required to transmit an I frame than P and B frames. It can be clearly seen that in cases D–G, by appropriately tuning the TXOP limit parameter, the QFTD for the video frames can be significantly reduced. Overdimensioning the TXOP limit parameter causes the AC queue to seize too much bandwidth, which results in a deterioration in performance for the other competing traffic streams.

Table 3.3 summarizes the mean loss rate, packet delay, and QFTD for the different test cases averaged over all video bit rates for the different test cases. The reduction in QFTD (RQFTD) from the worst-case scenario, case B, is quantified as follows:

$$RQFTD = \left({QFTD_{CaseB} - QFTD_{Case}} \middle/ {QFTD_{CaseB}} \right) \qquad (3.10)$$

As expected, the reference best case, case A, exhibits the best performance as the video stream does not have to share the medium with other streams. It is expected that the loss rate is higher for case B because there is a greater buffer occupancy at the AP, as both the video traffic and background traffic share the same transmission buffer, which leads to packets being dropped at the incoming buffer. In all other cases, the loss rate is negligible because video can tolerate a small degree of packet loss. The mean packet delay is obtained by averaging over all packets. It can be seen that the mean packet delay is related to the QFTD.

In case D it can be seen that by using the mean frame size to dimension the TXOP limit parameter, the QFTD is reduced 67 percent, while in case E using the mean plus one standard deviation reduces the QFTD by 72 percent. There is a small performance gain in using the mean plus one standard deviation to dimension the TXOP limit parameter, as it reduces the QFTD by less than 3 ms, as in case E. Similarly, in cases F and G, the QFTD is reduced by 67 and 68 percent, respectively.

From Table 3.3 it can be seen that there is a small difference in the mean QFTD for cases D and F and for cases E and G. The benefit in providing differentiated services for the constituent frame types in cases F and G can be seen on examination of the QFTD for the individual frame types. Figure 3.12 shows the mean QFTD for the individual I, P, and B frame types. By comparing cases D and F, it can be seen that by providing differentiated service to the individual frame types, the mean QFTD for I frames is reduced by 5 ms, while the mean QFTD for B frame is increased by 6 ms. A similar effect can be seen in cases E and G: the I frame QFTD

Table 3.2 TXOP Test Cases

Case A: Only the video stream is transmitted through an IEEE 802.11b AP. This represents the best-case scenario.	**Case A:** Video only
Case B: The video stream and 5 Mbps of background traffic are transmitted through an IEEE 802.11b AP. This represents the worst-case scenario as both the video and background traffic packets are put into the same queue and must wait for their turn in accessing the medium.	**Case B:** Video + 5 Mbps
Cases C, D, and E: The video stream is transmitted through the AC_VI queue and 5 Mbps of background traffic is transmitted through the AC_BK queue and the AC_BK queue has a TXOP limit = 0. In case C both AC queues have IEEE 802.11b settings and a TXOP limit = 0. In case D the AC_VI queues have a TXOP limit parameter value that is related to the mean number of packets required to transmit the video frame ($\overline{N}$) averaged over all frames (ALL) irrespective of frame type, i.e., $TXOP_{\overline{(N+\sigma)}_{ALL}}$. In case E the AC_VI queue has a TXOP limit that is related to the mean number of packets plus one standard deviation ($\overline{N}+\sigma$) averaged over all frames (ALL) irrespective of frame type, i.e., $TXOP_{\overline{(N+\sigma)}_{ALL}}$.	**Case C:** TXOP = 0 **Case D:** $TXOP_{\overline{N}_{ALL}}$ **Case E:** $TXOP_{\overline{(N+\sigma)}_{ALL}}$
Cases F and G: The I, P, and B frames of the video stream are transmitted through the AC_VO, AC_VI, and AC_BE queues, and the background traffic is transmitted through the AC_BK queue with a TXOP limit = 0. The AC queues used for the video frames are configured with a TXOP limit parameter that is related to the number of packets for each frame type, where the subscripts I, P, and B refer to the I, P, and B video frames, respectively. In case F, the TXOP limit parameter is related to the mean number of packets ($\overline{N}$) for each frame type, i.e., $TXOP_{\overline{N}_I}$, $TXOP_{\overline{N}_P}$ and $TXOP_{\overline{N}_B}$. In case G the TXOP limit parameter is related to the mean plus one standard deviation of the number of packets ($\overline{N}+\sigma$) for the different frame types.	**Case F:** $TXOP_{\overline{N}_F}$ **Case G:** $TXOP_{\overline{(N+\sigma)}_F}$

is reduced by 2 ms, while the QFTD for B frames is increased by 5 ms. By providing differentiated service to the constituent frame types, the end-to-end video frame transmission delay for I or P frames can be reduced. I and P frames have a higher priority and a greater impact on the end-user perceived QoS over B frames.

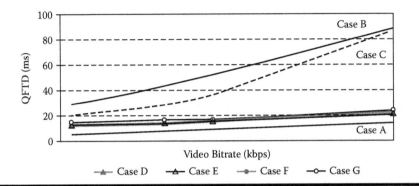

Figure 3.11   Mean QFTD with video bitrate.

Table 3.3   Experimental Results for Varying TXOP

Case	TXOP limit	Mean packet delay (ms)	QFTD (ms)		RQFTD
			Mean	Standard deviation	
A	—	3.43	8.42	3.61	0.84
B	—	26.38	52.85	25.42	0.00
C	0	19.15	38.14	32.97	0.28
D	$TXOP_{\overline{N}_{ALL}}$	6.70	17.28	2.25	0.67
E	$TXOP_{(\overline{N}+\sigma)_{ALL}}$	6.52	14.98	3.20	0.72
F	$TXOP_{\overline{N}_F}$	7.97	17.21	3.33	0.67
G	$TXOP_{(\overline{N}+\sigma)_F}$	8.16	16.85	5.11	0.68

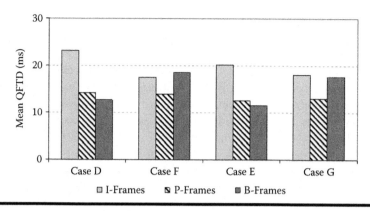

Figure 3.12 Mean QFTD for I, P, and B frames.

3.7 Conclusions

In this chapter, we have experimentally demonstrated the challenges for delivering multimedia streaming applications over IEEE 802.11. The end-to-end delay for multimedia applications is important because if a packet is delayed past its playout time, the packet is effectively lost, which ultimately affects the end-user perceived quality of the video stream. Video is a frame-based media, whereby frames are generated at a particular frame rate, which manifests itself as periodic bursts of packets at the AP. The AP must gain access to the medium to send each packet in the burst. Because each packet must wait for the packets in the queue ahead of it to be transmitted, the end-to-end delay steadily increases until all packets in the burst have been transmitted. This behavior results in the end-to-end delay for consecutive packets belonging to a single video frame rising and falling in a sawtooth manner that is related to the number of packets required to transmit the video frame and the size of those packets. The height of this sawtooth corresponds to the delay required to transmit the entire video frame, QFTD. We consider the QFTD to be of particular importance because a video frame cannot be correctly played out until all or most of the packets belonging to the video frame have been received. The difference in delay between consecutive packets in a burst is defined as the Interpacket Delay (IPD) and is related to the packet size. We have shown how the IPD and QFTD are affected differently, with a contention and downlink load saturation at the AP.

We demonstrated that the QoS-enabling features of IEEE 802.11e can be used to provide differentiated service to video streaming applications to reduce the QFTD. The periodic packet bursts that characterize video streaming applications can be exploited to reduce the transmission delay for video frames through tuning of the TXOP limit parameter in IEEE 802.11e. We have shown that the distribution of video frame sizes can be used to efficiently dimension the TXOP limit parameter such that 60 percent of video frames are capable of being transmitted within a single TXOP interval to transmit the complete burst of packets corresponding to a single video frame. We showed that by using the mean video frame size to dimension the TXOP limit parameter, the transmission delay for the video frame is reduced by 67 percent under heavily loaded conditions. By differentiating between the constituent video frame types through transmitting the I and P frames through the VI AC queue and the B frames through the BE AC queue, there is a performance improvement in terms of reducing the frame transmission delay for the I frames at the cost of increasing the frame transmission delay for the B frames. Furthermore, by providing prioritized access to the different frame types, we can reduce the likelihood of packets relating to I or P frames being lost because these frames have a higher priority and a greater impact on the end-user QoS over B frames.

Further research is being conducted to provide prioritized access to the audio streams and video streams and increase the number of parallel multimedia sessions that can be supported through an appropriate tuning of the AIFSN, CWmin, and CWmax settings in conjunction with the TXOP limit parameter.

References

[1] Insight Research Corp. 2006. Streaming media, IP TV, and broadband transport: Telecommunications carriers and entertainment services 2006–2011. http://www. insight-corp.com/reports/IPTV06.asp (accessed April 2006).

[2] N. Cranley and M. Davis. 2005. Performance analysis of network-level QoS with encoding configurations for unicast video streaming over IEEE 802.11 WLAN networks. Paper presented at WirelessCom 2005, Maui, HI.

[3] Q. Zhang, W. Zhu, and Y.-Q. Zhang. 2005. End-to-end QoS for video delivery over wireless Internet. *Proc. IEEE* 93(1).

[4] Q. Ni. 2005. Performance analysis and enhancements for IEEE 802.11e wireless networks. *IEEE Network* 19:21–27.

[5] IEEE Standard 802.11e. 2005. IEEE standards for local and metropolitan area networks: Specific requirements: Wireless LAN medium access control (MAC) and physical layer (PHY) specifications: Medium access control (MAC) quality of service enhancements. Part 11, Amendment 8.

[6] M. Li, M. Claypool, R. Kinicki, and J. Nichols. 2003. Characteristics of streaming media stored on the Internet. WPI-CS-TR-03-18 (May), Computer Science Technical Report Series, Worchester, MA: Worcester Polytechnical Institute.

[7] J. Liu. 2000. Signal processing for Internet video streaming: A review. Paper presented at Proceedings of SPIE Image and Video Communications and Processing.

[8] C. Krasic, J. Walpole, and W.-C. Feng. 2003. Quality adaptive media streaming by priority drop. Paper presented at Proceedings of NOSSDAV'03. 1–4 Monterey, CA.

[9] G. Conklin, G. Greenbaum, et al. 2001. Video coding for streaming media delivery on the Internet. *IEEE Transactions on Circuits and Systems for Video Technology* 11(3).

[10] R. Koenen. *MPEG-4 overview. ISO-IEC JTC1/SC29/WG 11: N4668 coding of moving pictures and audio.* MPEG-4 WG11, version 21.

[11] Apple Computer. QuickTime file format. http://developer.apple.com/techpubs/quicktime/qtdevdocs/PDF/QTFileFormat.pdf.

[12] Quicktime 6.3 + 3GPP. http://developer.apple.com/documentation/QuickTime/QT6_3/chap1/chapter_1_section_1.html (accessed April 2006).

[13] Hint track format. http://developer.apple.com/documentation/QuickTime/REF/Streaming.29.htm#pgfId=19901.

[14] WinDump. http://mirrors.wiretapped.net/security/packet-capture/wincap/windump (accessed April 2006).

[15] NetTime. http://nettime.sourceforge.net/ (accessed April 2006).

[16] S. B. Moon, P. Skelly, and D. Towsley. 1999. Estimation and removal of clock skew from network delay measurements. In *Proceedings of IEEE InfoComm'99*, March, 227–34.

[17] Helix Streaming Server. https://www.helixcommunity.org/ (accessed April 2006).

[18] T. Kuang and C. Williamson. 2001. RealMedia streaming performance on an IEEE 802.11b wireless LAN. In Proceedings of IASTED Wireless and Optical Communications (WOC) Conference (July), Alberta, Canada: Banff, pp. 306–11.

[19] Darwin Streaming Server. http://developer.apple.com/darwin/projects/streaming/.

[20] QuickTime streaming. http://developer.apple.com/documentation/QuickTime/RM/Streaming/StreamingClient/StreamingClient.pdf (accessed April 2006).

[21] Cisco Aironet 1200. http://www.cisco.com/en/US/products/hw/wireless/ps430/index.html (accessed April 2006).

[22] J. Jun, P. Peddabachagari, and M. Sichitiu. 2003. Theoretical maximum throughput of IEEE 802.11 and its applications. In *Proceedings of the Second IEEE International Symposium on Network Computing and Applications (April)*, Washington, DC.

[23] N. Cranley and M. Davis. 2006. The effects of background traffic on the end-toend delay for video streaming applications over IEEE 802.11b WLAN networks. Paper presented at 17th Annual IEEE International Symposium on Personal, Indoor and Mobile Radio Communications (PIMRC), Helsinki, Finalnd.

[24] T. Suzuki, N. Fukushi, A. Noguchi, and S. Tasaka. 2006. Effect of TXOP-bursting and transmission error in application-level and user-level QoS in audio-video transmission with IEEE 802.11e EDCA. Paper presented at 17th Annual IEEE International Symposium on Personal, Indoor and Mobile Radio Communications (PIMRC), Helsinki, Finland.

[25] E. Kim and Y.-J. Suh. 2004. ATXOP: An adaptive TXOP based on the data rate to guarantee fairness for IEEE 802.11e wireless LANs. In *IEEE 60th Vehicular Technology Conference (VTC2004)*, vol. 4, pp. 2678–82.

[26] M. van der Schaar, Y. Andreopoulos, and Y.-Z. Hu. 2006. Optimized scalable video streaming over IEEE 802.11 a/e HCCA wireless networks under delay constraints. *IEEE Transactions on Mobile Computing* 5:755–68.

[27] B.-S. Kim, S.-W. Kim, Y. Fang, and T. F. Wong. 2005. Two step multipolling MAC protocol for wireless LANs. *IEEE JSAC in Communications* 23(6).

Chapter 4

Performance Modeling and Analysis of IEEE 802.11e Contention Free Bursting Scheme under Unsaturated Traffic

Jia Hu, Geyong Min, Mike E. Woodward, and Wei Guo

Contents

4.1　Introduction

Wireless local area networks (WLANs) based on the IEEE 802.11 standard [16] have become ubiquitous over recent years, especially encouraged by the success of the Internet and the proliferation of portable devices, such as laptop computers and personal digital assistants. Accompanying the rapid deployment of WLANs, academic and industrial communities have carried out in-depth research activities by analytical or numerical means to gain insights on the key performance attributes, such as system capacity and quality of service (QoS), perceived by network users. The architecture of IEEE 802.11 includes the definitions of the physical (PHY) layer and the Medium Access Control (MAC) sublayer. The original MAC sublayer employs a mandatory contention-based channel access function called Distributed Coordination Function (DCF), which is based on the Carrier Sense Multiple Access with Collision Avoidance (CSMA/CA) protocol [19]. The standard also specifies an optional polling-based channel access function called Point Coordination Function (PCF). However, PCF is rarely implemented in commercially available WLANs and has received little attention due to its complexity and inefficiency [25].

With the rapid growth in the popularity of multimedia applications such as Voice-over-IP and video conferencing, the demand for high bandwidth and differentiated QoS in WLANs is increasing dramatically [11, 25]. To support MAC-level QoS, an enhanced version of the IEEE 802.11 MAC protocol [16], namely, IEEE 802.11e [17], has been standardized. The IEEE 802.11e MAC employs a channel access function called Hybrid Coordination Function (HCF) [17], which comprises the contention-based Enhanced Distributed Channel Access (EDCA) and the centrally controlled Hybrid Coordinated Channel Access (HCCA). EDCA is the QoS mechanism that will be supported by most WLAN vendors, whereas HCCA is difficult to find on the market due to its high cost and complexity associated with actual implementation. As an extension to the basic DCF mechanism of the legacy 802.11, EDCA provides a priority scheme by differentiating the interframe space and contention window size [17]. Moreover, it develops a new burst transmission mechanism [17] (referred to as contention free bursting [CFB] or transmission opportunity [TXOP] bursting) to improve the system efficiency. This innovative operation amortizes the contention overhead by allowing a station to transmit multiple frames consecutively in a burst after it gains the channel and thus improves the channel utilization.

Analytical performance evaluation of DCF has been extensively studied in recent years [1–3, 5–10, 13, 14, 20, 21, 26, 27, 29, 30, 31, 33, 34, 35–37, 41, 43, 44]. As a promising mechanism to enhance the performance of DCF, the burst transmission scheme has recently attracted significant research attention [12, 15, 22, 23, 39, 40, 42]. However, most existing studies [15, 22, 32, 39, 40, 42] are based on the assumption of saturation traffic loads, which implies that all stations in the network always have frames to transmit, and thus excludes any need to consider queuing or traffic models for performance analysis. In contrast, realistic network conditions are nonsaturated, as very few networks are in a situation where all nodes have frames to send all the time. Therefore, it is important to remove the assumption of saturation traffic to obtain a realistic and deep understanding of the performance of the CFB scheme. To this end, this chapter aims to present a new analytical model for the CFB scheme that can be used to gain an in-depth insight into its performance under more realistic traffic conditions. We adopt a Markov model to analyze the backoff procedure under unsaturated traffic conditions and then model the transmission queue at each station as a bulk service queueing system. The analytical model is validated through comparing the numerical results with those obtained from extensive ns2 simulation experiments.

The rest of this chapter is organized as follows. In Section 4.2, we present an overview of the legacy DCF and the CFB scheme followed by a detailed survey of the related work. Section 4.3 describes an analytical model for the CFB scheme with unsaturated traffic. Section 4.4 validates the model and reports the results of performance evaluation. Finally, we conclude this chapter in Section 4.5.

4.2   Background

4.2.1 Medium Access Control (MAC)

In DCF [16], a station with a pending frame to transmit first senses the channel. If the channel is detected idle for a distributed interframe space (DIFS), the station transmits the frame. Otherwise, if the channel is sensed busy (either initially or during the DIFS), the station defers until the channel is detected idle for a DIFS, and then generates a random backoff time. The backoff time is uniformly chosen in the range [0, $W_i - 1$], where W_i is current contention window (*CW*) size, i, $i \in [0, m]$, is the backoff stage, and m represents the maximum number of backoff stages. At the first transmission attempt, *CW* is set to the minimum value, $CW_{min} = W_0$. After each unsuccessful transmission, *CW* is doubled, up to a maximum value $CW_{max} = W_m = 2^m W_0$. It remains at the value CW_{max} until it is reset to CW_{min} after the successful frame transmission, or if the unsuccessful transmission attempts reach a retry limit. The backoff time is downcounted by one for each slot (an interval of fixed duration specified in the protocol [16]) when the medium is idle, halted when the medium becomes busy, and resumed when the medium is idle again for a DIFS. A station transmits a frame when its backoff timer reaches zero. Other stations that hear the transmission of the frame set their Network Allocation Vector (NAV) to the expected period of time in which the channel will be busy, as indicated in the duration identity (ID) field of the frame. This is called the *virtual carrier sensing* mechanism. If either the virtual carrier sensing or physical carrier sensing [19] indicates that the channel is busy, the station commences the backoff procedure. Upon the successful reception of the frame, the destination station sends back an acknowledgment (ACK) frame immediately after a short interframe space (SIFS) interval. If the source station does not receive the ACK within a specified ACK timeout interval, it retransmits the frame according to the given backoff rules. Each station maintains a retry counter that is increased by one after each retransmission. If the retry counter reaches the retry limit, then the frame is discarded. The above-mentioned procedure is referred to as the basic access method. Hidden terminal problems [38] occur when a station is unable to detect a potential competitor for the medium because they are not within transmission range of each other. To combat the hidden terminal problems, DCF also defines an optional four-way handshake scheme (RTS/CTS) whereby the source and destination exchange request-to-send (RTS) and clear-to-send (CTS) messages before the transmission of the actual data frame.

CFB is a new scheme introduced in EDCA to improve the system efficiency. As shown in Figure 4.1, a station gaining the channel transmits the frames available in its buffer successively provided that the duration of transmission does not exceed a certain threshold, referred to as the TXOP limit [17]. Each frame is acknowledged by an ACK after an SIFS interval. The next frame is transmitted immediately after it waits for an SIFS upon receiving this ACK. If the transmission of any frame fails, the burst is terminated and the station contends again for the channel to retransmit the failed frame. The CFB scheme is an efficient way to improve the channel utilization of wireless MAC protocols because the contention overhead is shared by all the

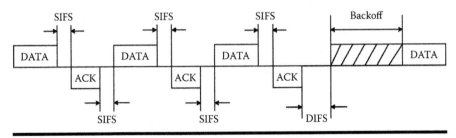

Figure 4.1   Mechanism of the CFB scheme.

frames transmitted in a burst. Moreover, it enables service differentiation between multiple traffic classes by virtue of various TXOP limits [15]. Another advantage of using TXOP is that in multirate WLANs the access time to the medium can be more fairly distributed by allocating the faster stations with the larger TXOP limit. Thus, the slow stations no longer severely degrade the performance of the higher-rate stations [8].

4.2.2 Related Work

Performance analysis of DCF has come under much scrutiny in recent years. Cali et al. [6, 7] have presented the analysis of the saturated throughput of the p-persistent CSMA/CA protocol by modeling the backoff counter value as a geometric distribution. Their study has shown that it is possible to maximize the throughput by tuning the backoff window size at runtime. Bianchi's well-known analytical model [1, 2] has adopted a bidimensional discrete-time Markov chain to derive the saturation throughput for the DCF, assuming the ideal channel conditions (i.e., no hidden terminals and capture effects [38]). Many subsequent studies have built upon Bianchi's work. For instance, Ziouva and Antonakopoulos [44] have improved Bianchi's model by taking account of the busy medium conditions for invoking the backoff procedure. Based on their model, the performance measures in terms of throughput and average service time have been calculated. Wu et al. [41] have modified Bianchi's model to deal with the retry limit. They have also proposed a new scheme called DCF+, which is more suitable than DCF for the Transmission Control Protocol (TCP). Kumar et al. [21] have studied the fixed-point formulation based on the analysis of Bianchi's model and showed that the derivation of transmission probability can be significantly simplified by removing the Markovian assumptions. They have also applied their model to obtain the throughput of TCP-controlled file transfer under some working scenarios.

The queueing delay becomes the dominant factor of the total delay. However, the vast majority of the saturation models have only considered throughput, access delay, and service time, but neglected the queueing delay. For example, Carvalho and Garcia-Luna-Aceves [9] linearized Bianchi's model and performed the analysis of average service time as well as jitter. Chatzimisios et al. [10] have adapted Bianchi's model to handle retry limit and calculate the access delay (the delay seen by the frame at the head of the queue) on saturation state. Zenella and Pellegrini [43] have derived a closed-form probability-generating function for the service time in saturated networks, including the basic access scheme and RTS/CTS.

The aforementioned saturated models assume that all stations in the network always have frames to transmit and thus exclude any need to consider queuing dynamics or traffic models for performance analysis. However, realistic network conditions are nonsaturated, as very few networks are in a situation where all nodes have frames to send all the time. Therefore, it is important to develop analytical models under unsaturated conditions. There are several studies focused on modeling DCF under nonsaturated working conditions. Tickoo and Sikdar [36] have presented a G/G/1 queueing model for DCF and improved this model in [37] by providing a better approximation to the probability that the transmission queue is not empty. After deriving the saturation transmission probability through the average-value analytical model [35], Tickoo and Sikdar [36,37] obtained the transmission probability of finite loads by weighting the saturation transmission probability with the probability of the transmission queue being nonempty. Medepalli and Tobagi [26] have developed a unified analytical model where the unsaturated transmission probability is also derived by weighting the saturation transmission probability with the probability of a nonempty M/M/1 queue. The key features of their model include the ability to handle hidden and exposed terminals, directional antennas, multiple channels, and arbitrary traffic matrices. Ozdemir and McDonald [29] have obtained the unsaturated performance metrics based on the M/G/1/K queueing model, where the service time distribution is modeled by a Markov-modulated general distribution. Moreover, they have removed the fundamental assumption of Bianchi's model that every frame collides with a constant and independent probability regardless of the number of retransmissions it suffered. Miorandi et al. [27] have taken a processor-sharing view of DCF to evaluate the performance of Hypertext Transfer Protocol (HTTP) traffic over IEEE 802.11. They have proved that setting the TCP's advertised window size to a small value leads to insensitivity of mean file transfer times to the file size distribution.

None of the above-mentioned nonsaturated models have taken into account the case of heterogeneous stations. Cantieni et al. [8] have introduced an unsaturated model considering the multirate capabilities where the MAC buffer is modeled by an M/G/1 queue. In addition, they have invented a new fairness metric for general CSMA/CA multirate networks. Malone et al. [24] have presented an extension of Bianchi's model to a nonsaturated environment in the presence of heterogeneous loads. Their model has captured several important features of unsaturated operation, for instance, predicting the maximum throughput. To investigate the effect of buffering on resource allocation, Duffy and Ganesh [13] have further extended the model developed in [24] for stations with large buffers and Poisson arrivals.

As a promising mechanism to be incorporated in the MAC protocol of the next-generation WLANs [42], analytical modeling of burst transmission has drawn considerable research efforts. Vitsas et al. [40] presented an analytical model for the CFB scheme under saturation traffic loads to evaluate the performance measures of throughput, mean frame delay, and frame drop probability. Furthermore, the fairness problem was discussed for both DCF and burst transmission cases. Tinnirello and Choi [39] have analyzed and compared the system efficiency of the CFB and block ACK (BACK) schemes under the saturation condition. Through the analysis, they have shown that access mode and ACK policy have a great impact on the overall system throughput, given a frame size and a TXOP limit. Li et al. [22] have investigated the saturation throughput of the BACK scheme under noisy channel conditions. Their results have indicated that the BACK scheme is very effective in high-rate wireless networks, and the number of frames in each block needs to be discussed before transmission to provide better efficiency. Fei et al. [15] have proposed an analytical model to evaluate the throughput of different access categories (ACs) as a function of different TXOP limits. It has been observed that the choice of TXOP limit for different ACs can lead to increased throughput for some ACs but reduced throughput for others.

For the simplicity of performance analysis, existing models of burst transmission schemes have primarily focused on the analysis of system throughput only and ignored some realistic factors, such as unsaturated traffic loads and finite buffer capacity. As stated earlier, it is of merit for a study to remove the assumption of saturation traffic to obtain a realistic and deep understanding of the performance of the CFB scheme. Another weakness of existing models is that the important QoS performance metrics, such as end-to-end delay and loss probability, have not been investigated. Distinguishing from existing work, this chapter presents an analytical model that can be used to gain a thorough insight into the QoS performance of the CFB scheme under more realistic working conditions.

4.3   System Model

In this section, we present the fundamental methodology and components to develop the analytical model for burst transmission in WLANs under unsaturated traffic conditions. The transmission queue at each station is modeled as a bulk service queueing system where the arrival traffic follows a Poisson process with rate λ (frames/second). The service time of the queueing system is the MAC layer service time, which is defined as the time interval from the instant that a burst starts to contend for the channel, to the instant that the transmission of the burst is successful. To calculate the service time, the binary exponential backoff (BEB) algorithm for the frame transmission is modeled by a two-dimensional discrete-time Markov chain with the unsaturated condition taken into account.

4.3.1 Modeling of BEB Procedure

Consider a scenario of n homogeneous stations under unsaturated Poisson traffic. Each frame reaching the head of the queue needs to perform the BEB procedure before the actual transmission starts. Using the similar notations of Bianchi's model [2], let $s(t)$ and $b(t)$ denote the stochastic processes representing the backoff stage and the backoff time counter for the station at time t, respectively. A discrete and integer timescale in which t and $t + 1$ correspond to the start of two successive slot times [2] is adopted in the model. The virtual time adds by one slot when the station either decrements its backoff counter or attempts a transmission. These time slots have real-time intervals of variable lengths, as it may include a channel contextion, a collision, or an idle slot time size σ.

Assuming that the collision probability p is independent of the number of retransmissions that a frame has suffered [2], the bidimensional process $\{s(t), b(t)\}$ can be modeled as a discrete-time Markov chain. To analyze the backoff procedure under unsaturated traffic, we introduce a new state, namely, IDLE, representing the case that the transmission queue at a station is empty. Let b_i denote the probability that a station is at state IDLE, p_0 represent the probability that a station has no frame waiting for transmission while completing the backoff process, and p_e be the transition probability from the IDLE state to the backoff states at level 0 (i.e., $s(t) = 0$), as shown in Figure 4.2.

Let $P\{i_1, k_1|i_0, k_0\}$ be the short notation of one-step transition probability, i.e., $P\{i_1, k_1|i_0, k_0\} = P\{s(t + 1) = i_1, b(t + 1) = k_1|s(t) = i_0, b(t) = k_0\}$. The one-step transition probabilities of the Markov chain modeling the backoff procedure can be written as

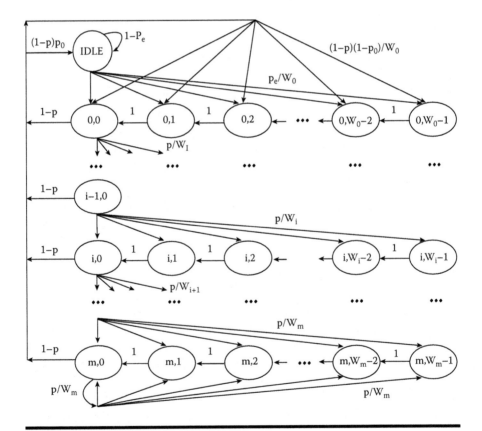

Figure 4.2   Markov chain model under unsaturated traffic conditions.

$$
\begin{cases}
P\{i,k \mid i,k+1\}=1 & k \in [0, W_i-2], i \in [0,m] & \text{(1.a)} \\[4pt]
P\{i,k \mid i-1,0\}= p/W_i & k \in [0, W_i-1], i \in [1,m] & \text{(1.b)} \\[4pt]
P\{m,k \mid m,0\}= p/W_m & k \in [0, W_m-1] & \text{(1.c)} \\[4pt]
P\{0,k \mid i,0\}=(1-p)(1-p_0)/W_0 & k \in [0, W_0-1], i \in [0,m] & \text{(1.d)} \\[4pt]
P\{IDLE \mid i,0\}=(1-p)p_0 & i \in [0,m] & \text{(1.e)} \\[4pt]
P\{0,k \mid IDLE\}= p_e/W_0 & k \in [0, W_0-1] & \text{(1.f)}
\end{cases}
\qquad (4.1)
$$

These equations account, respectively, for: (4.1a) the backoff counter is decremented; (4.1b) the backoff stage increases after an unsuccessful transmission; (4.1c) the backoff stage reaches the maximum value, and thus does not increase after an unsuccessful transmission; (4.1d) the station returns to backoff stage 0, in the case of having at least one pending frame in the transmission queue after a successful transmission; (4.1e) the station reaches state IDLE because its transmission queue is empty after a successful transmission; (4.1f) the station leaves state IDLE and starts backoff from stage 0 because it has an arrival frame.

Let $b_{i,k}$ and b_{IDLE} be the stationary probabilities of the Markov chain. First, the steady-state $b_{i,0}$ have the following relations:

$$
\begin{cases}
b_{i,0} = p^i b_{0,0} & i \in [0, m) \\[10pt]
b_{m,0} = \dfrac{p^m}{1-p} b_{0,0} & i = m
\end{cases}
\qquad (4.2)
$$

Next, the balance equation in the IDLE state is given by

$$
p_e b_{IDLE} = p_0(1-p)\sum_{i=0}^{m} b_{i,0} \qquad (4.3)
$$

Owing to the Markov chain regularities, for each $k \in [0, W_i-1]$, we have

$$
b_{i,k} = \frac{W_i-k}{W_i}
\begin{cases}
(1-p)(1-p_0)\sum_{j=0}^{m} b_{j,0} + p_e b_{IDLE} & i=0 \\[10pt]
pb_{i-1,0} & i \in (0, m) \\[10pt]
p(b_{m-1,0}+b_{m,0}) & i=m
\end{cases}
\qquad (4.4)
$$

Using equations (4.2) and (4.3), equation (4.4) can be simplified as

$$b_{i,k} = \frac{W_i - k}{W_i} b_{i,0} \quad i \in [0,m] \quad k \in [0, W_i - 1] \tag{4.5}$$

Therefore, with equations (4.2) and (4.5), $b_{0,0}$ can be finally determined by imposing the following normalization condition:

$$1 = \sum_{i=0}^{m} \sum_{k=0}^{W_i-1} b_{i,k} + b_{IDLE} = \sum_{i=0}^{m} \frac{b_{i,0}(W_i+1)}{2} + b_{IDLE} \tag{4.6}$$

from which

$$b_{0,0} = \frac{2(1-2p)(1-p)(1-b_{IDLE})}{(1-2p)(W+1) + pW(1-(2p)^m)} \tag{4.7}$$

As any transmission occurs when the backoff counter reaches zero, regardless of the backoff stage, the probability that a station transmits in a randomly chosen slot time, τ, can be expressed as

$$\tau = \sum_{i=0}^{m} b_{i,0} = \frac{b_{0,0}}{1-p} \tag{4.8}$$

The probability, p, that a transmitted frame encounters a collision, is the probability that at least one of the remaining stations transmits in a time slot:

$$p = 1 - (1-\tau)^{n-1} \tag{4.9}$$

Equations (4.8) and (4.9) represent a nonlinear system with the unknown variable b_{IDLE}, which can be solved through a numerical method. As a special case, when the stations are working under saturated traffic with $b_{IDLE} = 0$, the model reduces to Bianchi's [2]. The unknown b_{IDLE} will be derived in Section 4.3.3.

4.3.2 Analysis of Service Time

It is worth noting that only the head-of-burst (HoB) frame needs to contend for the channel. We define the service time as the duration from the instant that an HoB frame begins to contend for the channel to the instant that the entire data burst is acknowledged following successful transmission. The service time includes two parts: the channel access delay and burst transmission delay. The former is the time interval from the time that the frame reaches the head of its transmission queue until it wins the contention and is ready to be transmitted. The latter is the time duration of successfully transmitting a burst. Let $E[S_i]$, $E[A]$, and $E[B_i]$ denote the mean service time, channel access delay, and burst transmission delay, respectively, where i represents the actual number of frames transmitted in a burst. $E[S_i]$ can be written as

$$E[S_i] = E[A] + E[B_i] \tag{4.10}$$

Given that a frame is successfully transmitted after j collisions ($j \geq 0$) its channel access delay consists of the delay from j unsuccessful transmissions and delay from ($j + 1$) backoff stages; thus, the average channel access delay is given by

$$E[A] = \sum_{j=0}^{\infty} \left(\left(jT_c + \sigma' \sum_{i=0}^{j} \frac{W_i - 1}{2} \right) p^j (1-p) \right) \tag{4.11}$$

where $p^j(1 - p)$ is the probability that the frame is successfully transmitted after j collisions, jT_c is the collision time that the frame experiences during the previous transmission attempts before the ($j + 1$)-th transmission attempt, ($W_i - 1$)/2 is the average number of time slots that the tagged station defers in the i-th backoff stage, and σ' is the average length of a slot time when the remaining ($n - 1$) stations contend for the channel.

Let P_{tr} be the probability that at least one station among the other ($n - 1$) stations transmits in a considered slot. P_{tr} is given by

$$P_{tr} = 1 - (1-\tau)^{n-1} \tag{4.12}$$

The probability, P_s, that there is a successful transmission among the other ($n - 1$) stations given that a transmission occurs on the channel can be written as

$$P_s = \frac{(n-1)\tau(1-\tau)^{n-2}}{P_{tr}} \tag{4.13}$$

The average length of a slot time is obtained by considering the fact that a successful transmission occurs in a slot time with probability $P_{tr}P_s$, the slot time is empty with probability ($1 - P_{tr}$), and a collision happens with probability $P_{tr}(1 - P_s)$. Thus, σ' can be written as

$$\sigma' = (1 - P_{tr})\sigma + P_{tr}P_sT_s + P_{tr}(1 - P_s)T_c \tag{4.14}$$

where σ is the duration of an empty slot time, T_s is the average time for the successful transmission of a burst, and T_c is the average collision instant.

Note that only the HoB frame can be involved in the collision using a burst transmission mechanism. T_c is given by

$$T_c = DIFS + L + H + SIFS + ACK \tag{4.15}$$

where L is the average time required for transmitting the frame payload, and H is the average time required for transmission of the frame header. The average time for the successful transmission of a burst, T_s, is given by

$$T_s = \frac{\sum_{i=1}^{K} E[B_i]P_i}{1 - P_0} \tag{4.16}$$

where K denotes the maximum number of frames that can be transmitted in a TXOP limit, the denominator $(1 - P_0)$ means that the occurrence of burst transmission is conditioned on the fact that there is at least one pending frame in the transmission queue, P_i ($1 \le i \le K$) is the probability of having i frames in the burst, and $E[B_i]$ is the burst transmission delay. $E[B_i]$ is dependent on the number of frames within a burst and can be given by

$$E[B_i] = DIFS + i(L + H + 2SIFS + ACK) - SIFS \tag{4.17}$$

4.3.3 Queueing Model

The transmission queue at each station can be modeled as a continuous-time M/$G^{[1,K]}$/1/N queueing system [18], where the superscript $[1, K]$ denotes that the actual number of frames transmitted in a burst ranges from 1 to K, and N represents the buffer size at each station.

As mentioned above, the server becomes busy when a frame reaches the head of the transmission queue. The server becomes free after a burst of frames are acknowledged by the destination following successful transmission. The service time is dependent on the actual number of frames transmitted in a burst and is modeled by an exponential distribution function with mean $E[S_i]$. Thus, the service rate, μ_i, can be given by

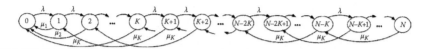

Figure 4.3   The M/G[1,K]/1/N queue state transition rate diagram.

$$\mu_i = \frac{1}{E[S_i]} \tag{4.18}$$

Figure 4.3 illustrates the state transition rate diagram of the queuing system where each state denotes the number of frames in the system. The transition rate from state i to $i + 1$ ($0 \le i \le N - 1$) is the arrival rate λ of the Poisson process. A transition out of state i to $i - K$ ($K \le i \le N$) implies that the burst transmission of K frames is complete and the transition rate is μ_K. The change from state i to 0 ($1 \le i \le K - 1$) denotes that all i frames in the system are transmitted within a burst and the transition rate is μ_i. The generator matrix $\mathbf{Q}$ of the Markov chain is given by

$$Q = \begin{bmatrix} -\lambda & \lambda & & & & & & & & \\ \mu_1 & -\lambda-\mu_1 & \lambda & & & & & & & \\ \mu_2 & & -\lambda-\mu_2 & \lambda & & & & & & \\ \vdots & & & \ddots & \ddots & & & & & \\ \mu_{K-1} & & & & -\lambda-\mu_{K-1} & \lambda & & & & \\ \mu_K & & & & & -\lambda-\mu_K & \lambda & & & \\ & \ddots & & & & & \ddots & & & \\ & & \mu_K & & & & & -\lambda-\mu_K & \lambda & \\ & & & \mu_K & & & & & -\mu_K \end{bmatrix} \tag{4.19}$$

We utilize the Gaussian elimination algorithm [4] to calculate the equilibrium probability P_i that there are i frames in the system. Following this algorithm, the last column of matrix $\mathbf{Q}$ is replaced by the unit vector $1 = [1, 1, \ldots , 1]^T$ to obtain the parameter matrix $\mathbf{A}$. Then $\mathbf{A}$ is split into upper triangular matrix $\mathbf{U}$ and lower triangular matrix $\mathbf{L}$. The steady-state probability vector $\mathbf{P} = (P_i, i = 0, 1, \ldots , N)$ of the Markov chain satisfies the following equations:

$$\begin{cases} \mathbf{YL} = \mathbf{B} \\ \mathbf{PU} = \mathbf{Y} \end{cases} \tag{4.20}$$

where vector $\mathbf{B} = [0, 0, \ldots 0, 1]$ and $\mathbf{Y}$ is the intermediate solution vector. Following the algorithm presented in [4], we can solve these equations and yield the steady-state vector **P**. Thus, we obtain the probability b_{IDLE}, which equals the probability P_0 that the transmission queue is empty.

4.3.4 Performance Measures

The end-to-end delay is the duration from the time epoch a frame enters the transmission queue of the station to the time epoch the frame is removed from the transmission queue. It is equivalent to the queueing delay plus service time. By virtue of Little's law [18], the end-to-end delay, $E[D]$, is given by

$$E[D] = \frac{E[N]}{\lambda(1-p_b)} \tag{4.21}$$

where $E[N] = \sum_{i=0}^{N} iP_i$ is the average number of frames in the queueing system. $\lambda(1-p_b)$ is the effective arrival rate to the transmission queue due to the fact that the arriving frames are dropped if the finite buffer is found full.

The loss probability, p_b, which is the probability of an arriving frame from the upper layer finding the finite buffer of the MAC layer full, is equal to P_N. Given the loss probability p_b, the normalized system throughput S can be computed by

$$S = \frac{n\lambda E[P](1-p_b)}{C} \tag{4.22}$$

where n is the number of stations, $E[P]$ is the average frame payload length, and C is the channel data rate.

4.3.5 Implementation of the Model

Due to the interdependent relationships among τ, p, P_0, we resort to the following iterative algorithm to calculate the system performance metrics:

Step 1: Initialize the steady-state probability vector **P**.
Step 2: Calculate τ, p through the Markov chain model.
Step 3: Given τ and p, calculate the burst service time $E[S_j]$.
Step 4: Given $E[S_j]$, calculate the steady-state probability vector **P** using the M/$G^{[1,K]}$/1/N queuing model. This will result in a new value of P_0.
Step 5: Repeat steps 2, 3, and 4 until P_0 converges.

Table 4.1 System Parameters

Frame payload	8,000 bits
MAC header	224 bits
PHY header	192 bits
ACK	112 bits + PHY header
Channel data rate	11 Mbit/s
Basic rate	1 Mbit/s
Slot time	20 μs
SIFS	10 μs
DIFS	50 μs
CWmin	32
CWmax	1,024
Buffer size	50 frames

4.4 Numerical Results

4.4.1 Model Validation

To validate the analytical model, we compare the numerical performance results achieved through the analysis against those obtained from the ns2 [28] simulation experiments. There are ten stations located in a rectangular grid with dimension 100×100 m. The parameters used in the analytical model and simulation are summarized in Table 4.1.

Figure 4.4 depicts the throughput, end-to-end delay, and loss probability, respectively, versus the normalized total offered load for different TXOP limits. Note that the case $K = 1$ represents the legacy DCF scheme. It is shown that the analytical results match those obtained from simulation at a good level. More specifically, Figure 4.4[(a), (c)] demonstrates that our model possesses excellent accuracy to predict the system throughput and loss probability. Figure 4.4(b) reveals that there are exact agreements on end-to-end delay between the simulation results and those predicted by the analytical model subject to the light and medium loads as well as saturated region. However, some discrepancies appear as the network is transited from the region of medium loads to the saturated region. The overestimate is mainly due to the approximations that have been used to make the development of the model feasible, such as the assumption that the service times are independent stochastic variables. Nevertheless, it can be concluded that the analytical model produces accurate results in stable conditions, which are the conditions of interest in most performance evaluation studies of network protocols. The simplicity and reasonable accuracy of the analytical model make it a practical and cost-effective performance evaluation tool for the burst transmission mechanism in 802.11e MAC.

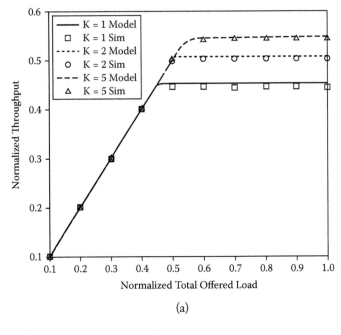

(a)

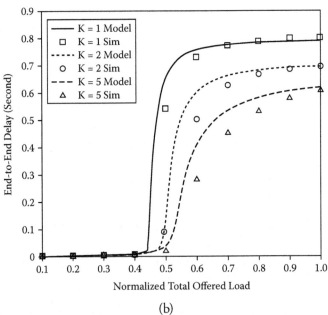

(b)

Figure 4.4 **Performance metrics versus normalized total offered load with varying TXOP limit.**

4.4.2 Performance Evaluation

Figure 4.4(a, b) shows that the CFB scheme can improve the system throughput and reduce the end-to-end delay when the traffic load is moderate and high. We also observe that under the legacy DCF scheme, the throughput becomes saturated and the delay increases sharply when the total offered load reaches 0.45. After the load exceeds 0.6, the end-to-end delay increases gradually. A similar phenomenon can be found for the CFB scheme. Also, the maximum throughput increases as the TXOP limit increases. It can be seen from Figure 4.4(c) that the loss probability is negligible when the traffic load is low. We also note that the loss probability increases dramatically as soon as the network becomes saturated. For the CFB scheme, the loss probability is lower than that of legacy DCF and drops as the TXOP limit increases.

4.4.2.1 Impact of Buffer Size

To investigate the impact of the buffer size on the performance of the CFB scheme, Figure 4.5 depicts the throughput, end-to-end delay, and loss probability against the offered load with the varying buffer sizes of 5, 10, and 20 frames. The TXOP limit is fixed at five, and the network has ten stations. It is observed that the larger buffers achieve a higher throughput and lower loss probability while leading to an increasing delay. Furthermore, when the buffer size goes beyond ten frames, the improvement of throughput and loss probability is not significant, and in contrast, the delay keeps growing. For the case of delay-insensitive data traffic, throughput and loss probability are the most important performance metrics of interest, so we can set a large buffer size. While for the case of delay-sensitive traffic like voice and video, a large buffer leads to high delay, which may be intolerable for these inelastic applications. Thus, a small buffer is preferable in this case.

4.4.2.2 Optimal TXOP Limit

As observed in the previous section, the system performance improves when the CFB scheme is incorporated in the wireless MAC protocol, but the percentage of performance enhancement decreases as the TXOP limit increases. Also, the short-term fairness becomes weak as the TXOP limit increases [40]. Thus, it is interesting to find the optimal TXOP limit that can provide the better performance.

Figure 4.6 depicts the performance result of the CFB scheme under saturated traffic conditions, with the TXOP limit increasing from one to ten frames, by varying the number of stations. On the one hand, we observe that a TXOP limit of six is good enough for the case of ten stations, while the larger TXOP limit introduces the negligible performance improvement in terms of throughput, delay, and loss probability (merely 1 percent). On the other hand, when the network size

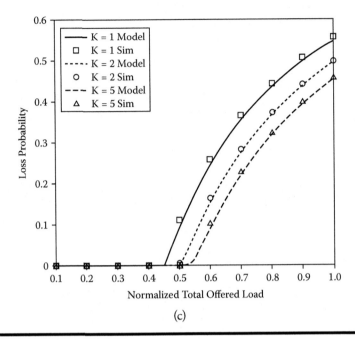

Figure 4.4 (continued)

increases, the optimal TXOP limit should be larger. For instance, in the case of 50 stations, the throughput increases 45 percent when the TXOP limit increases from six to ten. Another important observation is that the enhancement of system performance becomes more noticeable as the network size grows. Specifically, it introduces 24 percent enhancement of throughput, 29 percent decrease of delay, and 20 percent drop of loss probability, respectively, when the number of stations is ten, while these figures rise to 50 percent, 40 percent, and 30 percent, respectively, when the network size grows to 50. Thus, the results reveal that the CFB scheme is very efficient when the network size is large.

4.4.2.3 Number of Stations

As a case study we utilize the analytical model to predict the system performance behaviors and control the wireless channel access. Each station generates 550 Kbps FTP applications where the frame interarrival times follow an exponential distribution. The CFB scheme is incorporated in the wireless MAC protocol, the TXOP limit is set to 6, and the buffer size is fixed at 20. Figure 4.7 shows the predicted performance results as the number of stations increases. It can be seen that the throughput of each station stabilizes at 550 Kbps, the delay increases gradually, and the loss probability is negligible, respectively, when the network size grows up from five to ten. When the network size goes beyond ten, the throughput decreases

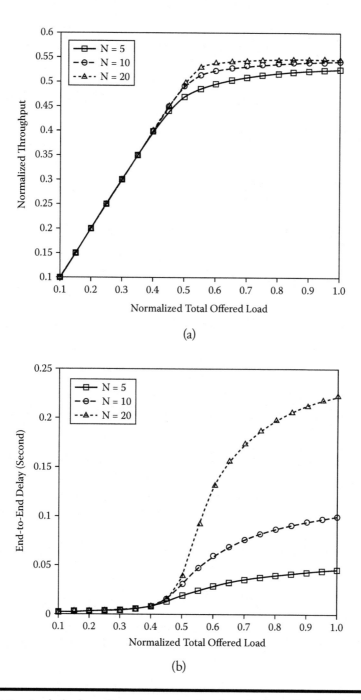

Figure 4.5   Analytical results versus normalized total offered load with varying buffer size.

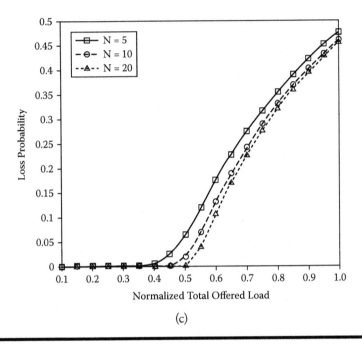

(c)

Figure 4.5 (continued)

because the system cannot support such a large number of stations. Meanwhile, the delay and loss probability climb up linearly as the network size increases. Therefore, we conclude that all the performance metrics of each station deteriorate clearly when the number of stations increases beyond ten, which is therefore the maximum number of users that the network can accommodate. Any extra access to the medium beyond the system capacity will result in excessive loss and delay of admitted users. Based on the predictions of this model, the desired performance requirements can be met for a specific QoS application by limiting the number of stations admitted in the network.

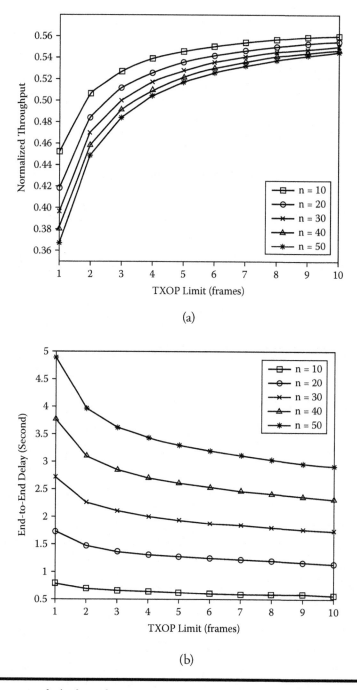

(a)

(b)

Figure 4.6   Analytical results versus TXOP limit with varying number of stations.

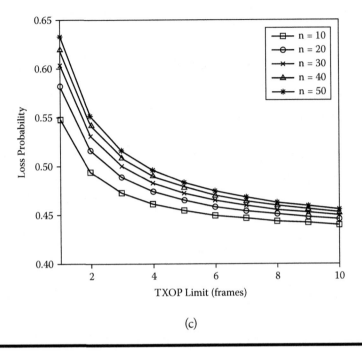

(c)

Figure 4.6 (continued)

4.5   Conclusions

The CFB scheme is a new method to alleviate the contention overheads and thus to improve the channel utilization of IEEE 802.11 WLANs. In this chapter, we have presented an analytical model to evaluate several important performance metrics of the CFB scheme, including throughput, end-to-end delay, and loss probability under unsaturated environments. First, the backoff procedure under the unsaturated traffic condition has been modeled by a bidimensional Markov chain where a new state representing a station with an empty transmission queue is introduced. Second, the service time of burst has been decomposed into two parts: the channel access delay and burst transmission delay, which were derived based on the analysis of the backoff procedure. The queueing dynamics have finally been obtained by modeling the transmission queue at each station as a bulk service queueing system. The results from the analysis have been compared with extensive ns2 simulations to validate the accuracy of the model. The model has been used to investigate the efficiency of the CFB scheme and evaluate the effects of buffer size, TXOP limit, and number of stations on the system performance. In addition, we have utilized this model as a tool to assess the maximum number of users that can be supported by the system with specific QoS requirements.

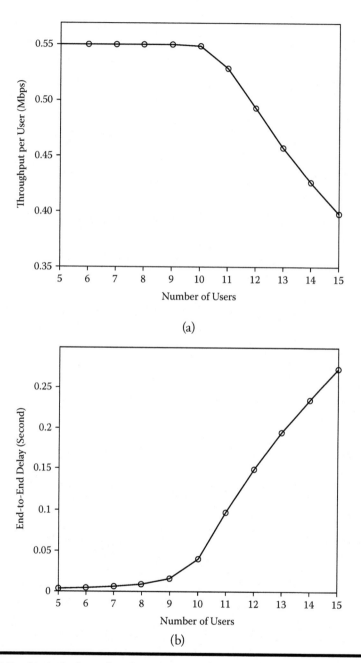

Figure 4.7   Analytical results of system capacity using 550 Kbps FTP traffic.

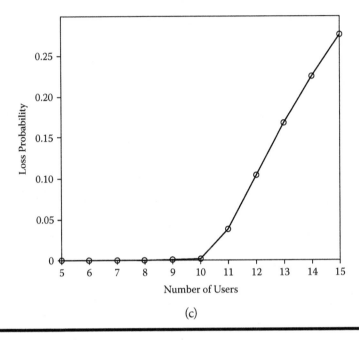

(c)

Figure 4.7 (continued)

References

[1] G. Bianchi. 1998. IEEE 802.11 saturation throughput analysis. *IEEE Communications Letters* 2:318–20.

[2] G. Bianchi. 2000. Performance analysis of the IEEE 802.11 distributed coordination function. *IEEE Journal on Selected Areas in Communications* 18:535–47.

[3] G. Bianchi and I. Tinnirello. 2005. Remarks on IEEE 802.11 DCF performance analysis. *IEEE Communications Letters* 9:765–67.

[4] G. Bolch, S. Greiner, H. Meer, and K. S. Trivedi. 1998. *Queueing networks and Markov chains: Modeling and performance evaluation with computer science applications.* New York: John Wiley & Sons.

[5] L. Bonini, M. Conti, and E. Gregori. 2004. Runtime optimization of IEEE 802.11 wireless LANs performance. *IEEE Transactions on Parallel and Distributed Systems* 15:66–80.

[6] F. Cali, M. Conti, and E. Gregori. 1998. IEEE 802.11 wireless LANs: Capacity analysis and protocol enhancement. In *Proceedings of the Annual Joint Conference of the IEEE Computer and Communications Societies (INFOCOM'98)*, 142–49.

[7] F. Cali, M. Conti, and E. Gregori. 2000. Dynamic tuning of the IEEE 802.11 protocol to achieve a theoretical throughput limit. *IEEE/ACM Transactions on Networking* 18:785–99.

[8] G. Cantieni, Q. Ni, C. Barakat, and T. Turletti. 2005. Performance analysis under finite load and improvements for multirate 802.11. *Computer Communications* 28:1095–109.

[9] M. Carvalho and J. J Garcia-Luna-Aceves. 2003. Delay analysis of IEEE 802.11 in single-hop networks. In *Proceedings of the IEEE International Conference on Network Protocols (ICNP'03)*, 146–55.

[10] P. Chatzimisios, A. C. Boucouvalas, and V. Vitsas. 2003. IEEE 802.11 packet delay: A finite retry limit analysis. In *Proceedings of the IEEE Global Telecommunications Conference (GLOBECOM'2003)*, 950–54.

[11] X. Chen, H. Zhai, X. Tian, and Y. Fang. 2006. Supporting QoS in IEEE 802.11e wireless LANs. *IEEE Transactions on Wireless Communications* 5:2217–27.

[12] S. Choi, J. del Prado, N. S. Shanker, and S. Mangold. 2003. IEEE 802.11e contention-based channel access (EDCF) performance evaluation. In *Proceedings of the IEEE International Conference on Communications (ICC'2003)*, 2:1151–56 1151–56.

[13] K. Duffy and A. J. Ganesh. 2007. Modeling the impact of buffering on 802.11. *IEEE Communications Letters* 11:219–21.

[14] P. E. Engelstad and O. N. Osterbo. 2006. Analysis of the total delay of IEEE 802.11e EDCA and 802.11 DCF. In *Proceedings of the IEEE International Conference on Communications (ICC'2006)*, vol. 2:552–59. 552–59.

[15] P. Fei, H. M. Alnuweiri, and V. C. M. Leung. 2006. Analysis of burst transmission in IEEE 802.11e wireless LANs. In *Proceedings of the IEEE International Conference on Communications (ICC'2006)*, 2:535–39 535–39.

[16] IEEE std802.11. 1999. Standard for wireless LAN medium access control (MAC) and physical layer (PHY) specifications.

[17] IEEE 802.11e/013.0. 2005. *Standard for wireless LAN medium access control (MAC) and physical layer (PHY) specifications: Medium access control (MAC) quality of service (QoS) enhancements.*

[18] L. Kleinrock. 1975. *Queueing systems: Theory.* vol. 1. New York: John Wiley & Sons.

[19] L. Kleinrock and F. A. Tobagi. 1975. Packet switching in radio channels. Part I. Carrier sense multiple access and their throughput-delay characteristics. *IEEE Transactions on Communications* 23:1417–33.

[20] Z. Kong, D. Tsang, B. Bensaou, and D. Gao. 2004. Performance analysis of IEEE 802.11e contention-based channel access. *IEEE Journal on Selected Areas in Communications* 22:2095–106.

[21] A. Kumar, E. Altman, D. Miorandi, and M. Goyal. 2007. New insights from a fixed-point analysis of single cell IEEE 802.11 WLANs. *IEEE/ACM Transactions on Networking* 15(3):588–601.

[22] T. Li, Q. Ni, and Y. Xiao. 2006. Investigation of the block ACK scheme in wireless ad-hoc networks. *Wireless Communications and Mobile Computing* 6:877–88.

[23] K. Lu, D. Wu, Y. Fang, and R. C. Qiu. 2005. Performance analysis of a burst-frame-based MAC protocol for ultra-wideband ad hoc networks. In *Proceedings of the IEEE International Conference on Communications (ICC'2005)*, 5:2937–41.

[24] D. Malone, K. Duffy, and D. J. Leith. 2007. Modeling the 802.11 distributed coordination function in non-saturated heterogeneous conditions. *IEEE/ACM Transactions on Networking* 15:159–72.

[25] S. Mangold, S. Choi, P. May, O. Klein, G. Hiertz, and L. Stibor. 2002. IEEE 802.11e wireless LAN for quality of service. In *Proceedings of the European Wireless Conference*, 1:31–39.

[26] K. Medepalli and F. A. Tobagi. 2006. Towards performance modelling of IEEE 802.11 based wireless networks: A unified framework and its applications. In *Proceedings of the Annual Joint Conference of the IEEE Computer and Communications Societies (INFOCOM'2006)*.

[27] D. Miorandi, A. A. Kherani, and E. Altman. 2006. A queueing model for HTTP traffic over IEEE 802.11 WLANs. *Computer Networks* 50(1):63–79.

[28] NS-2 network simulator. http://www.isi.edu/nanam/ns/.

[29] M. Ozdemir and A. B. McDonald. 2006. On the performance of ad hoc wireless LANs: A practical queuing theoretic model. *Performance Evaluation* 63(11).

[30] P. Raptis, V. Vitsas, K. Paparrizos, P. Chatzimisios, and A. C. Boucouvalas. 2005. Packet delay distribution of the IEEE 802.11 distributed coordination function. In *Proceedings of the International Symposium on a World of Wireless Mobile and Multimedia Networks (WOWMOM'05)*, 299–304.

[31] J. W. Robinson and T. S. Randhawa. 2004. Saturation throughput analysis of IEEE 802.11e enhanced distributed coordination function. *IEEE Journal on Selected Areas in Communications* 22:917–28.

[32] S. Selvakennedy. 2004. The impact of transmit buffer on EDCF with frame-bursting option for wireless networks. In *Proceedings of IEEE Local Computer Networks (LCN'04)*, 696–97.

[33] G. Sharma, A. Ganesh, and P. Key. 2006. Performance analysis of contention based medium access control protocols. In *Proceedings of the Annual Joint Conference of the IEEE Computer and Communications Societies (INFOCOM'2006)*, 1–12.

[34] Z. Tao and S. Panwar. 2006. Throughput and delay analysis for the IEEE 802.11e enhanced distributed channel access. *IEEE Transactions on Communications* 54:596–603.

[35] Y. Tay and K. Chua. 2001. A capacity analysis for the IEEE 802.11 MAC protocol. *Wireless Networks* 7:159–71.

[36] O. Tickoo and B. Sikdar. 2004. Queuing analysis and delay mitigation in IEEE 802.11 random access MAC based wireless networks. In *Proceedings of the Annual Joint Conference of the IEEE Computer and Communications Societies (INFOCOM'2004)*, 1404–13.

[37] O. Tickoo and B. Sikdar. 2004. A queueing model for finite load IEEE 802.11 random access. In *Proceedings of the IEEE International Conference on Communications (ICC'2004)*, 1:175–79.

[38] F. A. Tobagi and L. Kleinrock. 1975. Packet switching in radio channels. Part II. The hidden terminal problem in carrier sense multiple-access and the busy-tone solution. *IEEE Transactions on Communications* 23:1417–33.

[39] I. Tinnirello and S. Choi. 2005. Efficiency analysis of burst transmission with block ACK in contention-based 802.11e WLANs. In *Proceedings of the IEEE International Conference on Communications (ICC'2005)*, 5:345–60.

[40] V. Vitsas, P. Chatzimisios, A. C. Boucouvalas, P. Raptis, K. Paparrizos, and D. Kleftouris. 2004. Enhancing performance of the IEEE 802.11 distributed coordination function via packet bursting. In *Proceedings of the IEEE Global Telecommunications Conference Workshops (GLOBECOMW'2004)*, 245–52.

[41] H. Wu, Y. Peng, K. Long, S. Cheng, and J. Ma. 2002. Performance of Reliable Transport Protocol over IEEE 802.11 wireless LAN: Analysis and enhancement. In *Proceedings of the Annual Joint Conference of the IEEE Computer and Communications Societies (INFOCOM'2002)*, 2:599–607.

[42] Y. Xiao and J. Rosdahl. 2003. Performance analysis and enhancement for the current and future IEEE 802.11 MAC protocols. *ACM SIGMOBILE Mobile Computing and Communications Review (MC2R)* 7:6–19.

[43] A. Zenella and F. Pellegrini. 2005. Statistical characterization of the service time in saturated IEEE 802.11 networks. *IEEE Communications Letters* 9:225–27.

[44] E. Ziouva and T. Antonakopoulos. 2002. CSMA/CA performance under high traffic conditions: Throughput and delay analysis. *Computer Communication* 25:313–21.

Chapter 5

QoS Services in Wireless Metropolitan Area Networks

Haitang Wang, Bin Xie, and Dharma P. Agrawal

Contents

5.1   Introduction

The Internet and cellular networks are worldwide an indispensable part of our daily life because of their universal coverage, adequate flexibility, ease of use, and continuously declining cost. People are turning to the Internet and cellular network to communicate with others, make calls, send and receive e-mail, and check the latest news around the world. Other advance services through the Internet and the cellular network include buying or selling products online, downloading ring tones for cell phones, music and streaming video with a laptop or PDA, and engaging in chats by using Internet messengers or Short Message Services (SMS) via cell phones, etc.

In spite of all the above network services that are provided by the Internet (i.e., Ethernet, IEEE 802.11) and cellular network (i.e., Global System for Mobile [GSM] communication systems and time division multiple access [TDMA]), there still are a number of technical challenges that need to be addressed, especially for quality-of-service (QoS) purposes:

■ **QoS provisioning:** QoS refers to the ability of the network to provide a better service to the selected network. The Internet Protocol (IP) was originally designed to provide the same for all best-effort services, and thus it is not the ideal solution for integrated services, such as voice, video, and high-speed data, that exhibit widely different QoS requirements.

- **Low data rate:** The existing wide area network (WAN) technology, such as general packet radio service (GPRS), only offers an average throughput of 10 Kbps, which is far too slow from users' satisfaction point of view.
- **Costly infrastructure:** The conventional networks use copper, fiber-optic cable, and coaxial cable for link connection, and are all considerably expensive. In particular, in rural areas and developing countries, which may lack optical fiber or copper-wire infrastructures for broadband services, service providers are unwilling to install the expensive equipment for these regions with little profit potential.

To address the above limitations, the broadband wireless access (BWA) technology has recently developed and aims to provide high-speed wireless access over a wide area while satisfying QoS requirements for a variety of services. The BWA network is planned to provide broadband wireless connectivity for mobile users with at least equivalent access speed as cable modems. The goal of this chapter is to provide a better understanding of the QoS issues in the BWA network; the background of the BWA network and the QoS issues are as follows.

5.1.1 BWA Networks and IEEE 802.16 Wireless Metropolitan Area Networks

A BWA network is typically deployed in a metropolitan area, and the wireless access systems for the BWA, however, are usually owned by a service provider. The target subscriber is often an entire enterprise rather than an individual subscriber. The BWA network has the following four critical benefits:

- **Provision of high-speed mobile data and multimedia services:** A BWA network is intended to provide a wireless access alternative at a high data rate to Digital Subscriber Line (DSL) and cable network technology. It is designed to support a variety of services, including bandwidth-consuming services like data and multimedia services (i.e., Voice-over-Internet Protocol [VoIP], stream video, File Transfer Protocol [FTP], and World Wide Web [WWW]).
- **Ubiquitous wireless connectivity:** The BWA network has the ability to cover a wide geographic area without costly infrastructure to deploy cable links to individual sites. For example, one base station (BS) of a BWA network covers up to 30 miles, to which the ubiquitous broadband access should be led. Moreover, the IEEE 802.16e standard can support a mobility speed of up to 70–80 miles per hour and the asymmetrical link structure that enables the subscriber station (SS) to support various terminal devices, e.g., PDAs, cell phones, or laptops.

■ **Low cost:** The BWA network is different from a generic broadband network such as the Broadband Integrated Services Digital Network (BISDN), which refers to various network technologies (fiber or optical) implemented by Internet service providers (ISPs) and network service providers (NSPs) to achieve a transmission speed higher than 155 Mbps for the Internet backbone. The BWA network, on the other hand, is the wireless network that enables the delivery of last-mile wireless connectivity by which we can connect isolated customer buildings to an ISP or a wired backbone network without any pre-existing cables or telephoning networks. Therefore, the BWA network considerably reduces the investment cost that is caused by cable installation and expensive infrastructure devices. In addition, the BWA network allows fast deployment without the time-consuming installation of the various infrastructure devices. International Data Corporation has estimated that fewer than 10 percent of the office buildings in the United States are reachable by fiber cable. Therefore, BWA technology should offer a large market opportunity.

■ **Flexible deployment:** BWA technology provides highly flexible deployment architectures, including point-to-multipoint and point-to-point mesh networks for residential, business, and campus environments. The current target subscriber of the BWA is often an entire enterprise rather than an individual subscriber. However, it plans to enable network services for individual mobile users.

The IEEE standard for delivering integrated services (voice, video, high-speed data, WWW) over the BWA network is called IEEE 802.16 (its commercial version is referred to as WiMAX [Worldwide Interoperability for Microwave Access]). The IEEE 802.16 standard defines the wireless MAC and PHY layers of the BWA network. The IEEE 802.16 standard–based BWA network is officially known as IEEE 802.16 Wireless Metropolitan Area Networks (WirelessMANs). The European Telecommunications Standards Institute (ETSI) standardized the High Performance Radio Metropolitan Area Network (HIPERMAN) as an alternative to WiMAX, and South Korea has developed its own standard called WiBro. In the rest of the chapter, we focus further on the QoS issues in IEEE 802.16 wireless MANs.

5.1.2 QoS in IEEE 802.16 Wireless MANs

Currently, a variety of applications, especially multimedia services, are difficult to support by the traditional networks due to the limited capacity, i.e., available end-to-end throughput and restricted accessibility. For example, cellular networks offer high radio coverage but limited network capacity. On the other hand, a wireless LAN offers high-speed communication with the Internet, but has small coverage, which is not suitable for high-mobility users. One of the notable features of IEEE 802.16 wireless MAN is that it is designed to support a variety of services with QoS requirements, while providing high-bandwidth network accessibility and seamless roaming support for high-mobility users. Compared to conventional packet-switched networks (e.g., the original Internet), which were designed to provide best-effort service and do not provide full reliability in support of QoS-demand services, IEEE 802.16 wireless MAN aims at providing better service (better than best effort) to various classes of network traffic, which is called QoS.

One of main challenges for wireless MANs is to simultaneously provide QoS for various services that have very different QoS characteristics due to the following reasons. First, different services have different QoS requirements, such as minimum guaranteed rate, latency, blocking probability, maximum tolerant bit error rate (BER), etc. Next, the traffic of different services exhibits a wide range of diverse characteristics, such as various packet sizes, different interpacket arrive times, etc. Thereafter, enabling QoS in wireless networks is much more difficult than in wired networks because the characteristics of a wireless link are highly variable and unpredictable, on both a time-dependent basis and a spatial-dependent basis [1].

To cope with these issues, the QoS in wireless networks is usually managed at the Medium Access Control (MAC) layer. In the MAC layer of IEEE 802.16 wireless MANs, four types of scheduling services are defined, unsolicited grant service (UGS), real-time polling service (rtPS), non-real-time polling service (nrtPS) and best effort (BE), representing the various multimedia applications in wireless networks. Asynchronous Transfer Mode (ATM) constant-bit-rate (CBR) voice service and E1/T1 over ATM are the typical connections for the UGS, which is for real-time service flows having fixed-size data packets. The rtPS is well suited for connections that carry services such as variable bit rate video and audio. The nrtPS is suitable for the service of Internet access with minimum guaranteed rate, such as FTP. On the other hand, the intention of the BE service is to provide efficient services for WWW applications. The IEEE 802.16 MAC is defined as connection oriented to support QoS for various types of scheduling services. All services, including inherently connectionless service (packet service like IPv4, IPv6), are mapped to a connection in the IEEE 802.16 MAC. The bandwidth is allocated by a request/grant mechanism that is designed to handle four (UGS, rtPS, nrtPS, BE) types of scheduling service.

In the remainder of this chapter, we will elaborate on the QoS issues of IEEE 802.16 in detail. In Section 5.2, we illustrate the network architecture and operation modes. The PHY and MAC of IEEE 802.16 are illustrated in this section as well. The basic MAC mechanism defined by IEEE 802.16 is discussed specifically in relation to QoS provisioning. In Section 5.3, the mechanisms of the uplink and downlink frame transmission as well as the QoS-enhanced MAC protocols are discussed. Finally, Section 5.4 addresses some open issues in terms of IEEE 802.16 QoS.

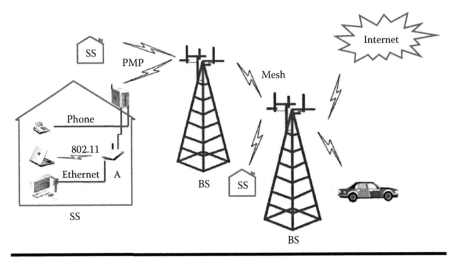

Figure 5.1 The architecture of an IEEE 802.16 wireless MAN.

5.2 QoS Support in IEEE Wireless MANs

5.2.1 Architecture of IEEE 802.16 Wireless MANs

The original IEEE 802.16 standard defines the wireless MAN air interface for the 10 to 66 GHz range, and IEEE standard 802.16-2004 [2], also known as 802.16d, has added the specification for the 2 to 11 GHz range. IEEE 802.16e-2005, also known as 802.16e, provides an improvement on the modulation schemes defined in the original IEEE 802.16 standard. IEEE 802.16e also provides mobility support up to the speed of 70 to 80 miles per hour. Figure 5.1 illustrates a general network architecture of the IEEE 802.16 wireless MAN. As defined in IEEE 802.16, the BWA setup is like a cellular system using the BS for serving multiple SSs up to a radius of several miles. Within an SS (e.g., building, house, small campus, etc.), a large number of end users with different broadband access requirements are present. An SS sends wireless traffic at a speed ranging from 2M to 155M bps from a fixed antenna on the top of a building to the BS. The BS receives transmissions from multiple SSs and sends traffic over extreme high speed wireless links to an Internet service provider (ISP) or public switched telephone network (PSTN) directly or via other BSs. As shown in Figure 5.1, the end user inside the building connects to the network with traditional in-building networks, such as Ethernet (IEEE 802.3) or wireless LANs (IEEE Standard 802.11) for data and video, and phone lines for voice. However, the final goal of the standard may eventually allow an efficient extension of the wireless MAN networking protocol directly to the individual user. The links from the BS to the home receiver and from the home receiver to the end user would likely use different physical layers, but design of the MAC could accommodate such a connection with needed QoS [3].

Using orthogonal frequency division multiplexing (OFDM), IEEE 802.16 wireless MAN allows high-speed data transmission over multiple broad frequency ranges. At the 10–66 GHz range, it requires a direct line of sight (LOS) between transmitter and receiver. The advantage of the LOS is to reduce the multipath distortion and therefore have an increased bandwidth. The IEEE 802.16 wireless MAN can provide a single-channel data rate theoretically up to 134 Mbps on both the uplink and downlink. With an enhanced orthogonal frequency division multiple access (OFDMA), IEEE 802.16e greatly improves non-line-of-sight (NLOS) performance. NLOS is the most appropriate technology available when obstacles such as trees and buildings are present. Due to this improvement provided by NLOS, stations can be mounted on homes or buildings rather than towers on mountains. NLOS in the 2–11 GHz range provides the advantage of reducing the installation cost by making under-the-eaves customer premises equipment (CPE) installation and making the problem of locating adequate CPE mounting locations relatively easy. The single-channel data rate theoretically can be up to 75 Mbps on both uplink and downlink. If multiple channels are used for a single transmission, the bandwidth can reach up to 350 Mbps, which is higher than the wireless LAN (up to 54 Mbps) and third-generation (3G) cellular technologies (up to 115 Kbps).

Two types of architecture are supported by IEEE 802.16 wireless MAN. One is point to multipoint (PMP) architecture, and the other is point-to-point (mesh) architecture. Figure 5.2 illustrates these two architectures. In the PMP mode, a BS connects to the backbone public networks, and SSs only connect to the BS, which provides SSs with last-mile access to the public network. The mesh architecture is optional for IEEE 802.16 standard. In this mode, as shown in Figure 5.2, an SS

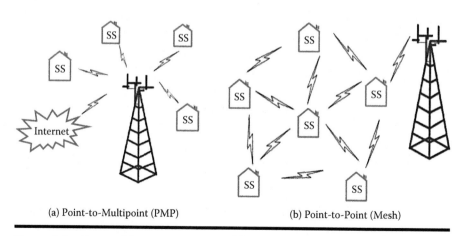

(a) Point-to-Multipoint (PMP) (b) Point-to-Point (Mesh)

Figure 5.2   Two architectures supported by IEEE 802.16 wireless MAN.

may connect to a number of other neighboring SSs, and traffic of an SS can be routed through other SSs before reaching the BS. Therefore, some SSs may also perform the functions of the BS in the mesh architecture.

5.2.2 Wireless MAN-SC PHY for 10–66 GHz Frequency Band

5.2.2.1 Frame Transmission: TDD and FDD

Both time division duplexing (TDD) and frequency division duplexing (FDD) techniques are supported in IEEE 802.16 wireless MAN-SC PHY for 10–66 GHz to allow for flexible spectrum usage. In an FDD system, the uplink and downlink channels are located on separate frequencies, but utilize a fixed duration frame. In a TDD system, the uplink and downlink transmissions occur at different times and share the same frequency. The PHY specification of IEEE 802.16 wireless MAN operates in a framed format. In the FDD, the uplink frame occurs concurrently with the downlink frame. In the TDD, as shown in Figure 5.3, a frame has a fixed length and is divided into many physical time slots. Each frame contains a downlink (DL) subframe and an uplink (UL) subframe, as shown in Figure 5.3. The downlink subframe comes first, beginning with information for frame synchronization and control. The TDD frame is adaptive, meaning that the bandwidth allocated to both directions can vary, depending on the traffic in both directions.

The downlink subframe in the TDD system includes three parts, as illustrated in Figure 5.4. The first part is frame start preamble, used for synchronization

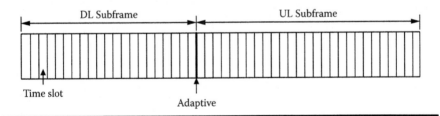

Figure 5.3   The TDD frame structure.

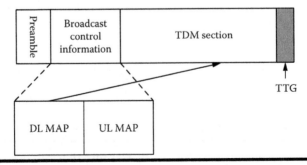

Figure 5.4   The downlink subframe structure.

and equalization. The second part is the broadcast frame control section, which contains the DL-MAP and UL-MAP, indicating the time slots at which the downlink and uplink data transmissions begin. The UL-MAP messages are contained in the downlink subframe to indicate the bandwidth allocation, starting no later than the next downlink frame. The UL-MAP contains an information element (IE) that implies the time slots in which SS can transmit during the uplink subframe. These are followed by the time division multiplexing (TDM) section carrying the downlink data. The transmit/receive transition gap (TTG) is the gap between the downlink and uplink subframes. This gap is a grace period for BS/SS to switch from transmit/receive to receive/transmit mode. In the downlink, the wireless channel is a broadcasting media. Each SS receives and decodes the control information of the downlink and looks for the data for that SS and discards the rest.

5.2.2.2 TDMA

In the uplink, time division multiple access (TDMA) is employed to avoid data collisions and fairly share the channel capacity operated in the 10–66 GHz frequency band. The noncollision is achieved by the BS that allocates a group of time slots to each SS. Upon receiving the UL-MAP, SSs will transmit their data according to the scheduled time slots. The structure of uplink subframe is shown in Figure 5.5. The uplink subframe of the IEEE 802.16 wireless MAN may include three parts: (1) the contention-based allocations for initial system access, (2) the contention-based allocation for broadcast or multicast bandwidth requests, and (3) the allocated time slots for individual SS packet transmission. Any order and quantity of these three parts may occur in the uplink subframe. The subscriber station transmission gap (SSTG) separates the transmissions of individual SSs, allowing for ramping down the previous data, followed by a preamble with which it allows the BS to synchronize to the new SS. The receive/transmit transition gap (RTG) is the guard time between the uplink subframe and subsequent downlink by which the BS/SS can switch its mode.

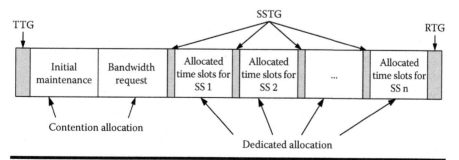

Figure 5.5 The uplink subframe structure.

5.2.3 MAC Support of IEEE Wireless MANs

5.2.3.1 Scheduling Services of IEEE 802.16 Wireless MANs

5.2.3.1.1 QoS Metrics

End users are expected to obtain services from IEEE 802.16 wireless MANs with different QoS guarantees such as end-to-end throughput, latency, and delay jitter. In this section, we first introduce these QoS parameters to evaluate the QoS performance from different angles.

Traffic rate: Defines the information rate of the service. The traffic rate is expressed in amount of data per second that can be successfully transported over a link. The rate is usually measured in bits per second. As defined in IEEE 802.16, *maximum sustained traffic rate* indicates the peak information rate of the service. *Minimum reserved traffic rate* specifies the time-average minimum amount of data to be transported on behalf of the service flow. End-to-end throughput is another similar QoS parameter that is defined as the amount of data in bits per time unit successfully delivered from one user to the end user.

Latency: The amount of time it takes for a packet to travel from the source to the destination. The *maximum latency* is defined in IEEE 802.16 to be the maximum amount of time it takes for a packet from the reception by the BS or SS on its network interface to transmission out by its RF interface.

Delay jitter: The fluctuation or variation of end-to-end delay from one packet to the next packet within the same packet stream/connection/flow. *Tolerated jitter* specifies the maximum delay variation for the connection.

5.2.3.1.2 QoS Scheduling Services

The IEEE 802.16 wireless MAN supports various types of services ranging from high QoS-guaranteed services to best-effort services. In particular, these services can be divided into the following four categories (Table 6.1): unsolicited grant service (UGS), real-time polling service (rtPS), non-real-time polling service (nrtPS), and best effort (BE). For example, the issue of VoIP with QoS provisioning is the major problem to ensure a quality VoIP service. In this section, we present the traits of these scheduling services, and then we investigate the correspondent mechanisms that are used for each service type to request bandwidth for achieving the QoS requirements specified by multimedia applications.

5.2.3.1.2.1 UGS

UGS is designed to support real-time services, which have strict delay requirements. The data flow of UGS generates fixed-size data packets on a periodic basis, such as

Table 5.1   Scheduling Services and Corresponding Characteristics

Scheduling type	Data packet size	Delay requirements	Request bandwidth
UGS	Fixed	Strict	Not needed
rtPS	Variable	Less stringent	Unicast, piggyback
nrtPS	Variable	Not specific	Unicast, piggyback, contention
BE	Variable	Not specific	Unicast, piggyback, contention

T1/E1 [2]. The state information regarding the state of the UGS service flow, such as whether the flow has overflowed the transmission queue or not, is passed from the SS to the BS. The SS is prohibited from using any contention request for UGS connection, and the BS does not provide any unicast request opportunity for the SS. Piggyback requests on the outgoing data unit are also prohibited. To provide the QoS for a UGS, the BS shall provide fixed-size grants to the service flow at periodic intervals based on the maximum sustained traffic rate. Any explicit request from the SS is not allowable, which eliminates the overhead and latency of bandwidth requests so as to meet the real-time requirement of the UGS. The size of allocations for the UGS flow shall be at least sufficient to hold the fixed-length data of the UGS flow. The BS has the ability to adaptively allocate additional capacity to the SS when a backlog in the transmission queue of the SS is detected.

The key QoS metrics for the UGS are the maximum sustained traffic rate, maximum latency, and tolerated jitter. In general, the minimum reserved traffic rate has the same value as the maximum sustained traffic rate. The BS shall not allocate more bandwidth than the maximum sustained traffic rate unless the transmission queue of this service flow is detected to overflow.

5.2.3.1.2.2   rtPS

rtPS is designed to support real-time application with less stringent delay requirements that generate variable-size data packets on a periodic basis, such as VoIP, videoconferencing, video on demand, and other multimedia applications. Because the size of the arriving packets is not fixed, rtPS connections are required to notify the BS of their current bandwidth requirements. As a result, this service causes a higher control overhead than that of UGS. The BS provides periodic request opportunities for the SS to meet flow's real-time demands and to allow the SS to specify the size of desired bandwidth grants. This service supports the variable size of allocations for the purpose of achieving the optimum efficiency of data transport. In order for the service to work appropriately, the SS is allowed to use unicast request issued by the BS for connection. However, it is prohibited from using any other request opportunities, such as contention or piggyback requests. The BS may issue unicast request opportunities even if the prior requests have not been fulfilled yet, to ensure the real-time requirements of the service.

The key QoS metrics for the rtPS are the maximum sustained traffic rate, minimum reserved traffic rate, and maximum latency.

5.2.3.1.2.3 nrtPS

nrtPS is designed to support delay-tolerant data streams that generate periodic variable-size data packets and require a minimum data rate. The applications include FTP, corporate database transactions, and those applications that have transmission rate limitations. [3]. The nrtPS is similar to the rtPS but has larger intervals for unicast requests to ensure that the service flow gets request opportunities, and during network congestion, the connections may utilize random access transmission opportunities for the SS to send their bandwidth request.

5.2.3.1.2.4 BE

The nrtPS and BE services are both designed for applications that do not have any specific delay requirement. The main difference is that the nrtPS connections reserve a minimum amount of bandwidth, which can boost performance of bandwidth-intensive applications [1], such as FTP. However, BE is intended for supporting data streams for which no minimum transmission rate is required and therefore may be handled on a space-available basis [3]. For example, the Web-browsing application can tolerate high delay jitter. The bandwidth for the connections of the BE service varies within a wide range. Thus, it is allowed to burst up to the maximum link bandwidth when the traffic of other services is not present. The SS is allowed to use contention request opportunities as well as unicast request opportunities for the BE service flow. The interval of unicast request opportunities should be longer than the nrtPS, and the availability of dedicated opportunities is subject to network load.

5.2.3.2 Request and Grant Mechanism

The IEEE 802.16 MAC is defined as connection oriented to support QoS for the above scheduling services. Every service, including an inherently connectionless service (packet service like IPv4, IPv6), is mapped to a connection. Each downlink connection has a packet queue at the BS. The BS downlink scheduler selects the next transmitted packets from those queues for the transmission to SSs in the next frame. The packet selection from the queue is based on the QoS demands of each connection and the current status of queues. On the other hand, each uplink connection has a packet queue at the SSs. The SSs uplink scheduler selects the next transmitted packets from those queues for the uplink transmission to BS based on the QoS requirements, the current status of the queues, as well as the grants from the BS.

5.2.3.2.1 Bandwidth Request

The required bandwidth is time varying (increasing or decreasing) from most services. After a connection is admitted into the network, the SS shall send a request to the BS to ask for a bandwidth allocation for all connections except for incompressible UGS connections. The bandwidth needs of the incompressible UGS connections remain constant once the connections are established. The bandwidth requirements of other service connections may change depending on traffic load. In most cases, the bandwidth request message is issued by the SS to BS during the process of uplink allocation. However, the request cannot be issued in the initial ranging interval. A bandwidth request message may come as a stand-alone bandwidth request or as a piggyback request. Two types of bandwidth request are defined. One is incremental, and the other is aggregate. Incremental requests indicate that the connection still needs the quantity of bandwidth requested besides its current perception of the bandwidth need. On the contrary, the aggregate bandwidth request states the total quantity of bandwidth needed by the connection. The aggregate bandwidth request shall be sent periodically. And piggyback bandwidth requests shall always be incremental.

5.2.3.2.2 Polling

In addition to piggybacking the request on the transmitting data unit, polling is also a mechanism by which the BS allocates bandwidth to the SSs, as shown in Figure 5.6. SSs send bandwidth request messages using the bandwidth specially allocated by the polling process of the BS. The bandwidth grant can be allocated to an individual SS by unicast polling or to a group of SSs by multicast or broadcast polling. The unicast polling is polled by the BS to an individual SS. The bandwidth request is not accomplished in the form of an explicit message, but is contained as a series of IEs in a UL-MAP. If the SS does not need the allocated bandwidth, the allocation will be padded and thus wasted. As we stated before, the SS with an active incompressible UGS connection of sufficient bandwidth shall not be polled individually. It is noted that the polling

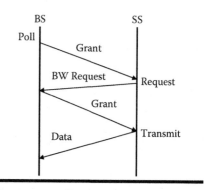

Figure 5.6 Illustration of unicast polling.

is normally done on a per-SS basis rather than a per-connection basis. On the other hand, if enough bandwidth is not available for unicast polling, SSs may be polled in a multicast group fashion or in a broadcast poll process. The bandwidth allocated for sending out requests is multicast or broadcast to a group of SSs, and any SS belonging to the polling group can request bandwidth in any request interval that is allocated in the UL-MAP. The disadvantage is that the collision occurs when two or more SSs send out bandwidth requests at the same time. To reduce collision, only SSs who need bandwidth are allowed to request. In addition, content resolution algorithms are used to reduce possible collision.

5.2.3.2.3   Grants

The BS controls the access to the channel in the uplink direction by grant transmission opportunities for SSs. Unlike a bandwidth request, which is related to individual connections, each bandwidth grant is on a per-SS basis. In other words, the bandwidth grant is allocated in an aggregated grant to the SS. In this case, the SS may receive a shorter transmission grant than it actually needs. When this happens, the SS may request again or discard the data packets. The grant decision is done by the BS scheduler based on the QoS parameters as well as the current packet queues status. The unsolicited grant needs no request for bandwidth. A fixed amount of bandwidth on a periodic basis is granted by the BS to the UGS connection.

5.3   QoS-Related Design Issues of IEEE Wireless MANs

The QoS for IEEE 802.16 wireless MAN applications cannot be achieved in a straightforward manner, as illustrated in Section 5.2 and should be addressed from many other aspects. The design factors for QoS management addressed in this section include admission control, packet scheduling, and buffer management.

5.3.1 QoS Management

The IEEE 802.16 MAC, defined as connection oriented, is designed to support different QoS for various services. The QoS of such networks can be managed in three ways: *connection admission control*, *packet scheduling*, and *buffer management*. The connection admission control is used to limit the number of connections/flows admitted into the network so that each individual connection/flow can get its desired QoS. For simplicity, we use the term *flow* with the same meaning as *connection* in the following subsections. It is noted that connection admission control only exists in a connection-oriented network. Connectionless networks, such as the

Internet, do not have any connection admission control mechanisms. Connection admission control is an important part of proactive congestion control in a connection-oriented network. After the connection is admitted and set up, the network needs to decide which packet gets transmitted first on the output link if multiple packets on ongoing connections arrive at the transmitter. The connection admission control is made in terms of the connection-level QoS requirement, such as the connection blocking probability and the average number of connections in the system. The determination of transmission sequence is accomplished by a process of packet scheduling by which the promised QoS of each flow can be achieved. In other words, packet scheduling deals with the packet-level requirements, such as the packet end-to-end delay and delay jitter. Furthermore, buffer management achieves the QoS control by stipulating the buffer size and determining which packet to be dropped if the buffer is overflowed. In a connection-oriented network, QoS cannot be efficiently achieved only by a simple packet scheduling once the network load is extremely high. An efficient connection admission control is required to control the connections in an appropriate number. In other words, a fair admission control on the connection level and an efficient packet scheduling as well as an optimal buffer management are essential to QoS provisioning for the IEEE 802.16 wireless MAN. However, none of them are addressed in the IEEE 802.16 standard. In the following, we explain QoS management from these three aspects in detail.

5.3.1.1 Connection Admission Control

The connection admission control consists of the actions taken by the network during the connection setup phase to decide whether an incoming connection can be accepted in the network. The decision is made based on whether the QoS requirement of the incoming connection violates the QoS of existing connections. Therefore, before a decision is made, the network has to evaluate the QoS attributes of both the incoming connection and the existing connections. A connection can be accepted only if sufficient resources (bandwidth) are available to establish the connection with its required QoS, while the promised QoS (minimum transmission rate, maximum delay) of existing connections in the network must not be significantly affected by the new connection. Note that the promised QoS for a connection may not be the same as the actually received QoS. The received QoS may be better than the promised QoS when network resources are sufficient. Connection admission control has to predict the fraction of the network resources that will be consumed by the traffic generated by each service. And the strategy and policy of connection admission control shall vary according to the type of services—UGS, rtPS, nrtPS, or BE—and also depend on the stochastic nature of the traffic of services.

5.3.1.2 Packet Scheduling

Packet scheduling is a part of traffic control in the networks and is referred to as the decision process used to choose which packet should be sent out first. In the connection-oriented network, connection admission control is deemed to resource reservation at the connection level; packet scheduling, on the other hand implements fair resource allocation in the packet level. The general packet scheduling algorithms include first in, first out (FIFO), round-robin, fair queuing, weighted fair queuing, etc. In FIFO, packets are forwarded in the same order in which they arrive at the transmitter. Round-robin and fair queuing are used for best-effort scheduling, and weighted fair queuing can be used as the QoS provision scheduling strategy.

5.3.1.3 Buffer Management

Like network bandwidth, buffers are another network resource whose consumption should be controlled. The buffer management is to regulate the occupancy of a finite buffer queue. The buffer management makes the decision to admit or drop an incoming packet into the queue according to the state information, such as the content of the buffer queue, the flow to which the packet belongs, the number of packets in the flow current in the buffer queue.

In the following sections, we focus on the existing approaches for IEEE 802.16 wireless MAN with regard to connection admission control and packet scheduling.

5.3.2 Existing Admission Control Schemes for IEEE 802.16 Wireless MANs

As illustrated above, connection admission control is where the BS admits a limited number of flows in the network so that each individual flow obtains its desired QoS. When the BS receives a connection request for setting up a new connection, it has to determine whether to accept or reject it. The decision is usually made based on the long-term bandwidth requirement of the connection and the current network state, such as available bandwidth. The long-term bandwidth requirement refers to the estimation of the average bandwidth requirement for the whole connection transmission. It may be different from the actual granted bandwidth for every service data unit (SDU) transmission. Similar to most admission controls, the BS sets aside a certain amount of bandwidth for the service flow, which supports efficient granting control in the MAC layer, i.e., packet scheduling at the packet level. The admission control performs a trade-off between accepting a request for connection and the resultant QoS degradation of ongoing connections. Basically, the connection request will be rejected if the QoS of the ongoing connections is predicted to be degraded to lower than their minimum QoS requirements due to acceptance of the new connection.

5.3.2.1 Dynamic Admission Control
Based on Scheduling Services

The UGS application, like E1/T1, is most commonly used by people for daily communication. On the contrary, non-UGS connections, such as rtPS, nrtPS, and BE flows that support variable-bit-rate stream video, FTP, or Hypertext Transfer Protocol (HTTP) applications, are mostly used for entertainment and getting information from the network. Due to this reason, blocking a new UGS flow may cause more serious problems than blocking a new non-UGS flow from the viewpoint of an end user. For example, a phone call, which is a UGS service, is usually related to an important commercial or personal conversation. With this consideration in mind, Wang et al. [4] developed a dynamic connection admission control scheme that gives UGS connection a higher priority than a non-UGS connection. Thus, the request for a UGS connection request is accepted without restriction if the required bandwidth is available. On the contrary, the request for non-UGS is accepted only when the total used bandwidth is not greater than the predetermined value. Before the illustration of the approach proposed by Wang et al. [4], we present the denotations as follows:

B: Total bandwidth allocated for an SS

U: Bandwidth exclusively reserved for UGS connections

b_{UGS}: Bandwidth required by a UGS connection

b_{rtPS}: Bandwidth required by an rtPS connection

b_{nrtPS}^{min}: The minimum bandwidth required for the nrtPS connection

b_{nrtPS}^{max}: The maximum bandwidth required for the nrtPS connection

δ: Amount of degraded bandwidth for every degradation step that is introduced by admitting a new connection

l_{nrtps}^{n}: Current degradation level

The required bandwidth must be satisfied to meet QoS requirements of UGS and rtPS connections. Due to the property of nrtPS flow, the required bandwidth of an nrtPS flow may vary within the range of $[b_{nrtPS}^{min}, b_{nrtPS}^{max}]$. If sufficient bandwidth is available (i.e., light connection load), each nrtPS flow can be transmitted at a higher rate. As the number of ongoing connections increases, the ongoing nrtPS flows can give up some bandwidth to the new connections to have more UGS, nrtPS, or rtPS connections in the system. This is called connection degradation model. The degradation is performed stepwise, and moreover, all the nrtPS connections in the system maintain the same degradation level, whereupon the current reserved bandwidth for each nrtPS connection is $b_{nrtPS}^{max} - l_{nrtps}^n \delta$, which satisfies $b_{nrtPS}^{max} - l_{nrtps}^n \delta \geq b_{nrtPS}^{min}$. The maximum degradation step is $l_{nrtps}^{max} = (b_{nrtPS}^{max} - b_{nrtPS}^{min}) / \delta$.

The process of the dynamic connection admission control scheme is as follows:

■ When a request for a UGS connection arrives at the BS, if the bandwidth currently set aside for all ongoing connections plus b_{UGS} is less than or equal to the total bandwidth allocated for an SS B, the request is accepted. The b_{UGS} bandwidth for its lifetime transmission is reserved for this incoming connection. Otherwise, the request for the UGS connection is rejected.

■ When a request for an, rtPS connection arrives at the BS, if the total bandwidth set aside for all ongoing connections plus b_{rtPS} is less than or equal to $B - U$, the connection is admitted and the BS sets aside b_{rtPS} bandwidth for the connection. Otherwise, the BS degrades the bandwidth set aside for all ongoing nrtPS connections, until the total bandwidth set aside for all ongoing connections plus b_{rtPS} is not greater than $B - U$. If the currently set aside bandwidth plus b_{rtPS} is still greater than $B - U$ and the maximum degradation step l_{nrtps}^{max} of nrtPS connections has been reached, the request for rtPS connection is rejected; otherwise, it is admitted by reserved b_{rtPS} bandwidth.

■ When a request for an nrtPS connection arrives at the BS, if the total bandwidth already set aside for all ongoing connections plus $b_{nrtPS}^{max} - l_{nrtPS}^n \delta$ is less than or equal to $B - U$, the connection is admitted into the system. The bandwidth reserved for this nrtPS connection is $b_{nrtPS}^{max} - l_{nrtPS}^n \delta$. If not admitted, the BS degrades the bandwidth set aside for all ongoing nrtPS connections until the current total bandwidth for all ongoing connections plus the bandwidth for the new nrtPS connection is not greater than $B - U$. If this can be reached, the nrtPS connection is admitted with $b_{nrtPS}^{max} - l_{nrtps}^{n'} \delta$, where $l_{nrtps}^{n'}$ $(l_{nrtPS}^n \leq l_{nrtps}^{n'} \leq l_{nrtPS}^{max})$ is the updated degradation level of all nrtPS connections after having a new connection. Otherwise, the nrtPS connection is rejected.

■ When the request for the BE connection arrives at the BS, the request is always admitted, but the BS will not set aside any bandwidth for such a connection. In 802.16 MAC, the BE connections get the transmission opportunity only when the connections of other services do not transmit packets. Generally, BE connections do have long idle periods, which is referred to as think time, and data in each transmission is relatively small, especially in the uplink direction. Therefore, the QoS of the BE can be easily satisfied.

In the dynamic connection admission control and bandwidth allocation scheme, UGS flows are given the highest priority and the performance of UGS flows have been significantly improved by using this approach. Furthermore, both the bandwidth utilization and the average number of connections in the system are increased by the above defined bandwidth borrowing and degradation. However, in this approach the packet-level QoS requirements, i.e., end-to-end delay and delay jitter, are not taken into consideration when making the decision of connection admission control. In the next section, we introduce another scheme for connection admission control that considers the QoS requirements from both the connection level and the packet level.

5.3.2.2 Optimization-Based Connection Admission Control

The connection admission control approach proposed by Niyato and Hossain [5] aims to maximize the level of satisfaction on the received QoS (bandwidth, average delay) for different service types in IEEE 802.16 wireless MANs. The level of QoS satisfaction is evaluated by utility functions, and an optimization problem is formulated and solved to obtain the optimal amount of allocated bandwidth for the ongoing connections and newly arriving connection if it is admitted.

The utility function represents the satisfaction level of the offered QoS for a service type: the higher the utility of the service, the higher the satisfaction of the QoS obtained for the service. The utility for connection i depends on the amount of allocated bandwidth, average delay, transmission rate, and traffic model of the service it belongs to. Specifically, the utility function for the UGS connection is represented by

$$U_{UGS}(b_i) = \begin{cases} 1, & b_i \geq b_{UGS}^{(req)} \\ 0, & otherwise. \end{cases}$$

The utility function for the UGS connection equals 1 if the amount of allocated bandwidth (b_i) for connection i is higher than or equal to the required bandwidth ($b_{UGS}^{(req)}$):

$$U_{BE}(b_i) = \begin{cases} 1, & b_i \geq 1 \\ 0, & otherwise. \end{cases}$$

The utility function for a BE connection is equal to 1 if the connection is admitted into the network. The utility functions for rtPS and nrtPS connections are represented by modified sigmoid functions [6] of the packet-level performance metrics. $d(\overline{\gamma}, \lambda, b_i)$ and $\tau(\overline{\gamma}, \lambda, b_i)$ denote the average delay and transmission rate, respectively, as the functions of protocol data unit (PDU) arrival rate (λ) and average signal-to-noise ratio (SNR) ($\overline{\gamma}$) when the amount of allocated bandwidth is b_i. The utility functions for the rtPS and nrtPS are

$$U_{rtPS}(b_i) = 1 - \frac{1}{1 + \exp(-g_{rt}(d(\overline{\gamma}, \lambda, b_i) - d_i^{(req)} - h_{rt}))},$$

$$U_{nrtPS}(b_i) = \frac{1}{1+\exp(-g_{nrt}(\tau(\overline{\gamma},\lambda,b_i)-\tau_i^{(req)}-h_{nrt}))},$$

where g_{rt}, g_{nrt}, h_{rt}, and h_{nrt} are the parameters of sigmoid functions; g_{rt} and g_{nrt} determine the sensitivity of the utility function to the delay or throughput requirement; and h_{rt} and h_{nrt} represent the center of the utility function.

The optimization problem of maximizing the sum of the utilities for all connections is formulated to allocate the total bandwidth of the C unit among M (both ongoing and incoming) connections:

Maximize:

$$\sum_{i\in O_{UGS}}(U_{UGS}b_i^{(up)})+U_{UGS}b_i^{(do)})+$$

$$\sum_{i\in O_{rtPS}}(U_{rtPS}b_i^{(up)})+U_{rtPS}b_i^{(do)})+$$

$$\sum_{i\in O_{nrtPS}}(U_{nrtPS}b_i^{(up)})+U_{nrtPS}b_i^{(do)})+ \tag{5.1}$$

$$\sum_{i\in O_{BE}}(U_{BE}b_i^{(up)})+U_{BE}b_i^{(do)})$$

Subject to:

$$b_i^{(up)}=b_{UGS}^{(up,req)}, b_i^{(do)}=b_{UGS}^{(do,req)} \quad \text{for } i\in O_{UGS}, \tag{5.2}$$

$$d(\overline{\gamma},\lambda_i^{(up)},b_i^{(up)})\le d_i^{(up,req)},$$
$$d(\overline{\gamma},\lambda_i^{(do)},b_i^{(do)})\le d_i^{(do,req)} \quad \text{for } i\in O_{rtPS}, \tag{5.3}$$

$$\tau(\overline{\gamma},\lambda_i^{(up)},b_i^{(up)})\le \tau_i^{(up,req)},$$
$$\tau(\overline{\gamma},\lambda_i^{(do)},b_i^{(do)})\le \tau_i^{(do,req)} \quad \text{for } i\in O_{nrtPS}, \tag{5.4}$$

$$b_i^{(up)}=b_i^{(do)}=1 \quad \text{for } i\in O_{BE}, \tag{5.5}$$

$$b_{min}\le b_i^{(up)}, b_i^{(do)}=b_{max} \quad \forall i, \tag{5.6}$$

$$\sum_{\forall i} b_i^{(up)} + \sum_{\forall i} b_i^{(do)} \leq C, \tag{5.7}$$

$$\sum_{i \in O_{UGS}} b_i \leq \Gamma_{UGS}, \sum_{i \in O_{rtPS}} b_i \leq \Gamma_{rtPS},$$

$$\sum_{i \in O_{nrtPS}} b_i \leq \Gamma_{nrtPS}, \sum_{i \in O_{BE}} b_i \leq \Gamma_{BE}, \tag{5.8}$$

where O_{UGS}, O_{rtPS}, O_{nrtPS}, and O_{BE} denote the set of UGS, rtPS, nrtPS, and BE connections, respectively; $b_{UGS}^{(up,req)}$ and $b_{UGS}^{(do,req)}$ respectively stand for the bandwidth required by a UGS connection for uplink and downlink transmissions; b_{min} and b_{max} are the minimum and maximum amounts of bandwidth that is allowed to be allocated to a connection; and Γ_{UGS}, Γ_{rtPS}, Γ_{nrtPS}, and Γ_{BE} are thresholds representing the amount of bandwidth reserved for UGS, rtPS, nrtPS, and BE connections, respectively. Equations (5.2)–(5.5) represent the QoS constraints for UGS, rtPS, nrtPS, and BE connections. All the connection-level parameters, Γ_{UGS}, Γ_{rtPS}, Γ_{nrtPS}, and Γ_{BE}, as well as packet-level parameters can be calculated by the optimizing-based scheme and queuing analysis presented in [6].

The connection admission control is based on results of the above optimization problem. If the problem is unsolvable, the incoming connection is blocked; otherwise, the connection is accepted.

In the approach, QoS constraints are taken into consideration not only in the connection level but also in the packet level. However, there are still blemishes in this approach. First, how to choose the utility functions is not interpreted clearly. Second, the assumption that the connection interarrival time and connection holding time both follow the exponential distribution does not well match with the practical integrated traffic in MANs. In particular, the exponential distribution is not well suited for traffic that exhibits self-similarity, such as the traffic of rtPS and nrtPS connections, as illustrated in the following section.

5.3.2.3 Traffic Self-Similarity–Based Admission Control

A traffic model is a stochastic process that can be used to predict the behavior of real traffic streams [7]. Work has been done on analyzing some QoS metrics using an approximate mathematical model by assuming that the connection arrival process is a Poisson process and the duration of the connection follows the exponential distribution. In the telecommunication network, these two assumptions hold for voice traffic. However, for broadband-consuming traffic, such as high-speed video and data traffic, the connection duration does not simply follow exponential distribution. Many recent works have indicated that wide area network traffic is better modeled as self-similar statistical processes [8]. In [4], authors further investigated the traffic models for IEEE 802.16 wireless WANs and proposed an admission control scheme to guarantee the blocking probability of flows for different service types.

The UGS has been characterized by constant-bit-rate traffic-like voice. For voice traffic, a Poisson process is used to model the connection arrival process with fixed hourly rates within $1-h$ periods, which means that the interarrival time between two UGS connections is exponentially distributed. The connection duration (holding time) is also exponential distribution. In the packet level, the packet interarrival time and the packet length are constant, which depends on the bit rate. Figure 5.7 shows such a traffic process above the packet level for the traffic of UGS connections.

rtPS and nrtPS services are mostly used for variable-bit-rate (VBR) video and FTP. Both types of traffic are high-speed fractal (self-similar) traffic characterized by bursts on many timescales. Such bursts are caused due to downloading large files or long periods of VBR video activities. In a large network, modeling rtPS and nrtPS traffic as a Poisson process fails to capture the dynamic property of the traffic. The accurate model for these types of traffic should capture the long bursts [9]. The Poisson Pareto burst process (PPBP) is a random process that can be used to model the burst property of such traffic [7]. PPBP is basically a Poisson process with a certain rate of Pareto-distributed overlapping bursts. Figure 5.8 shows the process above the packet level for rtPS and nrtPS. Like UGS, the arrival process of rtPS and nrtPS connections can be modeled by a Poisson process within $1 - h$ intervals. Each of these arrivals reflects an individual user starting a new session. The arrival of overlapped bursts that are initiated whenever the user transfers files is a Poisson process with rate k. The period representing the length of a burst has a Pareto distribution. During the burst duration period, the packet arrival process is constant with rate r. The Pareto distribution with shape parameter α and location parameter β has the following cumulative distribution function:

$$F(x) = P[X \leq x] = 1 - (\frac{\beta}{x})^{\alpha}, \qquad 1 < \alpha < 2, \beta > 0, x \geq \beta$$

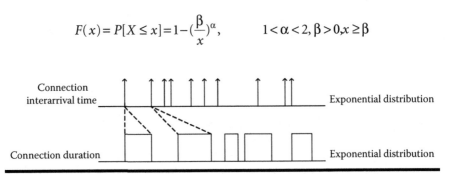

Figure 5.7 UGS traffic arrival process above the packet level.

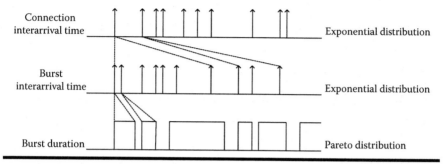

Figure 5.8   High-speed fractal traffic arrival process above the packet level.

By [9], the mean amount of work arriving within an interval of length t is ktr $\alpha\beta/(\alpha - 1)$, and the variance of the amount of work arriving in an interval of length t is

$$
\sigma^2(t) = \begin{cases} 2r^2\lambda t^2(\dfrac{\alpha\beta}{2(\alpha-1)} - \dfrac{t}{\beta}), & 0 \le t \le \beta \\[4mm] 2r^2\lambda\{\dfrac{\beta^3\alpha}{6(3-\alpha)} - \dfrac{\beta^2 t\alpha}{2(2-\alpha)} - \dfrac{\alpha\beta t^{3-\alpha}}{(1-\alpha)(2-\alpha)(3-\alpha)}\}, & t > \beta \end{cases}.
$$

The BE service is mainly about WWW traffic. WWW traffic has also been proven to be self-similar [10]. Figure 5.9 illustrates the BE traffic arrival process above the packet level. Every connection of WWW traffic represents a new request, and the connection arrival is a Poisson process, as shown in Figure 5.9. The distribution of a document transfer size is heavy-tailed with a Pareto distribution.

It has been proved [7, 11] that as the number of independent sources contributing to aggregate flows increases, the traffic weakly converges to a Gaussian process, which is as λ increases to a large number, the Gaussian process can be used to represent the traffic model of integrated services. The reason behind this is that in the IEEE 802.16 wireless MAN, from BS's view the traffic is aggregated by a large number of end users from SSs. Thus, the aggregated traffic process at BS can be approximated as a Gaussian model no matter what traffic model is used for each type of traffic.

Therefore, a method of determining the amount of bandwidth required for each type of traffic to ensure the given upper-bound blocking probabilities of each type of scheduling service above the packet level is formalized by the Chernoff bound-based approach [4] according to the above traffic models. An admission control and bandwidth allocation mechanism above the packet level to minimize the blocking

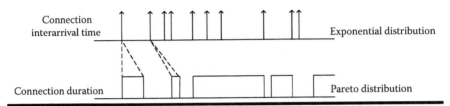

Figure 5.9   BE traffic arrival process above the packet level.

probability of each type of service is proposed. In this scheme, the total bandwidth (*C*) is partitioned for four types of scheduling services as C_i (i = 1, 2, 3, 4). C_i is calculated by a binary search algorithm with which the total bandwidth is fairly divided into four parts and each part corresponds to a type of multimedia service. The partition should achieve the goal that the upper bound connection/burst blocking probability of each type of service is minimized. If a new connection of UGS or BE arrives at the SS, it sends a request to the BS for its permission. It is admitted and reserved for bandwidth if the aggregated bandwidth of the same service, including for this connection, is less than C_i (i = 1, 4). Otherwise, it is blocked. If a new connection of rtPS or nrtPS arrives at the SS, the connection is always admitted. However, no bandwidth is reserved for this connection at this time. When the burst of connection arrives, the BS needs to decide whether to admit the burst based on the residual bandwidth and partitioned bandwidth. The burst is blocked when the total reserved bandwidth for the same service is larger than C_i (i = 2, 3).

Given a total bandwidth for an SS, this approach [4] can guarantee the minimum blocking probability for all scheduling services. The binary computation is not complex and computed only once offline by the BS before partitioning the bandwidth. Due to the self-similarity of rtPS and nrtPS traffic, the admission control is considered not only in the connection level but also in the burst level.

5.3.3 Existing Packet Scheduling Schemes or IEEE 802.16 Wireless MANs

This section provides insight into the current packet scheduling schemes proposed for IEEE 802.16 wireless MANs. Packet scheduling is to schedule the packet to be transmitted on the output link according to the QoS requirement of the connection. In the IEEE 802.16 wireless MAN, upon accepting a new connection the BS may poll this new connection and give an opportunity for the user to request a desired bandwidth or grant a bandwidth directly without negotiation, such as UGS connections. For the other services, the connection should send its bandwidth request to the BS for receiving bandwidth grants (e.g., time slots for transmitting data) from the BS. The grants are the result of uplink packet scheduling at the BS and will be included in the UL-MAP field in the downlink subframe. Efficient packet scheduling should meet different packet-level QoS requirements of different service types, such as the end-to-end delay requirements for real-time services.

5.3.3.1 Discriminating-Based Uplink Scheduler

In the proposed packet scheduler [13], all request packets and data packets are first pushed into distinguished transmission queues (type I, type II, and type III), as shown in Figure 5.10.

5.3.3.1.1   Type I Queue

Two types of traffic are pushed into the type I queue. The first type is the periodic generated data packets of UGS flows, and the second type is the periodic upstream unicast requests for rtPS and nrtPS flows. The data packets of UGS flows are generated at the time of $t_i = t_0 + i * interval$, where *interval* is the nominal grant interval for such a flow. Each generated data packet of the UGS flow has a marked delivery deadline of $t_i + jitter$, where *jitter* is the tolerated grant jitter for such a flow. The unicast requests of rtPS and nrtPS are generated in the same manner as the above UGS data packets. However, in this case the *interval* for the request means the nominal polling interval and the *jitter* is the tolerated polling jitter. The scheduler provides a strict semi-preemptive priority to the requests and UGS data packets in such a way that they are scheduled for transmission before their corresponding deadlines. If the newly arrived type I request or data packets can be transmitted before their deadlines without the need to preempt the delivery of current ongoing packets, the undergoing packets in type II or type III queues will be transmitted without inter-

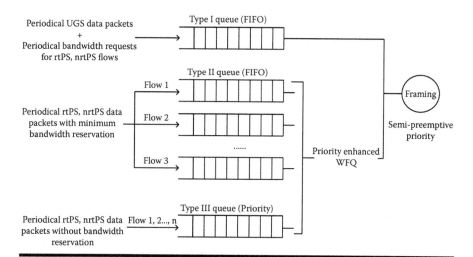

Figure 5.10   Scheduling architecture for uplink transmission.

ruption. On the contrary, if the type I request or data packets are approaching their deadlines, the scheduling of the packet for type II or type III has to be interrupted. The preemption may result in fragmentation of the low-priority data packet. If the fragmentation cannot be avoided, the size of non-UGS data packets should be small enough so as to satisfy the tolerated jitter of UGS flows.

5.3.3.1.2   Type II Queue

As shown in Figure 5.10, the type II queue contains multiple queues that are used for rtPS or nrtPS flows having minimum bandwidth reservation. Priority-enhanced policy such as weighted fair queuing (WFQ), self-clocked fair queuing (SCFQ), or start-time fair queuing (SFQ), may be employed to schedule the packet transmission of the type II queue. For example, a WFQ weight is assigned to each type II queue to reflect the minimum bandwidth reserved for the corresponding service flow. If the two data packets from two different type II queues have identical WFQ virtual finish times, then the first data packet to be scheduled is the one that has higher priority. This makes the higher-priority packet incur less delay.

5.3.3.1.3   Type III Queue

The data packets of service without minimum bandwidth reservation are kept in the type III queue. The type III queue is shared by all these service flows (rtPS, nrtPS, BE). Like type II queues, the priority-enhanced WFQ can be employed to handle type II and type III queues. The weight of the type III queue is calculated by subtracting all the reserved bandwidth for type II queues, UGS flows, and contention minislots (time slots allocated for random access requests for nrtPS and BE) from the aggregate output link capacity. The data packets of the rtPS, nrtPS, and BE flows without minimum bandwidth reservation are aggregated into a type III queue to reduce the complexity of the priority-enhanced WFQ algorithm. This is because usually the number of flows with bandwidth reservation is smaller than the number of flows without bandwidth reservation. Moreover, the scheduling policy within type III queues is not FIFO, which is used in both type I and type II queues. It is nonpreemptive priority to the contrary. That is, all data packets are sorted according to their priorities in the type III queue.

5.3.3.2 *Dynamic Priority Downlink Transmission Scheduling*

The packet scheduling architecture proposed in [12] gives the real-time packet the same priority as BE packets when the deadline condition can be satisfied for the real-time packet. Different from most existing packet scheduling schemes, which always assign a higher priority to real-time packets, it reduces the cost of BE services.

5.3.3.2.1   Traffic Priority

Two levels of priority, low and high, are defined in the proposed packet scheduling architecture [13]. The priorities of real-time packets and BE packets default to low level, and the priority of real-time packets may change to high level in the following two cases.

The first case is that the estimated data volume (in bits) in the received buffer for the mobile user is smaller than the predefined thresholds. Using a video service as an example, the downlink received buffers are stored in the video packet buffer of each receiver for retrieving periodically and playing back. If the buffer goes to empty, the continuous video playing may be interrupted. To avoid this, the base station estimates the remaining data volume of video buffer based on the following relation:

$$Q_i(t) = \max\{Q_i(t-1) + D_i(t-1) - R_i L_f, 0\},$$

where $Q_i(t)$ denotes the estimated data volume in the video buffer at the receiver of end user i at frame t, $D_i(t-1)$ represents the amount of video traffic that user i has successfully received in frame $t-1$ from BS, R_i is the mean playback rate of video service at the receiver of end user i, and L_f is frame duration. Two thresholds, Q_{TH1} and Q_{TH2}, are used to decide the priority level of real-time service. If $Q_i(t) > Q_{TH1}$, the priority of the real-time packets to be transmitted to the receiver of user i is set to low. If $Q_{TH2} < Q_i(t) \leq Q_{TH1}$, the priority of the first real-time packet to be transmitted to the receiver of user i is set to high. If $Q_i(t) \leq Q_{TH2}$, the priority of the first two packets in the real-time transmission queue to the receiver of user i is set to high.

The second case is that the priority of real-time packets changes when a handoff of user i happens. In the process of handoff, mobile user i migrates its service from a BS to the neighboring BS. The transmission is interrupted when the connection is released from the old BS while the new connection to the new BS has not been established yet. Such transmission suspension causes performance degradation of real-time service. To reduce the transmission gap, the scheduler at BS assigns a high priority to the real-time packet for the handoff to user i.

5.3.3.2.2   Packet Scheduler

In general, the packet scheduler first assigns the bandwidth grants to the high-priority packets. If the number of total high-priority packets is larger than the number of packets that can be transmitted in a frame by all subchannels of the BS, the BS transmits the high-priority packets according to some kind of predefined scheduling policy. For example, the best channel first (BCF) is based on the principle that the transmission queues for the receivers of different users are sorted in order of downlink channel qualities. The proportional fair (PF)–based approach transmits the high-priority packets according to a transmission priority policy [12]. In the

PF priority policy, the transmission sequence considers two factors: the bandwidth grant for each user and the number of packets of each user in the transmission queue. The amount of bandwidth grants is proportional to the number of packets in the transmission queue. If the number of total high-priority packets is smaller than or equal to the number of packets that can be transmitted in a frame by all subchannels of the BS, the BS transmits the entire high-priority packet according to a predefined scheduling policy, such as BCF and PF. The remaining available bandwidth grants are allocated to the low-priority packets. The scheduling policy for the low-priority packets may also use BCF or PF or other, more suitable scheduling policies.

The packet scheduling architecture only gives high priority to the real-time packet when its deadline is close or during the handoff process. Otherwise, both real-time packets and BE packets have the same priority level. This architecture improves the performance of BE and guarantees the delay requirement of real-time service. However, these improvements come at the cost of high computational overhead, which again imposes the performance requirement on the SS and BS hardware. Moreover, this packet scheduling architecture is only designed for downlink transmission scheduling, without the consideration of the uplink.

5.4 Open Issues for QoS in IEEE 802.16 Wireless MANs

Essentially, the QoS in the IEEE 802.16 wireless MAN enables the network to provide better service to a variety of flows. This chapter introduced the IEEE 802.16 wireless MAN architecture and the standardized MAC protocols that are highly involved in the network QoS design. It then addressed several QoS enhancement techniques, including connection admission control, packet scheduling, and buffer management. However, the QoS in the IEEE 802.16 wireless MAN is still largely open to research, including QoS-differentiated admission control and bandwidth allocation, and mobility technologies. The key open issues include:

- **QoS-differentiated admission control and bandwidth allocation:** The practical traffic in the IEEE 802.16 wireless MAN may be huge and exhibits saliently spatial and temporal properties. For example, the traffic at a commodity and the traffic at the business area may exhibit different time-varying properties. Thus, a general stochastic process may not capture such differences. Furthermore, QoS-differentiated connection admission control schemes have to be developed under different performance criteria, such as reliability to network size and adaptability to real-time connections.
- **Wireless QoS:** The wireless QoS provision is a complex issue in wireless networks due to the error-prone nature of wireless channels. The QoS for

an application involves each layer of the protocol stack and crosses different heterogeneous network environments. Some advanced features such as cross-layer design and end-to-end QoS negotiation can be employed to improve the QoS design of the IEEE 802.16 wireless MAN. The access network (i.e., SS and BS) needs to be aware of these characteristics of the wireless links in the communication path to provide QoS.

■ **Seamless mobility:** Guaranteeing QoS for various service types plus mobility is another problem. To ensure QoS guarantees for mobile users, a seamless handoff mechanism is needed in networks. This becomes especially critical in the case of delay- or jitter-sensitive traffic such as voice and video. The seamless handoff for the 802.16 wireless MAN is the ability for a service to cross the different BSs without degradation to the service.

References

[1] Cicconetti, L. C., et al. 2006. Quality of service support in IEEE 802.16 networks. *IEEE Network Magazine* 20:50.

[2] IEEE 802.16 Working Group. 2004. *IEEE standard for local and metropolitan area networks: Air interface for fixed broadband wireless access systems.* IEEE Standard 802.16-2004 (revision of IEEE Standard 802.16-2001), Part 16.

[3] Eklund, R. B. C., et al. 2002. IEEE standard 802.16: A technical overview of the Wireless MAN™ air interface for broadband wireless access. *IEEE Commun. Mag.* 40:1.

[4] Wang, H., He, B., and Agrawal, D. P. 2007. Above packet level admission control and bandwidth allocation for IEEE 802.16 wirelessMAN. *Simul. Model. Pract. Theory* 15:366.

[5] Niyato, D., and Hossain, E. 2006. A queuing-theoretic and optimization-based model for radio resource management in IEEE 802.16 broadband wireless networks. *IEEE Trans. Comput.* 55:1473.

[6] Xiao, M., Shroff, N. B., and Chong, E. K. P. 2001. Utility-based power control in cellular wireless systems. *Proc. IEEE INFOCOM* 1:412.

[7] Zukerman, M., Neame, T. D., and Addie, R. G. 2003. Internet traffic modeling and future technology implications. *Proc. IEEE INFOCOM* 1:587.

[8] Leland, W. E., et al. 1994. On the self-similar nature of Ethernet traffic (extended version). *IEEE/ACM Trans. Networking* 2:1.

[9] Addie, R. G., Zukerman, M., and Neame, T. D. 1998. Broadband traffic modeling: Simple solutions to hard problems. *IEEE Commun. Mag.* 8:88.

[10] Crovella, M., and Bestavros, A. 1995. *Explaining World Wide Web traffic self-similarity.* Revised Technical Report TR-95-015, Computer Science Department, Boston University.

[11] Intanagonwiwat, C., et al. 1999. On the weak convergence of long range dependent traffic process. *J. Stat. Planning Inference* 80:155.

[12] Jeon, W. S., and Jeong, D. G. 2006. Combined connection admission control and packet transmission scheduling for mobile Internet services. *IEEE Trans. Veh. Technol.* 55:1582.

[13] Hawa, M., and Petr, D. W. 2002. Quality of service scheduling in cable and broadband wireless access systems quality of service. In *Proceedings of the Tenth IEEE International Workshop on Quality of Service*, vol. 1, 247.

Chapter 6

Soft QoS Support for Mobile Ad Hoc Networks Based on End-to-End Path Probing and IEEE 802.11e Technology

Carlos T. Calafate, Juan Carlos Cano,
Pietro Manzoni, and Manuel Pérez Malumbres

Contents

6.1 Introduction

Digital wireless communication systems offer new possibilities and pose new engineering challenges with respect to legacy cabled networks. In the field of digital wireless communications itself we have different degrees of complexity according to the requirements, the technologies used, and the type of network involved.

Mobile ad hoc networks (MANETs) are an exciting field of research that, compared to infrastructure-based wireless networks, demands several technological improvements to adapt to on-off radio link conditions, distributed Medium Access Control (MAC) protocols, frequent topology changes, and, at a higher level, resource sharing among peers. Achieving quality-of-service (QoS) support in MANETs is a very complex task. Therefore, any solution usually requires cross-layer operation to be successful; this includes medium access, routing, and admission control, among others. One of the most important constraints to achieve QoS support in MANETs is related to radio interference and channel access. Because MANETs usually make use of the ISM frequency bands, several distinct technologies may cause radio interference among themselves; obviously, under these conditions, it is very difficult to achieve strict QoS guarantees.

MANETs typically rely on IEEE 802.11 [1] technology, which employs a distributed channel access algorithm. Therefore, a mechanism such as the one provided in annex E of the IEEE 802.11 standard is mandatory to classify and prioritize the different QoS traffic types.

The responsiveness of routing protocols is much more important in MANETs than in wired networks due to the presence of mobility; in fact, real-time multimedia data flows require highly responsive routing protocols to avoid long disconnection periods. At the routing layer, QoS-based route provision can also be enforced. Works such as [2–4] are proposals for QoS support at the routing layer.

A key element in any QoS framework is the presence of admission control. However, in MANETs, lack of ownership of those terminals along a path may make it impossible to do per-hop resource assessment and reservation. For those situations, a more flexible solution that avoids strict resource reservation is required, while offering soft QoS guarantees to real-time data flows.

This chapter is organized as follows: In Section 6.2 we present two well-known QoS frameworks for MANET environments—INSIGNIA [5] and SWAN [6]—which are an adaptation of the Integrated Services (IS) [7] and Differentiated Services (DS) [8] models, respectively, to MANET environments. As an alternative to these two proposals, we proceed by studying a novel QoS framework that offers soft QoS support by combining end-to-end path probing with IEEE 802.11e technology. An overview of the IEEE 802.11e technology is presented in Section 6.3, including a performance evaluation of IEEE 802.11e in both static and mobile environments. Section 6.4 is dedicated to DACME, a novel admission control system for MANET environments that enables real-time multimedia communication among peers. Distributed Admission Control for Manet Environments (DACME) is based on distributed admission control techniques and imposes very few requirements on MANET nodes. In fact, intermediate MANET stations only need to have IEEE 802.11e–capable interfaces and to handle packets according to the type-of-service (TOS) field in their Internet Protocol (IP) header. By imposing few requirements to intermediate stations, DACME is able to accommodate different paradigms of user cooperation in MANETs. The core of DACME is a probe-based mechanism that periodically measures end-to-end path conditions. These probe-based measurements are used by DACME agents to decide whether to admit traffic from an application based on its QoS requirements and the estimated available resources. By employing a series of mathematical adjustments, the proposed distributed admission control technique is able to offer reliable end-to-end measurements for bandwidth, delay, and jitter; moreover, the time spent in that process is relatively low (typically less than 50 ms). The performance of DACME in MANET environments, presented in Section 6.5, is assessed through a series of simulation-based experiments. In Section 6.6 we offer a summary of the chapter's contents, pointing out the main conclusions of the analysis made.

6.2 State of the Art on QoS Architectures for MANETs

The Internet was initially created to handle best-effort traffic alone. This means that there is no sort of resource reservation and so all users compete for bandwidth. This is one of the reasons why the Internet Protocol (IP) is connectionless, requiring no setup "signaling" for admission control. Later, enhancements in terms of available bandwidth and terminal's capabilities created the need for supporting new services in the Internet. These new services, though, performed poorly due to the best-effort policy. There was therefore a need to enhance the Internet to perform resource reservation in a fashion similar to that of telephony networks. The Resource Reservation Protocol (RSVP) [9] was created to fulfill this need as part of the Internet's Integrated Services (IntServ) architecture [7]. RSVP follows a receiver-based model because it is the responsibility of each receiver to choose its own level of reserved resources, initiating the reservation and keeping it active. The actual QoS control, though, occurs at the sender's end. The sender will try to establish and maintain resource reservations over a distribution tree. If a particular reservation is unsuccessful, the correspondent source is notified.

The Integrated Services architecture proved to be complex and required too many resources, suffering from scalability problems. The Differentiated Services (DiffServ) architecture [8] emerged as a more efficient alternative. In the latter, service level agreements (SLAs) are achieved between different domains. One of the main virtues of the Differentiated Services architecture is that it drops the traditional concept of signaling because it no longer requires the reservation of resources in all the network elements involved. The strategy consists in performing admission control on domain boundaries and then treating them in a differentiated manner inside the domain according to packet tagging on the domain borders, which is a much faster and lightweight process.

MANET environments differ greatly from the wired environments for which the DiffServ and IntServ models were created. The difference stems not only from the new problems encountered in MANETs (node mobility, frame collisions, variable conditions on the wireless channel, etc.), but also because MANETs do not follow the client/service provider paradigm inherent to both IntServ and DiffServ models. In MANETs the network is typically formed by users that cooperate, and except in situations where there is some centralized management entity (e.g., army), it relies on users' goodwill and limited resource sharing. So, new proposals were presented to achieve reliable QoS support in MANETs. The most representative ones are INSIGNIA [5] and SWAN [6], though in the literature others can be found (e.g., FQMM [10]). We will now expose the main characteristics of both INSIGNIA and SWAN proposals.

6.2.1 INSIGNIA

INSIGNIA is an in-band signaling system that supports fast reservation, restoration, and adaptation algorithms. With INSIGNIA all flows require admission control, resource reservation, and maintenance at all intermediate stations between source and destination to provide end-to-end quality-of-service support.

The INSIGNIA signaling system is designed to be lightweight in terms of the amount of bandwidth consumed for network control and to be capable of reacting to fast network dynamics such as rapid host mobility, wireless link degradation, intermittent session connectivity, and end-to-end quality of service conditions.

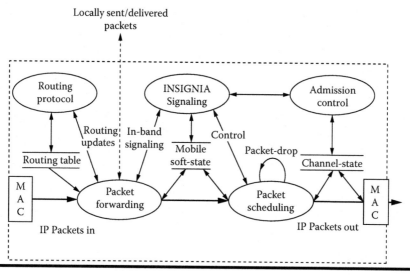

Figure 6.1 The INSIGNIA QoS framework. (Reprinted from Lee, S. B., et al., *J. Parallel Distributed Comput.* 60:374–406, 2000. Copyright 2000, with permission from Elsevier.)

In Figure 6.1 we offer an overview of INSIGNIA's architectural components, showing that the INSIGNIA framework is independent of the routing and MAC protocols used.

INSIGNIA relies on in-band commands, which are put inside the IP option field; these include service mode, payload type, bandwidth indicator, and bandwidth request fields.

When a node wants to perform a flow reservation, it activates a reservation mode bit (RES) in the IP option service mode field of a data packet and sends the packet to the destination. When a reservation is being established, each node along the path checks if it can offer the maximum QoS requested. In the event that all nodes can offer this maximum QoS requested, the destination will become aware of it by noticing that the bandwidth indicator is set to MAX. On the contrary, this field will be set to MIN, which means that all the packets sent by the source pertaining to the enhanced QoS traffic (payload type is Enhanced QoS [EQ]) will be degraded to best-effort traffic at the first bottleneck node.

When the destination of a reservation request receives a RES packet it sends a QoS report to the source node notifying it that the reservation succeeded. If during the reservation, or at a later time, the situation changes and the flow can no longer receive the requested QoS, the destination node can issue scaling/drop commands to the source node.

Mobility forces ongoing flows to be rerouted. A soft-state approach is used to manage resources in the presence of mobility. This means that as the route being used changes, new reservations along that new path are done automatically by a restoration mechanism, and old reservations are eliminated through a soft-state management mechanism.

When a node cannot meet the QoS requirements of a flow, it is downgraded to best effort, which means that no further nodes after that point will attempt to reserve resources for that flow. If later the bottleneck node becomes uncongested, it can allow the reservation to take place throughout the entire path, thereby achieving what is called degraded restoration.

One of the main drawbacks of INSIGNIA is that multi-hop wireless networks make the problem of selecting a path satisfying bandwidth requirements an Non-polynomial (NP)-complete problem, even under simplified rules for bandwidth reservation [11]; such a problem is commonly referred to as the coupled capacity problem. Such issues make deploying INSIGNIA on a real MANET environment a nontrivial task.

6.2.2 SWAN

SWAN is a QoS framework based on a stateless network model that aims at providing service differentiation in MANETs. One of the main advantages of SWAN is that it does not require the support of a QoS-capable MAC layer to provide service differentiation. Instead, it relies on rate control mechanisms that shape best-effort traffic at each node. SWAN's framework uses sender-initiated admission control and explicit congestion notification for real-time traffic to adapt to mobility and congestion conditions.

The SWAN model includes a number of mechanisms used to support rate regulation of best-effort traffic, as illustrated in Figure 6.2. The main elements of SWAN are a traffic classifier to differentiate real-time from best-effort traffic, a traf-

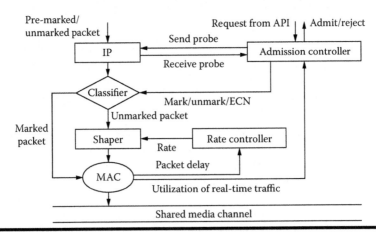

Figure 6.2   Architecture of the SWAN model. (From Ahn, G.-S., et al., *IEEE Tran. Mobile Comput.* 1:192, 2002. With permission.)

fic shaper to regulate the rate of best-effort data, a rate controller, and an admission controller.

The rate controller determines the departure rate of the shaper using an additive increase/multiplicative decrease (AIMD) rate control algorithm based on feedback from the MAC layer. This requires a constant monitoring of the actual transmission rate. When the difference between the shaping rate and the actual rate is greater than g percent of the actual rate, the rate controller adjusts the shaping rate to be g percent above the actual rate. This gap (i.e., g percent) allows the best-effort traffic to increase its actual rate gradually.

Concerning the admission controller, this element is responsible for allowing new connections to enter the MANET and also for estimating locally available bandwidth.

Local bandwidth estimations are obtained by measuring the rate of real-time traffic at each node, which requires calculating a running average of channel measurements to filter small-scale variations.

To admit new connections into the network, the source station sends a probing request packet to assess end-to-end bandwidth availability. Each node along the path will update the bandwidth value if its locally available bandwidth is lower than the one stated on the packet. The destination node sends a probing response packet back to the source node with the bottleneck field copied from the probing request message it received.

If the new flow is accepted, all of its packets are tagged as QoS packets. The network elements along that new path are unaware of the new flow, and they merely keep shaping best-effort traffic to offer good performance to QoS traffic.

The functioning of SWAN can be disturbed by mobility and false admission conditions. SWAN uses explicit congestion notification (ECN) to mark packets when a node finds that traffic is experiencing congestion. This is done by setting the ECN bits located on the IP header.

The use of ECN can be done in two different ways. The first one consists of marking all the packets flowing through a congested node.

When receiving ECN-marked packets, the destination node notifies this occurrence to the source. The source node waits for a random amount of time and initiates the reestablishment procedure; this avoids all sources probing the network at the same time.

Another solution would be for congested/overloaded nodes to randomly select a congestion set of real-time sessions and only mark packets associated with the set. This can be done using a hash function without keeping any per-flow state at the intermediate nodes. A congested node marks the congested set for a period of time T seconds and then calculates a new congested set. The main disadvantage of this scheme is that it requires some intelligence at intermediate nodes to manage the congested sets, as well as for determining if a flow is new or old, to correctly respond to false admission. The main advantage is that it enables a better utilization of resources than the source-based regulation technique.

One of the main drawbacks of SWAN is that its admission control mechanism requires all stations to keep track of the MAC's transmission delay of all packets to estimate available bandwidth; however, the association of a global estimate for transmission delay with a certain bandwidth in the link toward a specific target station is not straightforward, especially outside simulation scope. Also, the overhead introduced by the proposed shaping and measurement techniques can be significant for some mobile terminals where resources are scarce.

6.3 MAC Layer QoS Support

As an alternative to both the SWAN and INSIGNIA proposals, in this chapter we discuss a lightweight alternative that relies on IEEE 802.11e technology for traffic differentiation at the MAC layer, and a probe-based admission control system (DACME) that builds upon the services offered by IEEE 802.11e.

In this section we focus on the IEEE 802.11e technology, with a special emphasis on the ad hoc mode and its traffic differentiation capabilities in both static and mobile multi-hop networking environments.

6.3.1 IEEE 802.11e Technology

The IEEE 802.11e working group is extending the IEEE 802.11 MAC to provide QoS support. A subset of IEEE 802.11e, known as Wi-Fi Multimedia (WMM), is a Wi-Fi Alliance [12] interoperability certification. It is based on a draft standard of IEEE 802.11e.

The IEEE 802.11e standard [13] introduces the Hybrid Coordination Function (HCF), which defines two new medium access mechanisms to replace the legacy Point Coordination Function (PCF) and Distributed Coordination Function (DCF). These are the HCF Controlled Channel Access (HCCA) and the Enhanced Distributed Channel Access (EDCA).

Similarly to legacy IEEE 802.11 technology, the HCF may still break a superframe into a contention period (CP) and a contention free period (CFP), but now the HCCA is used in both periods, while the EDCA is used only during the CP. This new characteristic of HCF obviates the need for a CFP, because it is no longer required to provide QoS guarantees.

Because our focus is on ad hoc networks, we are only interested in the EDCA mode of operation. For more information on the HCF and the HCCA, please refer to [14].

Concerning IEEE 802.11e–enabled stations forming an ad hoc network, QoS support is achieved through the introduction of different access categories (ACs) and their associated backoff entities.

Contrary to the legacy IEEE 802.11 stations, where all the packets received by the MAC layer have the same priority and are assigned to a single backoff entity, IEEE 802.11e stations have four backoff entities (one for each AC) so that packets are sorted according to their priority. Each backoff entity has an independent packet queue assigned to it, as well as a different parameter set. In IEEE 802.11 legacy stations this parameter set was fixed: the interframe space equal to DCF Inter-Frame Space (DIFS) and the CWmin and CWmax parameters set to 15 and 1,023, respectively (for an IEEE 802.11a/g radio). With IEEE 802.11e the interframe space is arbitrary and depends on the access category itself (AIFS[AC]). We also have AC-dependent minimum and maximum values of the contention window (*CWmin*[AC] and *CWmax*[AC]). Moreover, IEEE 802.11e introduces an important new feature referred to as transmission opportunity (TXOP). A TXOP is defined by a start time and duration; during this time interval a station can deliver multiple frames consecutively without contention with other stations. This mechanism, also known as contention free bursting (CFB), increases global throughput through a higher channel occupation. An EDCA-TXOP (in contrast to an HCCA-TXOP) is limited by the value of TXOPLimit, which is a parameter defined for the entire QoS-Enhanced Basic Service Set (QBSS) and that also depends on the AC (TXOPLimit[AC]).

Table 6.1 presents the default MAC parameter values for the different ACs introduced by IEEE 802.11e. Notice that smaller values for the AIFSN, CWmin, and CWmax parameters result in a higher priority when accessing the channel; relative to the TXOPLimit, higher values result in larger shares of capacity, and therefore higher priority.

The relationship between AIFS[AC] and AIFSN[AC] is the following:

$$AIFS[AC] = SIFS + AIFSN[AC] \times aSlotTime, AIFSN[AC] = 2 , \qquad (6.1)$$

where SIFS is the shortest interframe space possible and *aSlotTime* is the duration of a slot. AIFSN[AC] should never be less than 2 in order not to interfere with AP operation.

Table 6.1 IEEE 802.11e MAC Parameters for an IEEE 802.11a/g Radio

Access category	AIFSN	CWmin	CWmax	TXOPLimit (ms)
AC_BK	7	15	1,023	0
AC_BE	3	15	1,023	0
AC_VI	2	7	15	3.008
AC_VO	2	3	7	1.504

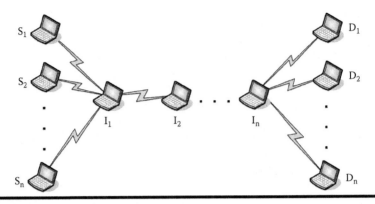

Figure 6.3 Static multi-hop scenario.

6.3.2 Performance of IEEE 802.11e in Static, Multi-Hop Environments

The QoS framework described in this chapter relies heavily on the performance and capabilities of the IEEE 802.11e technology. This technology, despite being designed targeting infrastructure-based wireless LANs, retains most of its effectiveness in multi-hop ad hoc networking environments, as we will now show.

To assess the performance and effectiveness of IEEE 802.11e in multi-hop environments, we first take as reference the static scenario shown in Figure 6.3. There are different source/destination pairs (S_i, D_i), and a variable number of intermediate nodes (I_i).

Using the static reference scenario, we start our analysis by observing the behavior in terms of throughput and end-to-end delay when varying the traffic load. Notice that, for each test, all the ACs and all stations acting as traffic sources are assigned a same packet generation rate.

Let us begin by observing the throughput decay as the number of hops is increased. Simultaneously, we assess if the share of bandwidth assigned to each AC in a single-hop situation remains the same as the number of hops increases. If traffic with lower priority obtains significantly higher bandwidth shares with increasing number of hops, we could conclude that the effectiveness of IEEE 802.11e is reduced.

In a first experiment we vary the number of hops from source to destination by varying the number of intermediate nodes (I_i). The number of source destination pairs is set to four, and all traffic sources generate a very high data rate so that their queues, for all four MAC access categories, are always full. Constant-bit-rate (CBR)/User Datagram Protocol (UDP) traffic is used, and the packet size is fixed at 512 bytes. The experiments here described were made using the ns-2 simulator [15].

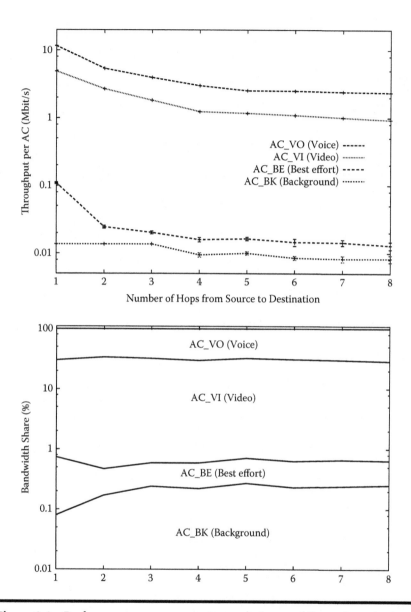

Figure 6.4 Performance per access category in terms of throughput (top) and bandwidth share (bottom) when varying the number of hops between source and destination. (From Calafate, C. T., et al., *IEEE MASCOTS Int. Symp.* **205, 2004. With permission.)**

Figure 6.4 shows the saturation throughput when varying the number of hops. As expected, the throughput for all traffic categories decreases as the number of hops increases according to the capacity decay (see [16] for more details). In terms of the total aggregate throughput, we observe that it drops from 16.6 Mbit/s (one hop) to 3.3 Mbit/s (eight hops). In terms of bandwidth share, Figure 6.4 shows the allocation of bandwidth to the different ACs. We observe that both voice (AC_VO) and video (AC_VI) traffic maintain a steady share of the available bandwidth, as desired. The best-effort (AC_BE) traffic slightly decreases its bandwidth share, while background (AC_BK) traffic increases it; nevertheless, it is always maintained very low.

To further validate the IEEE 802.11e technology in multi-hop environments, let us proceed by examining the stability of voice and video traffic when only best-effort and background traffics vary. The purpose is to assess the impact of non-real-time data packets (e.g., peer-to-peer file sharing) on real-time traffic, such as video or voice streaming. With this purpose we again fix the number of source destination pairs at four, and the number of intermediate nodes (I_i) is set to three. Source S_1 transmits nothing but voice traffic at a rate of 0.5 Mbit/s; likewise, source S_2 transmits solely video traffic, at a rate of 1 Mbit/s. Sources S_3 and S_4 transmit variable rates of best-effort and background traffic, respectively.

We find that neither video nor voice traffic throughputs are affected by increasing best-effort and background traffic loads.

In terms of delay, Figure 6.5 shows that voice traffic suffers from delay variations up to 70 percent, while for video traffic the delay variations can reach 91 percent.

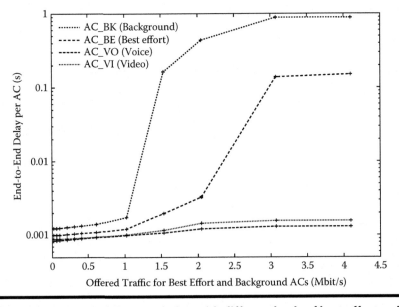

Figure 6.5   End-to-end delay variation with different loads of best-effort and background traffic. (From Calafate, C. T., et al., *IEEE MASCOTS Int. Symp.* 205, 2004. With permission.)

Nonetheless, the actual delay values can be considered low enough to support real-time applications adequately.

Overall, results show that the prioritization mechanism of IEEE 802.11e retains most of its effectiveness independently of the number of hops traversed by traffic or the load of best-effort and background traffic. As with legacy IEEE 802.11 networks, though, the impact of the number of hops on available bandwidth is considerable.

6.3.3 Impact of Station Mobility on QoS Performance

In MANET environments, node mobility is an important factor affecting QoS performance. Let us take, as an example, a typical mobile MANET environment, where 50 nodes move in a rectangular area sized 1900 × 400 m at a constant speed of 5 m/s according to the random waypoint mobility model. The routing protocol used in the tests is AODV [17], and all routing traffic is assigned the highest priority (AC_VO).

We compare the results obtained in such a scenario with the static scenario of Figure 6.3; in both cases the average number of hops from source to destination is four. A simple experiment with increasing load will put into evidence the impact of mobility on both throughput and delay. In both cases we vary the number of traffic source/destination pairs (S_i, D_i), and we set each traffic source to generate 0.2 Mbit/s (50 packets per second) on all MAC access categories.

Figure 6.6 shows the differences in terms of throughput between the static and mobile scenarios (ten random mobile scenarios were used). The results for the static scenario show that throughput values follow the line that represents offered traffic load quite closely before saturation. After saturation is reached, the throughput increase rate is no longer maintained, and it starts dropping after a certain point due to the contention mechanism inherent to IEEE 802.11.

Relative to mobile scenarios (see Figure 6.7), we observe that throughput values no longer follow the offered traffic load so strictly, though the points of maximum productivity for the different ACs are reached for a higher number of source stations. This is due to the higher degree of path diversity achieved in the mobile scenario [18]. So, while in the static scenario the maximum aggregated throughput is 4.1 Mbit/s (6 sources); in the mobile scenario this value is increased to 6 Mbit/s (14 sources).

In terms of delay, in the mobile scenario we observe that the minimum end-to-end delay values are higher than those in the static scenario. Moreover, the interval between the various ACs is not very high when there are only a few sources of traffic. This is due to mobility itself, which causes the routing protocol to react to route changes by buffering traffic on its queue. Similarly to what was found for throughput, now the end-to-end delay values do not reach saturation limits so quickly due to the expected traffic dispersion effect. In terms of traffic differentiation, we

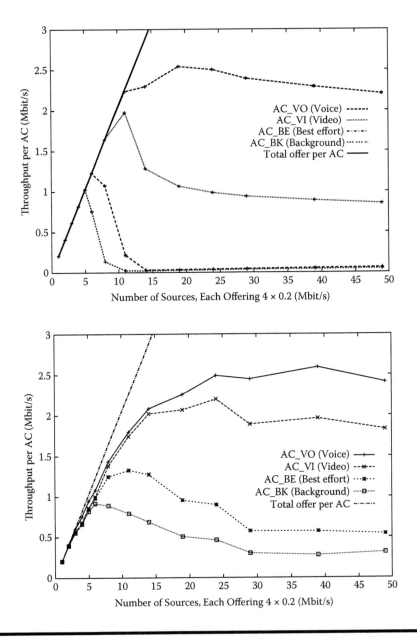

Figure 6.6 Throughput achieved in the static scenario (top) and on the mobile scenarios (bottom). (From Calafate, C. T., et al., *IEEE MASCOTS Int. Symp.* **205, 2004. With permission.)**

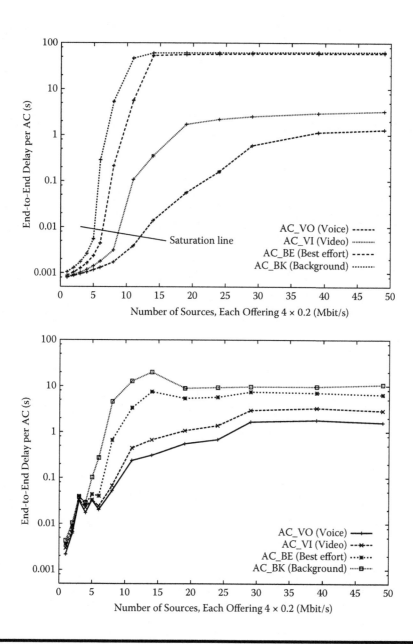

Figure 6.7 End-to-end delay achieved in the static scenario (top) and on the mobile scenarios (bottom). (From Calafate, C. T., et al., *IEEE MASCOTS Int. Symp.* 205, 2004. With permission.)

observe that in both scenarios the prioritization mechanism of IEEE 802.11e effectively offers better QoS to higher-priority traffic, and so we consider that the effectiveness of this mechanism in multi-hop environments is preserved.

6.4 DACME: Distributed Admission Control for MANET Environments

Having assessed the effectiveness of the IEEE 802.11e technology under mobility, we now focus on DACME.

DACME is a probe-based admission control mechanism that builds upon the IEEE 802.11e technology to achieve a full QoS framework for MANET environments. In terms of software requirements, only the source and destination of a QoS flow must have a DACME agent running. These DACME agents perform end-to-end QoS measurements according to the QoS requirements of multimedia streams. The remaining nodes will simply treat DACME packets as regular data packets, being unaware of the mechanism itself.

6.4.1 Overview and Architecture

In this section we will take a look at the different blocks that conform DACME's core.

Figure 6.8 shows the functional block diagram of a DACME agent. The main element of DACME is the QoS measurement module. This module is responsible for assessing QoS parameters on an end-to-end path. Another important element is the packet filter. Its purpose is to block all traffic that is not accepted into the MANET, and also to alter the IP type-of-service (TOS) packet header field in the packets of all accepted flows according to the QoS that has been requested.

An application that wishes to benefit from DACME must register with the DACME agent by indicating a connection identifier (*<source IP, source port, destination IP, destination port>*) and a Q_{spec}: *<U_{prio}, B_{avg}, D_{max}, J_{max}>*, which refer to the user-defined priority, the average data rate, the maximum delay, and the maximum jitter, respectively.

Once registration is completed successfully, the QoS measurement module is activated and will periodically perform path probing between source and destination. The purpose is to assess the current state of the path in terms of available bandwidth, end-to-end delay, and jitter. The destination agent, upon receiving probe packets, will update the destination statistics table, where it keeps per-source information of the packets received during the current probe. After receiving the last packet of a probe (or if a timeout is triggered) the destination agent will send a reply back to the source DACME agent. The QoS measurement module, upon receiving each probe reply, will update the state of the path accordingly. Once

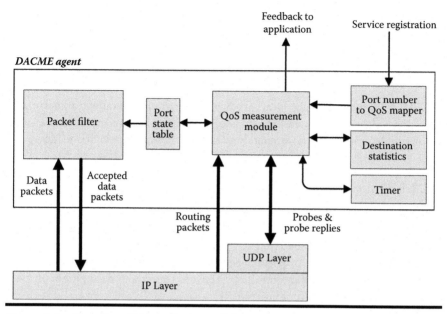

Figure 6.8 Functional block diagram of the DACME agent. (Reprinted from Calafate, C. T., et al., *Microprocessors Microsyst. J.* **2007. Copyright 2007, with permission from Elsevier.)**

enough information is gathered, it checks all the registered connections toward that destination and then decides when a connection should be accepted, preserved, or rejected, updating the port state table accordingly (with either accept or drop). If only part of the registered connections can be allowed, preference is given to those that have registered first. This module can also notify applications about service events using a call-back function, if requested at service registration.

6.4.2 Interaction with IEEE 802.11e

QoS parameters are typically set at the application level depending on the requirements of a particular application. The Internet Protocol (IP) supports traffic differentiation mechanisms in the sense that it allows tagging the packets according to QoS requirements, so that successive network elements can treat them adequately. This is achieved using the 8 bits of the type-of-service (TOS) header field in an IPv4 datagram header or the "traffic class" field in an IPv6 datagram header.

Within DACME's framework, actual QoS support is achieved when the packet filter configures the IP TOS header field of packets belonging to accepted data flows according to the data in the port state table. The IEEE 802.11e MAC must then map the service type defined in the IP TOS packet header field to one of the four MAC access categories available—voice, video, best effort, and background—according to Table 6.2.

6.4.3 End-to-End Path QoS Assessment through Probing

DACME's framework avoids local, per-node estimations of bandwidth, delay, and jitter. Instead, it relies on end-to-end measurements that are more accurate and simpler to determine. We now detail how each of these estimations is made.

6.4.3.1 Bandwidth Estimation

Relative to the support for bandwidth-constrained applications, DACME's framework relies on end-to-end bandwidth measurements, which consist of sending probes from source to destination. Each probe consists of back-to-back packets; the most appropriate number of packets per probe when operating in medium-sized MANET environments (about four hops between source and destination, on average) was found to be ten according to [19].

The DACME agent in the destination, upon receiving the probe, will obtain a measure of available end-to-end bandwidth and send it back to the source. It is calculated through the following expression:

$$B_{measured} = \frac{P_{size}}{AIT} \tag{6.2}$$

where P_{size} is the size of each probe packet in bits, and AIT is the average interarrival time for probe packets. AIT is defined as

$$AIT = \frac{\Delta t_{rec}}{n-1}, \tag{6.3}$$

where Δt_{rec} is the time interval between the first and the last packet arriving, and n is the number of packets received (not the number of packets sent).

To achieve more accurate results, this process can be repeated a certain number of times, though not too many times due to mobility-related impediments and also to avoid long start-up times. Optionally, the receiver can also indicate to the source the number of packets lost.

The DACME source agent, when receiving the probe reply packet, will collect the $B_{measured}$ values sent by the destination agent to be able to reach a decision on whether to admit the connection. The source agent must also correct the bandwidth estimation values before using them to reach a decision (see [19] for more details).

Table 6.2   User Priority to IEEE 802.11e Access Category Mapping (According to IEEE 802.1d Guidelines)

User priority	Designation	IEEE 802.11e access category	Common designation
1	BK (background)	AC_BK	Background
2	BK (background)	AC_BK	Background
0	BE (best effort)	AC_BE	Best effort
3	EE (video/excellent effort)	AC_BE	Best effort
4	CL (video/controlled load)	AC_VI	Video
5	VI (video)	AC_VI	Video
6	VO (voice)	AC_VO	Voice
7	NC (network control)	AC_VO	Voice

The strategy in which we propose to perform probabilistic admission control is described in Algorithm 6.1. Notice that, for applications with bandwidth constraints only, the decision to accept or block traffic will then be taken according to the value of the bandwidth flag alone.

This algorithm allows reducing the number of probes required to perform a decision to a value as low as two probes; it occurs often in those situations where it becomes quickly evident that the available bandwidth is either much higher or much lower than the requested one. If, after sending the maximum number of probes allowed, still no decision can reached, the chosen criterion consists of maintaining the previous path state. That way, if a connection is waiting for admission, it will remain blocked, and if it is active, it will remain active. Such a criterion aims at reducing the entropy in the MANET.

Algorithm 6.1: Probabilistic Admission Control Mechanism for Bandwidth-Constrained Applications

```
After receiving a bandwidth probe reply do {
  correct the bandwidth estimation using all available values
  if (there is a level of confidence of 95% that the
  available bandwidth is higher that the requested one)
  then set bandwidth flag to true
  else if (there is a level of confidence of 95% that the
  available bandwidth is lower that the requested one)
  then set the bandwidth flag to false
  else if (number of probes used is less than maximum
  allowed)
  then send a new probe
  else maintain the previous bandwidth flag value }
```

It should be noted that the DACME agent or the application itself should always reserve some extra bandwidth to cope with network bandwidth fluctuations, routing data, and probes from other sources. According to [20], typical values are between 0.5 and 1.5 Mbit/s. By reserving backup resources, the amount of QoS drops for incoming and outgoing video data is reduced, and more important, it avoids routing misbehavior.

If the application is only bandwidth constrained, the source will then notify it if the connection can currently be admitted or not. If the application also has requirements on end-to-end delay and delay jitter, the DACME source agent will perform more tests to assess the current end-to-end delay and delay jitter values. These topics will be handled in the next two sections.

6.4.3.2 Delay Estimation

When an application has bandwidth and delay requirements, or delay requirements alone, a DACME agent is required to offer a different measurement technique to handle this new constraint.

The technique used to measure end-to-end delay is similar to the measurements made by a ping application, with the difference that a new echo request packet is sent immediately after receiving an echo reply packet to reduce as much as possible the time spent while performing measurements. Also, the echo reply packet should have the same length and the same IP TOS field as the echo request one.

DACME requires at least three consecutive round-trip times to obtain a reliable measurement. Therefore, the technique we use to handle applications with delay requirements is the following: We start with several consecutive probe request/ probe reply rounds to assess the end-to-end delay. The value of the first round is discarded because it is used as a warm-up round to trigger routing and find end-to-end bidirectional paths. The results from the remaining probing rounds are averaged and stored. In case any of the packets are lost, the end-to-end path is considered to be broken and the traffic is blocked.

If the application is also bandwidth constrained, we then proceed to assess the available bandwidth following the strategy defined in the previous section. The only difference is that once code from Algorithm 6.1 is executed and a decision is taken relative to bandwidth, we must then proceed with Algorithm 6.2 to reach a decision based on end-to-end delay also. If the application is delay constrained alone, it will recur to Algorithm 6.2 immediately after the delay probing process ends.

The strategy followed in Algorithm 6.2 consists of rectifying end-to-end delay by finding worst- and best-case estimations in case the application is bandwidth constrained and the traffic is blocked. When traffic is flowing, or when the application is delay bounded only (which suggests that bandwidth requirements are minimal), there is no need to perform adjustments, and the measured value is directly used.

Algorithm 6.2: Probabilistic Admission Control Mechanism for Delay-Bounded Applications

```
Execute code from Algorithm 6.1 if appropriate. Then do {
 if (application is bandwidth-constrained && traffic is
currently blocked)
 then find worst- and best-case estimates for delay using
both delay and bandwidth measurements;
 else use the measured delay as the best- and worst-case
delay
 if (best-case delay > maximum delay allowed)
 then set delay flag to false
 else if (worst-case delay < 90% of the maximum delay
allowed)
 then set delay flag to true
 else if (application is bandwidth-constrained && number of
bandwidth probes used is less than maximum allowed)
 then send a new bandwidth probe
 else maintain the previous delay flag value }
```

We allow a small margin of uncertainty between 90 and 100 percent of the maximum delay requested to provoke hysteresis and so avoid frequent traffic fluctuations.

6.4.3.3 Jitter Estimation

In this section we complete our overview of DACME's QoS framework by analyzing support for jitter-constrained applications.

Relative to the jitter measurement process, the source must send packets with the same size, IP TOS field, and data rate as the application being served. According to [20], this process lasts for 250 ms.

The receiving end, aware of the source's packet sending rate by explicit notification, calculates the mean and standard deviation values for the absolute jitter and returns them to the source. Measurements made during the jitter measurement phase can also be used to obtain an estimate of the packet loss rate.

Jitter probes are only used if the application's traffic is blocked, and they are sent after delay and bandwidth probes if neither test denied the connection. In case that the traffic from the application is flowing through the network, there is no need to send jitter probes; this is because the destination agent can measure the jitter of the actual traffic and send it back to the source. Because most applications with jitter requirements are also delay bounded, the first probe reply packet of the delay measurement cycle is used to carry jitter information from destination to source. This avoids further probing if jitter requirements are not being met.

Independently of the method used to measure jitter (probes or actual traffic), once the source receives jitter statistics (absolute mean and standard deviation values), it will assess the compliance with the maximum value allowed using Algorithm 6.3.

Because jitter follows a normal distribution with a mean value of zero, about 95 percent of the cases fall between $\pm\sigma$. Therefore, in Algorithm 6.3, we accept traffic only if 95 percent of the packets have a jitter value lower than the maximum requested. We also introduce hysteresis by defining an interval $[1.9\sigma, 2\sigma]$, where the strategy consists of maintaining the previous state. As referred to earlier for bandwidth and delay, this aims at reducing fluctuations on traffic.

6.4.3.4 Timers

When designing an algorithm for a loss-prone network environment, we should always take care of handling losses in a clear and straightforward manner. In DACME this loss awareness is gained by recurring to timers, being a central element of both source and destination DACME agents.

Each source agent keeps a timer to be able to react in case a probe reply is never received. So, after sending a probe that may consist of one (in the case of delay probes) or more (in the case of bandwidth and jitter probes) packets, it sets the timer to go off after 500 ms. If no probe reply is received, causing the timer to be triggered, or in the case that the probing process is completed, the source will schedule a new probing cycle after 3 s ± 500 ms of jitter to avoid possible negative effects due to probe synchronization. This value offers a balance between the performance drop caused by poor reaction times and the overhead introduced by the probing process itself.

Focusing on bandwidth probes, the destination agent also maintains a timer to accommodate the possibility that not all the packets of a probe arrive. The purpose of maintaining a timer is to set a bound on the time consumed waiting for packets to arrive, thereby reducing blocking times at the source as much as possible. So,

Algorithm 6.3: *Probabilistic Admission Control Mechanism for Jitter-Bounded Applications*

```
After receiving a jitter reply do {
  if (2 × standard deviation < maximum jitter)
  then set jitter flag to true
  else if (1.9 × standard deviation > maximum jitter)
  then set jitter flag to false
  else maintain the previous jitter flag value }
```

when the destination receives the first packet of a bandwidth probe, it updates the current sequence number. When the second or the following packets are received, it continuously updates an internal timer, setting it to go off after

$$T = \frac{T_{last} - T_{first}}{N_{recv} - 1} \times (N_{rem} + \varepsilon) + \tau \qquad (6.4)$$

where T_{first} and T_{last} are the arrival times of the first and last packets received, N_{recv} is the number of packets currently received, N_{rem} is the number of packets that remain (not received yet), and ε is a fixed number of additional packets used to model a certain degree of tolerance. The purpose of the first part of the expression is to accommodate dynamically the observed network performance. With respect to constant τ, its purpose is to avoid malfunctioning in the presence of multipath routing protocols or other sources of intermittent delay variations.

Delay probes do not require a timer at the destination because the reply is immediate.

Concerning jitter probes, the strategy followed in terms of destination timers is very similar to the one proposed for bandwidth probes.

6.5 Performance Analysis

In this section we will assess the performance of DACME's QoS framework in supporting applications with bandwidth, delay, and jitter constraints. The experiments take place in a typical MANET environment, similar to the dynamic scenario described in Section 6.3.3. Concerning the data sources under study (regulated by DACME), these consist of four video streams and three voice streams. The video sources are simulated using CBR/UDP traffic at 1 Mbit/s using 512-byte packets. Voice sources are VoIP streams simulated using a Pareto on/off distribution with both burst and idle time set to 500 ms. The shaping factor used is 1.5, and the average data rate is 100 kbit/s.

In addition to DACME-regulated traffic sources, there are also four nonregulated background sources that generate negative–exponentially distributed traffic. For each of these sources, 50 percent of the total generated data belongs to the video AC, and the best-effort and background ACs receive a share of 25 percent each. Because the routing protocol makes use of the voice AC, too much voice traffic causes routing mechanisms to malfunction, and so that must be avoided for the results to be meaningful.

In the next section we will perform experiments to compare the performance obtained with and without DACME.

6.5.1 Performance under Bandwidth Constraints

Figure 6.9 shows the improvements—in terms of video goodput and voice packets dropped—obtained by using DACME. We compare these results to a solution where DACME is not used (turned off). We can observe that when DACME is not used, the average goodput for the different video sources drops steadily with increasing congestion. By using DACME, the average goodput is maintained higher (close to maximum). This occurs because sources are only allowed to transmit if the DACME agent verifies that the available bandwidth is enough.

In terms of end-to-end delay, Figure 6.10 shows the improvements in terms of end-to-end delay when increasing background traffic.

For the scenarios under analysis we see that the end-to-end delay values for both video and voice sources were lower with DACME active than with DACME turned off.

An interesting way to gain further insight into the benefits of DACME in MANET environments is to analyze the stability in terms of routing overhead or the lack of it.

Figure 6.11 shows the variation in terms of total routing packets when varying the amount of background traffic load. It shows that without the admission control mechanism offered by DACME, the routing protocol misbehaves due to congestion-related effects.

Relative to DACME's overhead, each source generates between 30 and 40 kbit/s of probing traffic, a very reasonable value taking into consideration that DACME follows a probe-based approach.

The results presented until now allow us to conclude that DACME allows achieving performance improvements, compared to a solution without DACME. In Figure 6.12 we show the results achieved in terms of percentage of admitted traffic. We can see that the main differences occur when the congestion levels are low. In these situations a quick reaction to newly found routes avoids blocking traffic for large periods if the new route can sustain the desired traffic rate.

It is interesting to notice that, as congestion increases, the amount of video traffic admitted decreases at a steady rate, contrary to voice traffic. This is due to the fact that video streams require much larger bandwidth shares.

6.5.2 Performance under Delay Constraints

In this section we continue the analysis of DACME by assessing the support for delay-bounded applications. For this study we set the aggregate background congestion to a moderate load to proceed with our experiments. The chosen value is 2.3 Mbit/s, and it is maintained throughout the simulations.

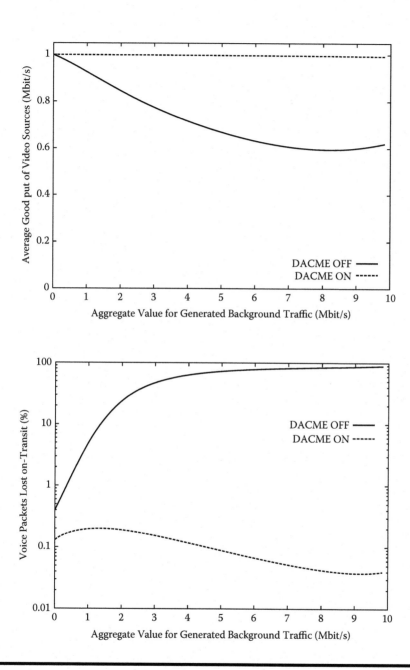

Figure 6.9   Improvements on video goodput (top) and voice drops (bottom) by using DACME.

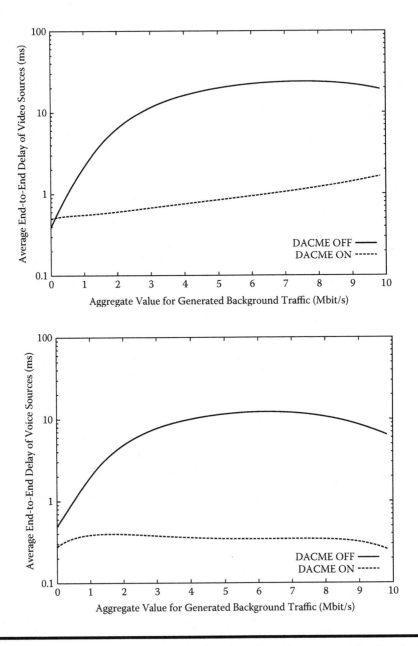

Figure 6.10   Average end-to-end delay values for video (top) and voice (bottom) sources.

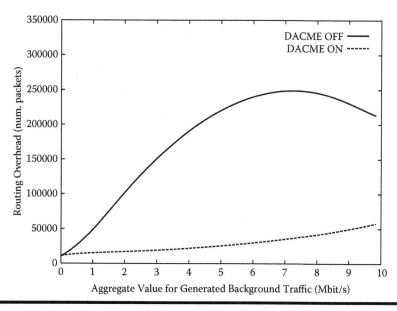

Figure 6.11   Routing overhead.

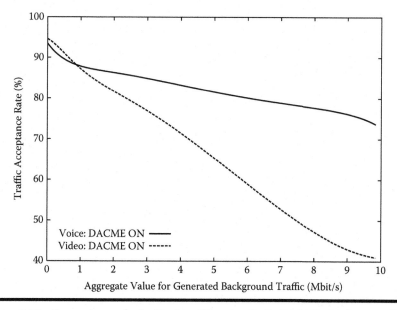

Figure 6.12   Percentage of admitted traffic using both DACME versions at different congestion levels.

The simulations made are similar to those performed for bandwidth-constrained applications. The only difference is that Q_{spec} includes bounds for delay also. In these experiments the maximum delay requested varies between 0.1 and 100 ms.

The lower bound is set to 0.1 ms because, for this delay requirement, no video or voice traffic is accepted into the MANET; the upper bound of 100 ms was chosen because this delay requirement can be met without difficulties.

In Figure 6.13 we present the traffic acceptance rate curves when varying the maximum delay settings. We observe that the impact of imposing delay requirements is more pronounced on video sources, being that the voice traffic only varies slightly; as expected, when demanding relatively high values for end-to-end delay (100 ms), the amount of traffic accepted for both video and voice sources is close to the one found when applying bandwidth constraints only.

We now proceed to measure the average end-to-end delay experienced by the video and voice sources. In Figure 6.14 we present the results found; we observe that the average end-to-end delay values are always well below the threshold defined, as desired. We find that the average end-to-end delay experienced by the video sources increases steadily with increasing delay thresholds; for the voice sources, though, we only appreciate slight variations. This phenomenon occurs because the MAC layer parameters associated with the voice access category do not allow much margin for such variations.

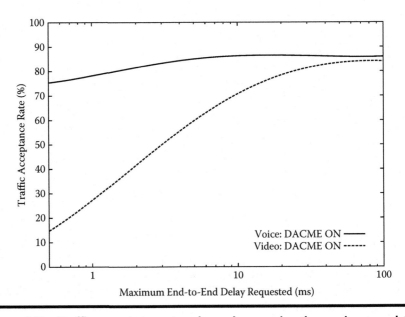

Figure 6.13 Traffic acceptance rate values when varying the maximum end-to-end delay requested.

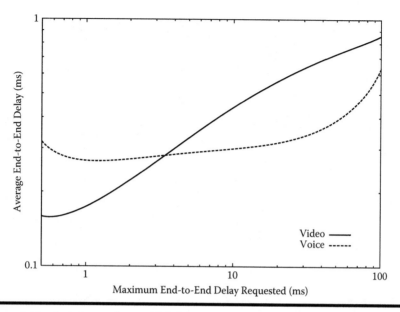

Figure 6.14 Average end-to-end delay variation when varying the maximum end-to-end delay requested for video and voice traffic.

If we now take into consideration the percentage of traffic that meets the predefined maximum value for end-to-end delay, we observe that voice traffic meets the requirements more strictly than video traffic (see Figure 6.15). These results also allow measuring the effectiveness of DACME's architecture in complying with the end-to-end requirements imposed. We find that although DACME agents only reassess the end-to-end delay values every 1.5 s when traffic is flowing, this strategy offers good results even when the scenario is characterized by an important degree of mobility: more than 80 percent of the accepted traffic meets the deadline always.

We proceed by analyzing the average overhead per source introduced by DACME. In Section 5.1 we found that, at the selected degree of congestion (aggregated background traffic of 2.3 Mbit/s), DACME's overhead was found to be 35 kbit/s (on average). Introducing additional probes to measure end-to-end delay does not have a significant impact on overhead. In fact, we find that when the requested end-to-end delay is low, DACME's overhead is slightly inferior to the one found without delay constraints. This effect occurs because sometimes the delay probes allow reaching a *deny flow* decision without requiring any measurements to decide about bandwidth.

Once we reach relatively high values for the requested end-to-end delay we find that the overhead, compared to the bandwidth-constrained solution, is increased by about 10 kbit/s.

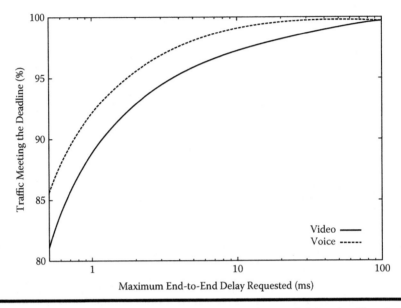

Figure 6.15   Percentage of traffic meeting the end-to-end delay deadline for video and voice traffic.

We will now proceed by doing a similar analysis in the scope of jitter-bounded applications.

6.5.3 Performance under Jitter Constraints

In the previous section we could appreciate the effectiveness of DACME to support applications with both bandwidth and end-to-end delay constraints. In this section we take a final step to evaluate the effectiveness of DACME in supporting applications with bandwidth, delay, and jitter requirements. With this purpose we maintain all the simulation parameters used in the previous section, fixing the value for the maximum end-to-end delay requested at 10 ms. The Q_{spec} now includes different requirements for the jitter, with values ranging from 0.1 ms up to a maximum value of 10 ms (equal to the maximum delay limit).

In Figure 6.16 we show the variation in terms of accepted video traffic as the maximum jitter allowed increases. Experiments show that 0.1 ms is a cutoff value for jitter and that when the maximum jitter allowed is 10 ms, the traffic acceptance rate is similar to the one found without jitter constraints for both video and voice MAC access categories.

We now proceed by analyzing the amount of video and voice traffic that meets the jitter requirements imposed. The results of Figure 6.17 show that when the jitter limits are too low, only about two-thirds of the traffic meets these limits. As

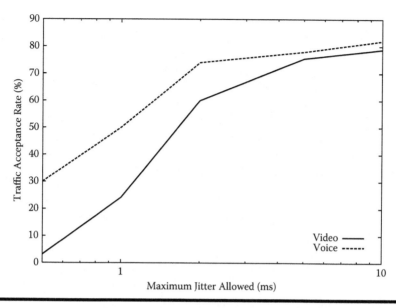

Figure 6.16   Traffic acceptance rate growth when increasing the maximum jitter allowed.

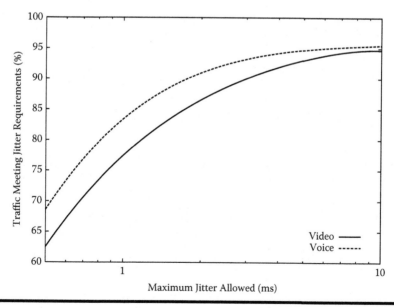

Figure 6.17   Percentage of traffic meeting the maximum jitter value requested.

we relax the jitter constraints, the percentage of traffic meeting the requirements increases significantly (up to 95 percent).

To conclude this section, we focus on DACME's overhead when jitter probes are also included. When the maximum jitter allowed is very low, the overhead generated by DACME is relatively high (about 110 kbit/s on average). This occurs because the DACME agent will be sending jitter probes every 3 s after the delay and bandwidth probes, which consume a considerable amount of bandwidth. As we increase the maximum jitter allowed, we find that DACME's overhead is reduced to about 75 kbit/s. As jitter constraints are relaxed, more traffic is admitted into the network; this enables performing jitter measurements based on actual traffic instead of probes, which explains the reduction observed.

6.6   Summary

The issue of quality-of-service support on MANETs is still in an early stage. Most of the proposals currently available only focus on improvements at a specific layer, not offering an overall QoS architecture.

In this chapter we analyzed a soft QoS architecture for MANET environments that enables real-time multimedia communication among peers. This architecture imposes few constraints on mobile terminals and mainly results from combining the traffic differentiation capabilities of the IEEE 802.11e technology with a novel admission control system—DACME—which is at the core of this architecture.

DACME is a probe-based distributed admission control mechanism built on top of IEEE 802.11e technology that relies on end-to-end probe measurements to support applications with bandwidth, delay, and, to a limited extent, jitter requirements. The probe-based measurements are used by DACME agents to decide whether to admit traffic from an application based on its QoS specification and the estimated available resources. Results show that probe-based admission control is able to offer reliable end-to-end measurements and that the time spent in that task is typically low.

We find that the probabilistic admission control technique used in DACME is effective in the presence of mobility and at different levels of congestion, and that delay and jitter constraints are met with a good level of accuracy.

Relative to the overhead introduced by the DACME's mechanism, it is usually in a range between 30 and 50 kbit/s, except when jitter support is also required, in which case it can reach values up to 110 kbit/s in the worst case.

Overall, DACME improves the performance experienced by users while simultaneously avoiding routing misbehavior and wasting MANET resources.

The soft QoS framework exposed in this chapter avoids strict resource reservations and local estimations of resource availability; therefore, it is expected to solve some of the problems encountered in previous proposals in this field.

Acknowledgments

This work was supported by the Ministerio de Educación y Ciencia (Spain) under grants TIN2005-07705-C02-01 and TIC2003-00339.

References

[1] IEEE 802.11 WG. 1999. *International standard for information technology—Telecom and Information exchange between systems—Local and metropolitan area networks—Specific requirements: Wireless medium access control (MAC) and physical layer (PHY) specifications.* ISO/IEC 8802-11:1999(E) IEEE Standard 802.11, Part 11.

[2] Lin, C. R., and Liu, J.-S. 1999. QoS routing in ad hoc wireless networks. *IEEE Journal on Selected Areas in Communications* 17:1426.

[3] Shigang, C., and Nahrstedt, K. 1999. Distributed quality-of-service routing in ad hoc networks. *IEEE Journal on Selected Areas in Communications* 17:1488.

[4] Xue, Q., and Ganz, A. 2003. Ad hoc QoS on-demand routing (AQOR) in mobile ad hoc networks. *Journal of Parallel and Distributed Computing* 63:154.

[5] Lee, S. B., et al. 2000. INSIGNIA: An IP-based quality of service framework for mobile ad hoc networks. *Journal of Parallel and Distributed Computing* 60:374.

[6] Ahn, G.-S., et al. 2002. Supporting service differentiation for real-time and best effort traffic in stateless wireless ad hoc networks (SWAN). *IEEE Transactions on Mobile Computing* 1:192.

[7] Braden, R., Clark, D., and Shenker, S. 1994. *Integrated services in the Internet architecture: An overview.* IETF RFC 1633.

[8] Blake, S. 1998. *An architecture for differentiated services.* IETF RFC 2475.

[9] Braden, R., et al. 1997. *Resource Reservation Protocol (RSVP)—Functional specification.* IETF RFC 2205, Version 1.

[10] Xiao, H., et al. 2000. A flexible quality of service model for mobile ad-hoc networks. In *IEEE 51st Vehicular Technology Conference Proceedings*, Tokyo, vol. 1, p. 445.

[11] Georgiadis, L., Jacquet, P., and Mans, B. 2004. Bandwidth reservation in multihop wireless networks: Complexity and mechanisms. In *International Conference on Distributed Computing Systems Workshops (ICDCSW'04)*, Hachioji-Tokyo, p. 762.

[12] Wi-Fi Alliance. http://www.wifialliance.com/.

[13] IEEE 802.11 WG. 2005. *Standard for information technology—Telecommunications and information exchange between systems—Local and metropolitan area networks—Specific requirements: Wireless LAN medium access control (MAC) and physical layer (PHY) specifications: Medium access control (MAC) quality of service enhancements.* IEEE 802.11e, Part 11, Amendment 8.

[14] Mangold, S., et al. 2003. Analysis of IEEE 802.11e for QoS support in wireless LANs. *IEEE Wireless Communications* 10:40.

[15] Fall, K., and Varadhan, K. 2000. *ns notes and documents.* The VINT Project. UC Berkeley, LBL, USC/ISI, and Xerox PARC.

[16] Li, J., et al. 2001. Capacity of ad hoc wireless networks. In *Proceedings of MobiCom*, 61.

[17] Perkins, C. E., Belding-Royer, E. M., and Das, S. R. 2003. *Ad hoc on-demand distance vector (AODV) routing.* Request for Comments 3561. MANET Working Group. http://www.ietf.org/rfc/rfc3561.txt. (accessed April 11, 2008)

[18] Grossglauser, M., and Tse, D. N. C. 2002. Mobility increases the capacity of ad hoc wireless networks. *IEEE/ACM Transactions on Networking (TON)* 10:477.

[19] Calafate, C. T., Manzoni, P., and Malumbres, M. P. 2005. Supporting soft real-time services in MANETs using distributed admission control and IEEE 802.11e technology. In *10th IEEE Symposium on Computers and Communications*, p. 217.

[20] Calafate, C. T. 2006. Analysis and design of efficient techniques for video transmission in IEEE 802.11 wireless ad hoc networks. Ph.D. dissertation, UPV.

Chapter 7

Quality of Service in Wireless Multi-Hop Ad Hoc Networks: A Cross-Layer Framework

Peng-Yong Kong, Dan Li, and Yan Zhang

Contents

This chapter proposes a general cross-layer framework for wireless multi-hop ad hoc networks to support proportional differentiation in end-to-end quality of service (QoS). In the framework, four mechanisms and three monitors in different layers of the protocol stack adaptively cooperate via information exchanged. With the proposed framework, a specific realization called proportionally differentiated multi-hop end-to-end delay (PDMED) is introduced for a Carrier Sense Multiple Access with Collision Avoidance (CSMA/CA)–based multi-hop network to provide a consistent and accurate proportional differentiation on the average end-to-end packet delay. PDMED requires a distributed scheduler to adapt to the information from a QoS monitor and dynamically adjusts the contention window of a flow based on its instantaneous deviation from the maximum normalized average end-to-end packet delay among neighboring flows. PDMED has been extensively evaluated through random event simulations using OPNET. The results confirm that it is capable of providing a consistent and accurate proportional differentiation in end-to-end packet delay, which is otherwise not achievable under various traffic conditions. A benchmark against IEEE 802.11e using video traces shows that PDMED is significantly more flexible in providing an accurate and controllable end-to-end proportional differentiation. We found that received signal-to-interference-and-noise ratio (SINR) is self-similar under the random waypoint mobility model. Hence, we proposed an improvement called PDMED+, which predicts the SINR based on the F-ARIMA process to improve the network throughput. PDMED+ adjusts the transmission time of a predicted fail packet to the time when the channel quality becomes good, so as to avoid occupancy of wireless channels by unsuccessful transmissions and to transmit the packet as soon as good channel quality is available. Simulation results from OPNET confirm PDMED+ can improve network throughput while continuing to maintain an accurate proportional differentiation on average end-to-end delay.

7.1   Introduction

Wireless networks have become increasingly popular in the network industry. They can provide mobile users with ubiquitous communication capability and information access regardless of locations. However, conventional wireless networks are often connected to a wired network and require a fixed wire-line backbone infrastructure. All mobile hosts in a communication cell can reach a base station on the wired network in one-hop radio transmission. In parallel with the conventional wireless networks, another type of wireless network model, based on radio-to-radio multi-hopping, has neither fixed base stations nor a wired backbone infrastructure. This is called wireless multi-hop ad hoc network, constituted of mobile nodes that act as both mobile host and mobile routers. The wireless multi-hop ad hoc network is expected to play an important role in civilian and military forums.

Naturally, a wireless ad hoc network is an autonomous system of mobile routers and their associated hosts, connected by wireless links forming an arbitrary graph. The routers are free to move randomly and organize themselves arbitrarily. Thus, the network's wireless topology may change rapidly and unpredictably. Such a network may operate in a stand-alone fashion or may be connected to the larger Internet.

Being self-organized and not relying on existing infrastructure, wireless multi-hop ad hoc networks have several salient and unique features [1]. First, their topologies are dynamic and change often rapidly because of unpredictable and arbitrary movement of nodes. Thus, node interconnectivity and link properties such as capacity and bit error rate cannot be predetermined. Next, the transmission medium has a bandwidth-constrained and time-varying capacity because of the unstable wireless link. In addition, distance between the ends of the link, obstacles in the environment, externally generated noise, and interference caused by other transmissions also make the capacity of the wireless communication reduced and apt to be highly variable. Finally, wireless ad hoc networks are only able to support power-constrained operation because of lightweight batteries to support portability. The limited power supply constrains the transmission range, data rate, communication activity, and processing speed of the devices. Without centralized administration, distributed operations on every node are also important characteristics of wireless ad hoc networks.

Given the features above, multi-hop ad hoc networks suffer from resource constraints and operation vulnerability. Therefore, providing quality-of-service (QoS) support in the network is a demanding task. Despite difficulty, providing QoS in a multi-hop ad hoc network is unavoidable because of rising popularity of multimedia applications and potential commercial usage of wireless ad hoc networks. Data with different timeliness requirements will be delivered through the networks. For example, real-time images need to be delivered immediately so that illegal intruders can be detected promptly. On the other hand, measured room temperature can be delivered with some delay to the control center.

Amid the challenging environment of wireless multi-hop ad hoc networks, among different QoS models, proportional differentiation is most suitable. This is because of its "tuning knob" feature that allows quantitative control of QoS spacing among different flows. Using this feature, we can delicately adjust the resource allocations between flows to achieve an optimized situation. Suppose that there are two real-time video flows that expect 0.05 s maximum end-to-end packet delay. If a packet is unable to reach its destination before exceeding the maximum delay, the packet may be dropped. The two flows are able to tolerate the packet drop ratios 0.01 and 0.05, respectively. When the network resource is so limited due to its timing-varying characteristics, such that both the packet drop ratio requirements cannot be met, then the tuning knob feature allows the "misery" to be proportionally distributed between the two flows subject to their original packet drop ratio requirement. This is fair but not achievable through the guaranteed service, relative differentiation service, and assured service models. There are numerous mechanisms across the protocol layers and timescales for QoS delivery in multi-hop ad hoc networks. Among these mechanisms are QoS routing protocols, admission control policies, resource reservation schemes, packet scheduling algorithms, QoS-capable MAC protocols, etc. Section 7.2 gives a comprehensive review on the existing QoS models and mechanisms.

To provide accurate proportional differentiation in end-to-end QoS, a single QoS mechanism can never be sufficient. Logically, a combination of these mechanisms have to work collaboratively to achieve the goal. For example, we may need a packet scheduling algorithm that transforms the QoS requirements into medium access priorities and works with a MAC protocol that provides the multiple priorities. Also, we need a channel monitor capturing the instantaneous channel quality so as to compensate its negative effects on the QoS schemes. Hence, Section 7.3 introduces a general framework in which different mechanisms from different protocol stacks can collaborate for the purpose of providing end-to-end QoS in wireless multi-hop ad hoc networks. Based on the framework, a specific realization call proportionally differentiated multi-hop end-to-end delay (PDMED) is designed and evaluated in Section 7.4.

In wireless multi-hop ad hoc networks, movement of nodes may lead to variation in signal strength and interference strength from other simultaneous transmissions. Commonly, the variations in wireless channels and node positions are tracked to improve channel utilization as well as hand-over and routing performance. In addition, we observe that in a wireless multi-hop ad hoc network with random waypoint mobility pattern, the received signal-to-interference-and-noise ratio (SINR) is self-similar. With this observation, we introduce in Section 7.5 an improvement to PDMED, namely PDMED+, that predicts the channel quality using the fractionally integrated autoregressive moving average (F-ARIMA) process. The predicted channel quality is used to adjust the transmission schedule for throughput improvement.

7.2 Related Work

To provide different guaranteed QoS to different types of applications, various distributed MAC protocols have been proposed in the literature. Specifically, these MAC protocols can provide different upper bounds in packet access delay. For example, [2] and [3] propose distributed time division multiple access (TDMA) protocols that can guarantee at least one collision free time slot for each node in a given duration. This guarantee is possible in the absence of a central controller by using a discrete mathematics mapping function to pseudo-randomly arrange the transmission and reception at each node. In the same spirit of bounding access delay, [4] proposes a distributed CSMA/CA MAC protocol that can guarantee access to a node by emulating a round-robin algorithm. This round-robin algorithm is enforced by making each node send a Black Burst, i.e., pulses of energy at the end of back-off, and the duration of Black Burst is proportional to the packet access delay. The node can only transmit its packet if the channel remains idle after its Black Burst. Otherwise, the node has to perform another back-off, which will increase the duration of its Black Burst.

While the two MAC protocols above are capable of providing guaranteed QoS in a distributed wireless ad hoc network, some forms of resource reservation are required. Due to unpredictable capacity, the reservation often means resource over-provisioning and thus makes the guaranteed QoS not scalable and efficient. Compared to guaranteed QoS, differentiated QoS is not to deliver a hard assurance in the perceived performance but to give different resources to different flows such that different levels of performance can potentially be achieved at the flows. This flexibility of differentiated QoS makes it suitable for wireless ad hoc networks with volatile capacity.

As a mechanism to provide differentiated QoS, prioritized channel access has been extensively studied. In [5], a MAC protocol is proposed such that different priorities are achieved by assigning different fixed Black Burst durations to different traffic classes. Within a priority class, a randomized initialization protocol is used to enforce a round-robin sequence of transmissions among distributed nodes so that collision can be avoided. While Black Burst is indeed a practical method to achieve prioritization, the priority is only local among all nodes within the region of one hop where there is no hidden node. In the presence of hidden nodes, a high-priority node may be marginalized compared to a node with lower priority. Hence, [6] proposes to tackle this misscheduling problem among all nodes within a region of two hops. According to [6], before sending a Black Burst at the end of a back-off, the high-priority node sends a busy tone, which will be echoed by its receiver. All low-priority nodes that hear the busy tone defer their transmissions.

Compared to Black Burst, differentiating back-off duration is another technique in providing different priorities. This technique has been adopted in [7] to provide QoS differentiation in IEEE 802.11 where a higher-priority node has a shorter back-off duration. It has been shown that this technique does not work well in a noisy environment with prevalent propagation impairments. Also, a shorter back-off duration cannot really provide a higher priority to TCP flow where its throughput is measured on an end-to-end basis. Under these conditions, [7] indicates that a better differentiation can be achieved by using a shorter distributed interframe spacing (DIFS) duration, instead of back-off duration, for a higher-priority node. This finding has also been reported in [8]. Further, [8] reveals that, while a combination of back-off duration and DIFS duration can provide good QoS differentiation, the differentiation can be dramatically affected by channel condition and number of active nodes. Specifically, when the number of nodes is large, an accurate differentiation is harder to achieve by merely controlling the back-off duration because of more frequent transmission collisions. On the other hand, with a smaller number of nodes, adjusting only DIFS duration is not efficient in achieving the desired differentiation due to a waste of transmission times.

In view of the individual deficiencies of both back-off duration and DIFS duration techniques, the IEEE 802.11 working group has made an effort to define a standard mechanism to use them collectively to achieve efficient QoS differentiations [9]. The effort yields IEEE 802.11e, which has been extensively studied in the literature [10, 11]. From the studies, controlling back-off duration is effective in introducing throughput differentiation, while adjusting DIFS duration amplifies the differentiation. The studies also show that IEEE 802.11e can provide differentiation when there is a fixed number of active nodes within a radio range in an idealistic channel even though the traffic load is at a saturated level. However, the differentiation is vulnerable to changes in the number of nodes and traffic load. This vulnerability is partly due to the definition of its differentiation, where a flow can choose one among a small number of service classes (or priorities) that best meet its QoS requirement, based on the assurance that the perceived QoS of higher classes will be better, or at least no worse than that of lower classes. This type of differentiation is called relative differentiation, compared to proportional differentiation, which offers predictable and controllable differentiations between different service classes [12].

A simple form of proportional differentiation in throughput has been termed fairness. Let g_i and ϕ_i be the throughput and proportional differentiation parameters, respectively, for node i. Then, unfairness may be expressed as follows:

$$\bar{\mathcal{F}} = \max_{\forall i,j}\left\{|\frac{g_i}{\phi_i} - \frac{g_j}{\phi_j}|\right\} \tag{7.1}$$

where a smaller $\bar{\mathcal{F}}$ means better fairness. To achieve good fairness, [13] has proposed a distributed fair MAC protocol to ensure a minimum fair share of medium to a node while maximizing the spatial channel reuse for throughput improvement. This is achieved by mapping the virtual clock of weighted fair queuing into the back-off duration of a contending node and by allowing a look-ahead window in the range of a virtual clock eligible for service. While [13] uses weighted fair queuing, similar works in achieving fairness by mapping the virtual clocks of other fair queuing models, such as start-time fair queuing and worst-case-fair fair queuing, into back-off duration have been reported in [14–17]. Unfortunately, all these works can only achieve proportional differentiation (fairness) locally or globally between two nodes over one hop. With multiple hops in a wireless network, we argue that the proportional differentiations should be achieved in an end-to-end manner across all hops, but not limited to a concatenation of local proportional differentiations at each hop.

To provide QoS across multiple hops, [18] has proposed a distributed packet scheduling algorithm for CSMA/CA-based MAC protocols to achieve an accurate transmission order, as if in a centralized scheduler that provides QoS differentiation. Based on the desired transmission order, the scheduling algorithm assigns to every packet an appropriate priority. With the priority of a head packet, each node can rank itself against all its neighboring nodes after overhearing their head packets' priorities, which are piggybacked on packet transmissions. According to the rank, a node will determine its back-off duration to achieve the desired transmission order. Although the algorithm is capable of ensuring an accurate transmission order in a multi-hop setting, it is for packet and not flow. Further, there is no end-to-end QoS across multiple hops.

To provide to a flow an end-to-end QoS across multiple hops, [19] has proposed a simple modification to the CSMA/CA MAC protocol so that DATA and ACK frames will carry piggybacked channel reservation for the next transmission, and thus no Request to Send (RTS)/Clear to Send (CTS) exchange is required except for the first packet of a traffic burst at the first hop. As such, as long as the first DATA frame manages to acquire the channel at the first hop, the subsequent packets are guaranteed channel access without further reservation delay in the absence of channel error. This scheme is able to provide a better QoS to a real-time flow, compared to a best-effort flow along a multi-hop path. However, it is not easy to support multiple real-time flows at the same time, especially when the different real-time flows have different QoS requirements.

In an effort to provide different QoS to different flows across multiple hops, [20] proposes a coordinated multi-hop packet scheduling algorithm that requires some modifications to and cooperations from the CSMA/CA MAC protocol. In [20], the end-to-end QoS requirement of a flow is transformed into an instantaneous priority by the packet scheduling algorithm. Here, a packet that has not been offered sufficient service in the previous hop will be given a higher priority in the future hops and vice versa. The priority of the current and next packets will be piggybacked onto RTS/CTS and DATA/ACK packets, respectively. Hence, all nodes within a hop know each other's instantaneous priorities, and only the node with the highest relative priority will contend for the channel, while the other nodes defer their transmissions. It is the mechanism of adjusting a packet's priority at a hop based on its experience in previous hops that enables end-to-end QoS across multiple hops. However, it is obvious that the opportunities of compensating a packet in downstream hops are limited by the number of downstream hops and the competition situations in downstream hops. These limitations make this scheme only capable of providing coarse QoS provision.

To have more adjustment space for QoS provision, [21] and [22] propose a framework to adjust network access for a packet of a flow according to the end-to-end performance of previous packets so as to compensate to the previous bias resource allocation promptly and achieve end-to-end assurances in multi-hop wireless networks. In the framework, dynamic class selection (DCS) gives a way to dynamically choose the priority for flows according to their instant end-to-end performances. Neighborhood proportional delay differentiation (NPDD) scheduler ensures the ability to differently allocate the access of the medium resources in a proportional ratio in queuing delay between flows in a node according to their priority. Medium access priority selection (MAPS) supports the priority order of packets of flows in a contending area in the MAC layer. Here, IEEE 802.11e is used to realize the priority in competing to access the medium. These algorithms also achieve end-to-end service assurance for flows via mapping end-to-end QoS targets into priority indexes. This similar service compensation mechanism has been adopted by [23] for the same goal. More aggressively, [23] intends to provide a guarantee in end-to-end packet delay through admission control. Because there is no intuitive way to compute the capacity of a multi-hop ad hoc network, the admission control is done using an admit-then-test method. Specifically, a flow with an end-to-end delay requirement is first admitted, and then its impact on the channel idle time is monitored. If the idle time becomes too short as a result of the new flow, another flow that has no end-to-end delay requirement is selected for rejection.

Among all the schemes above, none is capable of supporting proportional differentiations in end-to-end QoS. We have intensively studied the schemes that either adopt a proportional differentiation model or provide end-to-end QoS over multiple hops to investigate their potentials of providing proportional differentiation in end-to-end QoS. We summarized the findings in Table 7.1 The table shows the QoS goals of the existing schemes, their methods, and the problems that they may suffer in providing proportional differentiation in end-to-end QoS. Conclusively, the main problems are

1. By only providing guaranteed service, prioritized access, and proportional differentiation over one hop, a scheme is unable to provide proportional differentiation over multiple hops. A coordinated method to support proportional differentiation over multiple hops is necessary.
2. A mechanism that transforms end-to-end QoS of every flow into controllable parameters in medium access protocols on every hop is a must.
3. Too strict QoS provision methods, such as strict prioritized access or guaranteed service, may reduce the number of QoS-expected flows supported, and are not desirable for efficient utilization of network resources allocation.

7.3 Cross-Layer Framework for End-to-End QoS

In this section, we propose a cross-layer framework for end-to-end QoS in wireless multi-hop ad hoc networks. As illustrated in Figure 7.1, the framework consists of four mechanisms: traffic policing, centralized scheduler, distributed scheduler, and admission control. These mechanisms in turn are assisted by three monitors: QoS

Table 7.1 Existing Schemes for End-to-End QoS

Ref.	QoS goals	Methods	Problems
[16, 17]	Accurate fairness of network resources among all nodes	Approximate fair queuing by setting appropriate back-off time; use round number to control fairness	One-hop fairness; cannot support end-to-end QoS of flows over multiple hops
[18]	Fairness according to an ideal scheduling algorithm in multi-hop network	Priority medium access (nonpersistent CSMA/CA): hearing priorities from neighbor nodes; priority access to the node with the highest priority	Cannot support end-to-end QoS of flows with multiple hops
[19]	Guaranteed service for a real-time flow over multi-hop networks	Piggybacking channel reservation on DATA/ACK frames	Reserving network resource completely for a flow cannot support multiple flows with diverse QoS requirements
[20]	Service assurances for flows in multi-hop network	Priority medium access: hearing priorities from neighbor nodes; priority access to the node with the highest priority; coordination mechanisms to adjust priority of packets in downstream hops	Coarse QoS with quantitative control
[21, 22]	Service assurance of flows in multi-hop wireless networks	DCS-NPDD-MAPS framework; dynamically adjust flows' priorities according to end-to-end performances; proportional queuing delay between neighbors; IEEE 802.11e for prioritized medium access	Unable to quantitatively control the network resources between flows
[23]	Guarantee services flows in multi-hop wireless networks	Partitioning end-to-end requirement into a single hop; adjust contention windows according to the performance of satisfying the requirement in previous packets; admission control of low-priority traffics via monitoring congestion of channel	Unable to quantitatively control network resource allocation between flows; sacrificing low-priority flows to guarantee high-priority ones cannot efficiently utilize network resources

monitor, route monitor, and channel monitor. We will next describe these mechanisms and monitors as well as explain the interactions among them.

In Figure 7.1, the traffic policing is to ensure that the traffic arrival of a flow is in accordance with the declared traffic profile. For the arrived packets that have exceeded the profile, the traffic police will either discard them or mark them so the marked traffic can be discriminated when the need arises later.

The traffic profile component of traffic policing is also used in the other mechanism, i.e., admission control. Generally, admission control needs to derive the resource requirement of a flow based on the traffic profile before deciding if the flow should be admitted into the system. Normally, the flow is admitted only when the required resource is not more than the available resource in a route. Thus, routing is an integral part of the admission control and directly affects the admission deci-

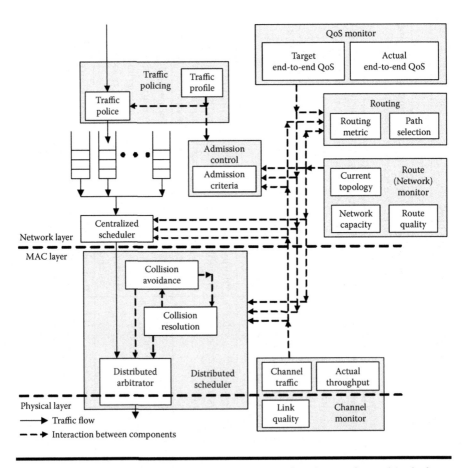

Figure 7.1   Cross-layer framework for proportional end-to-end QoS in wireless multi-hop ad hoc networks.

sion. The failure of admitting a flow to a route will prompt the routing component to search for another route before submitting the flow for admission decision again in an iterative approach. Despite rejecting a flow on a route due to insufficient resource, the available resource is not always known with certainty at the time of making the admission decision. This is due to the time-varying characteristics of link quality, which also affects the instantaneous actual end-to-end QoS and quality of a route. Thus, the admission control in the proposed framework needs to provide for and dynamically evaluate the impacts of the time-varying factors. As such, when the current route becomes unusable to a flow, the routing component may dynamically reroute the flow to another route that meets the flow's original performance requirements.

To dynamically evaluate the time-varying characteristics, the framework uses a channel monitor, a route monitor, and a QoS monitor. The channel monitor spans across both physical and MAC layers. In the physical layer, the channel monitor measures the link quality. In the literature, the link quality can be given in terms of bit error rate, received signal strength, signal-to-noise-and-interference ratio, etc. In the MAC layer, the channel monitor keeps track of the actual throughput as well as the channel traffic. In the framework, channel traffic is a general term that includes all received packets. It is based on these received packets, which may be erroneous or error-free, that other components in the framework may derive various information, such as the traffic load, actual QoS, current topology, etc.

Different from the channel monitor, which spans across the lowest two protocol layers, the route monitor appears only in the network layer. Here, the route monitor may quantify the route quality in terms of the effective end-to-end bit error rate, remaining time to a broken route, etc. Hence, the route quality is determined partly based on the mobility information and the qualities of its component links, which can be provided, among other things, by the channel monitor. In the proposed framework, the route monitor also keeps track of the current topology, which can be affected by mobility.

Similar to the route monitor, the QoS monitor is in the network layer, where the actual end-to-end QoS can be measured. The actual QoS can be compared against the target QoS, where an obvious difference suggests a failure in meeting performance requirements and thus triggers a sequence of activities in various mechanisms, such as rerouting and adjustments in transmission schedule.

Thus far, we have described traffic policing and admission control, which are two mechanisms working on different timescales. Specifically, traffic police must make a policing decision on each newly arrived packet, while admission control needs to decide on rerouting only after a sufficiently large number of packets have been transmitted or monitored such that the statistics collected by all the monitors are meaningful. Now, we introduce another mechanism: centralized scheduler. The centralized scheduler will decide among a set of the local flows which to serve after considering inputs from all the monitors. For example, while making a decision, the centralized scheduler needs to consider the target and actual end-to-end QoS of a flow that is provided by the QoS monitor.

In the centralized scheduler, the chosen flow will have its head packet sent from the network layer to its distributed scheduler in the MAC layer. Here, the distributed scheduler will decide which one from a set of neighboring nodes should transmit its packet to a physical medium using what parameters. These parameters, which include but are not limited to modulation scheme, carrier frequency, transmission power, packet length, etc., are decided by the distributed arbitrator, i.e., a component of the distributed scheduler. Note that the distributed arbitrator spans across two protocol layers because some of the transmission parameters it decides are physical layer parameters. All the transmission parameters are decided by the distributed arbitrator after taking into account the inputs from all the monitors, the collision avoidance function, and the collision resolution function. The two collision-related functions are needed as part of the distributed scheduler because collisions are likely to happen when the distributed arbitrator lacks perfect global information when making a transmission decision. While the distributed arbitrator works on a packet-by-packet basis, collision resolution and collision avoidance may or may not work on a packet timescale. As an example of collision avoidance, the CSMA/CA senses for the carrier and reserves the medium using RTS/CTS exchange for each packet. For the same purpose, TDMA uses a deterministic time slot allocation, which is performed only once for many packets.

Up to this point, we have described the mechanisms and monitors together with their interactions, as illustrated in Figure 7.1. We understand that the figure is not perfect because it does not show all the existing interactions. For example, the route quality in the route monitor is related to the link quality in the channel monitor, but this is not shown in the figure. We argue this is to avoid overcrowding the figure while keeping it conceptually correct. The key concept brought up by the framework is summarized as follows: In providing end-to-end QoS in a wireless multi-hop ad hoc network, we need the four mechanisms that are provided with feedback and dynamics by three monitors. These mechanisms and monitors operate across different protocol layers and timescales, and a change in any of the components will directly or indirectly affect the others. For the same purpose, to avoid overcrowding, the interactions between the four mechanisms are only shown indirectly through the monitors. For example, the admission control will affect the distributed scheduler by affecting actual QoS measured in the QoS monitor.

7.4 The PDMED Scheme

In this section, we present a realization of the proposed framework (see Figure 7.1), called PDMED [24], to provide an accurate proportional differentiation in end-to-end packet delay, which is also a performance metric for 802.11e. For ease of presentation, we assume that all the traffic flows are self-disciplined such that no traffic policing is required. We also assume the use of the CSMA/CA MAC protocol. This implies the collision avoidance function consists of RTS/CTS exchange and carrier sensing. Also, the collision resolution function is based on the paradigm that each flow has its own contention window size. Thus, collisions can be resolved by dynamically adjusting the contention window size, on which the back-off duration of a flow is determined. Let W_i be the contention window size of a flow i. Then, the back-off duration of a flow i, Δ, in terms of number of discrete intervals is decided as follows:

$$\Delta_i = U[0, W_i - 1] \tag{7.2}$$

where $U[x, y]$ is a function that generates random integer numbers within the range $[x, y]$. In (7.2), W_i is adjusted depending on the number of retransmission, m, the current flow i's packet has experienced such that $W_i = 2^m \times W_{min}$, where W_{min} is the minimum contention window size of all flows. While W_i increases with the number of retransmissions, it is upper bounded by W_{max}. The adoption of CSMA/CA also means that the centralized scheduler is implicit. Specifically, with CSMA/CA, only the local flow that has finished first counting down its back-off duration can contend for medium access with the other flows from neighboring nodes.

As a result of the few assumptions given above, the task of providing an accurate end-to-end proportional differentiation falls mainly on a distributed scheduler instead of the other three mechanisms. Thus, we will thereafter focus on designing the distributed scheduler and specifying how the QoS monitor, route monitor, and channel monitor should support the scheduler.

In designing the distributed scheduler, we let the QoS be defined in terms of average end-to-end packet delay. Thus, the target end-to-end QoS of the QoS monitor is to achieve proportional differentiation as follows:

$$\frac{d_i(t)}{\phi_i} = \frac{d_j(t)}{\phi_j}; \qquad \forall\, i, j, t \tag{7.3}$$

where ϕ_i has been defined earlier in (7.1), and $d_i(t)$ is the actual average end-to-end packet delay for flow i at time t. In practice, $d_i(t)$ must be measured at the destination node of flow i. From the expression above, the target QoS can be interpreted as achieving among all flows an equality in their normalized end-to-end packet delays, and the deviation of a flow i from the target QoS at time t can be quantified by $\beta_i(t)$ as follows:

$$\beta_i(t) = \max_{\forall j/i} \left\{ \frac{d_j(t)}{\phi_j} \right\} - \frac{d_i(t)}{\phi_i} \qquad (7.4)$$

From the equation, $\beta_i(t)$ is a positive real number where the smaller value means that it is closer to the QoS target, i.e., $\beta_i(t) = 0$. Thus, $\beta_i(t)$ is also used as the measurement for the actual QoS of flow i at time t.

To make $\beta_i(t)$ as close as possible to its target value 0, we propose to dynamically adjust the back-off duration of a flow based on its instantaneous deviation from the equality such that a flow with a relatively smaller $\beta_i(t)$ is given a shorter back-off duration to reduce its end-to-end packet delay. On the other hand, a flow with a relatively larger $\beta_i(t)$ is given a longer back-off duration to give way to transmissions from other flows with a smaller $\beta_i(t)$. However, there is no intuitive best-known method to perform the adjustment because of the following two problems: (1) The average end-to-end packet delay, $d_i(t)$, that is measured at the destination node is not readily available to the intermediate nodes and source node of the flow. (2) The normalized end-to-end packet delay of a flow is only known to the flow itself, but the computation of $\beta_i(t)$ requires the normalized delays of other contending flows.

Solving the two problems are the functions of the QoS monitor and channel monitor (refer to Figure 7.1), respectively. In the QoS monitor, a backward propagation scheme is proposed so that $d_i(t)/\phi_i$ computed at the destination node will be known by the flow's intermediate and source nodes. According to the backward propagation scheme, when a packet arrives at a flow i's destination node at time t, its average end-to-end delay is updated as follows:

$$d_i(t) = \frac{\tau_i(t) + (n(t)-1) \times d_i(t')}{n(t)} \qquad (7.5)$$

where $\tau_i(t)$ is the end-to-end delay of the packet arriving at time t, $n(t)$ is the total number of packets, including the newly arrived one, up to time t, and $d_i(t')$ is the previous average packet delay. Through the updating process, the destination node always has the latest value of normalized average end-to-end packet delay, i.e., $d_i(t)/\phi_i$. The latest value together with its respective flow identity will be piggybacked onto the MAC ACK frames that are transmitted in response to each successfully received MAC DATA frame of the flow. At the intermediate nodes, the piggybacked information will be extracted from the received MAC ACK frames and stored locally before being similarly piggybacked onto the upcoming MAC ACK frames of the flow. As such, the actual normalized end-to-end packet delay of each flow can be propagated from the destination node to the source node. We notice that there will be a time lag between the computation of an instantaneous normalized average end-to-end delay and its arrival at the intermediate and source nodes. In practice, the extent of the time lag depends on the number of hops. Through extensive simulations [24], we have found that the time lag has negligible effect on the QoS target for a small hop count.

In the channel monitor, a sniffer is proposed to read all the transmitted MAC ACK frames within a broadcast region. With the sniffer, each node can maintain a table containing the identities of all neighboring flows and their respective latest normalized average end-to-end delays. The table is updated each time a MAC ACK frame is received. With the up-to-date table, $\beta_{i,k}(t)$, i.e., the value of $\beta_i(t)$ (refer to (7.4)) at the k-th hop of flow i, can be computed as follows:

$$\beta_{i,k}(t) = \max_{\forall j \in I_{i,k/i}} \left\{ \frac{d_j(t)}{\phi_j} \right\} - \frac{d_i(t)}{\phi_i} \tag{7.6}$$

where $I_{i,k}$ is the set of flow i's neighboring flows at its k-th hop. Based on the computed $\beta_{i,k}(t)$, flow i can rank itself among all its neighboring flows. Specifically, the flow will be given the rank ℓ if its $\beta_{i,k}(t)$ is the ℓ-th highest among all the neighboring flows.

Let $r_{i,k}$ be the rank of flow i at its k-th hop when it has a packet to transmit there but senses a busy channel. In case no ranking can be performed, the default value for $r_{i,k}$ is unity. Also, let $W_{i,k} = 2^{m_{i,k}} \times W_{\min}$ be the flow's contention window size at its k-th hop when the packet is making the $m_{i,k}$-th retransmission attempt and $m_{i,k} = 0$ for a fresh packet. Then, instead of using the original CSMA/CA method in (7.2), the distributed scheduler will decide the flow's back-off duration, $\Delta_{i,k}$, as follows:

$$\Delta_{i,k} = \begin{cases} U[0, W_{\min} - 1] + I_{r_{i,k} \geq 2} \times \gamma_{i,k} \times W_{\min} & \text{if } m_{i,k} = 0 \\ U[0, \frac{W_{i,k} - 1}{h_i}] + W_{i,k} \times (\frac{h_i - k}{h_i} + r_{i,k} - 1) & \text{otherwise} \end{cases} \tag{7.7}$$

where h_i is the total number of hops for flow i and it is provided to the distributed scheduler by the route monitor in Figure 7.1. In (7.7), the term I_A is an indicator function defined as follows:

$$I_A = \begin{cases} 1 & \text{if A is true} \\ 0 & \text{otherwise,} \end{cases} \tag{7.8}$$

and $\gamma_{i,k}$ is a dynamic control parameter for flow i at its k-th hop. The control parameter has an initial value of unity, and it is dynamically adjusted only for a fresh packet at time t based on the actual normalized average end-to-end delay as follows:

$$\gamma_{i,k} = \begin{cases} \gamma'_{i,k}(t')+1 & \text{if} \quad 0<\beta_{i,k}(t')<\beta_{i,k}(t) \\ \gamma'_{i,k}(t')-1 & \text{if} \quad \beta_{i,k}(t)=0 \quad \text{and} \quad \gamma_{i,k}(t)>1 \\ \gamma'_{i,k}(t') & \text{otherwise} \end{cases} \quad (7.9)$$

where $\beta_{i,k}(t')$ and $\gamma'_{i,k}$ are the previous values of $\beta_{i,k}(t)$ and $\gamma_{i,k}$, respectively.

Comparing (7.7) and (7.2), we notice that the proposed distributed scheduler gives priority to a flow that is experiencing excessive normalized average end-to-end delay by allowing a smaller back-off duration. To ensure high responsiveness of the proposed mechanism, $\gamma_{i,k}$ provides an additional degree of freedom when ranking and prioritization alone are not sufficient to quickly bring down a high normalized delay. Also, the proposed method gives priority to a retransmitted packet compared to a fresh packet. This is to avoid the situation where multiple packets from the same flow are contending with each other arbitrarily. Among all the retransmitted packets, based on the heuristic disclosed in [25], the packet that is closer to the destination node will be given the priority to transmit so that the overall end-to-end delay can be reduced.

In PDMED, the message overhead is only the QoS performance value in the QoS monitor at the destination that is fed back to the source. The value is a floating point number of 2 bytes in the ACK frame. No other message and control packet is needed.

7.4.1 Performance Evaluation

The proposed PDMED has been extensively evaluated through random event simulations using OPNET [26]. The importance and effectiveness of its various components, namely, the back propagation, γ adjustment, and retransmission, have been confirmed [24].

After verifying PDMED's components, we now benchmark PDMED against IEEE 802.11e. For the purpose of simulation, the general static network topology as illustrated in Figure 7.2 is used first. In the network, there are only two flows: flow 1 ($S1 \rightarrow D1$) and flow 2 ($S2 \rightarrow D2$). From the figure, flow 1 and flow 2 have three and two hops, respectively. For the flows, their differentiation parameters are denoted by ϕ_1 and ϕ_2, respectively.

In the simulations, the raw bit rate of the communication channel is 1 Mbps. Also, refer to (7.2); W_{min} and W_{max} are fixed at 16 and 1,024 time slots, respectively. Here, the duration of each time slot Tslot = 50us. In addition, the delay of a packet

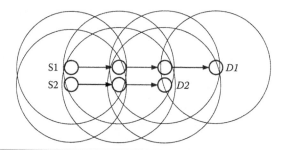

Figure 7.2 Topology of simulation scenario with different hops.

is the time elapsed since the packet's arrival at the MAC layer of its source node until the packet's subsequent arrival at the MAC layer of its destination node. These packets from their respective traffic sources are queued above, but not in, the MAC layer to avoid distortion in packet delay at a high traffic rate, when the delays of all flows increase exponentially, making any difference in their values not noticeable. Different ϕ_2/ϕ_1 ratios are achieved by fixing ϕ_1 at 1 while varying ϕ_2.

We use video traces from [27] that are coded by H.263 at 265 Kbps. Each of the coded video frames can be a few thousand bytes and thus potentially larger than the supported maximum MAC DATA frame payload size, i.e., 2,000 bytes. When this occurs, the oversized video frame is fragmented into multiple smaller frames of 2,000 bytes, with the final frame containing the residual bytes.

To begin with, we use the video trace from the movie *Jurassic Park*. Although both flow 1 and flow 2 use the same video trace, they have different time offsets. The offsets for flow 1 and flow 2 are 0 and 300 s, respectively. This means flow 2 starts playing the movie from its 300th second. Figure 7.3 shows the average end-to-end packet delay for different ratios of ϕ_2/ϕ_1. From the figure, the average packet delay for flow 1 equals that of flow 2 when $\phi_2/\phi_1 = 1$. Similarly, when $\phi_2/\phi_1 = 2$, the average delay of flow 2 is double compared to that of flow 1. This is a clear indication of an accurate proportional differentiation when the multi-hop ad hoc network is loaded with the actual video trace from a movie. As depicted in Figure 7.3, this accuracy in proportional differentiation is consistent when the evaluation is repeated using different video traces from other movies, e.g., *Silence of the Lambs* and *Star Wars*.

We benchmark PDMED against IEEE 802.11e, which is designed to provide QoS differentiation in a wireless ad hoc network. Different from PDMED, IEEE 802.11e achieves its goal by selecting an appropriate traffic class and setting different minimum and maximum contention window sizes, which are denoted by W_{min} and W_{max} in (7.2) for different flows within the selected traffic class. For this

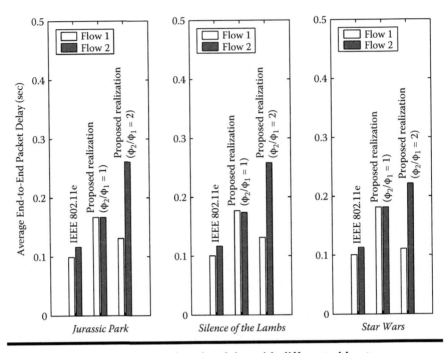

Figure 7.3 Average end-to-end packet delay with different video traces.

benchmark, we let both flows be from the same traffic class because PDMED does not have the concept of traffic classification and achieves its performance goal only by adjusting the contention window size of a flow. Unfortunately, there is no standardized method in IEEE 802.11e on how to set the contention window sizes to achieve its performance goal. Recall the finding in [11] that suggests that the one-hop average delay of an IEEE 802.11 flow is proportional to its minimum contention window size. Hence, we fixed the maximum contention window size at 1,024 time slots while setting the minimum contention window sizes for flow 1 and flow 2 to 16 and 32 time slots, respectively, to make flow 2's average end-to-end packet delay two times that of flow 1's.

For the evaluation described above, Figure 7.3 shows that IEEE 802.11e is not capable of providing an accurate proportional differentiation in end-to-end packet delay for all three movies. In the figure, despite failure in accurate proportional differentiation, IEEE 802.11e gives a lower average end-to-end delay. This is because IEEE 802.11e tends to have a smaller back-off duration than PDMED, especially when $\gamma\{i,k\}$ grows to a bigger value to provide accurate differentiation. For the same reason, Figure 7.4 shows that PDMED yields a lower throughput than IEEE 802.11e. The smaller throughput and higher delay are the cost incurred by PDMED in achieving the accurate proportional differentiation.

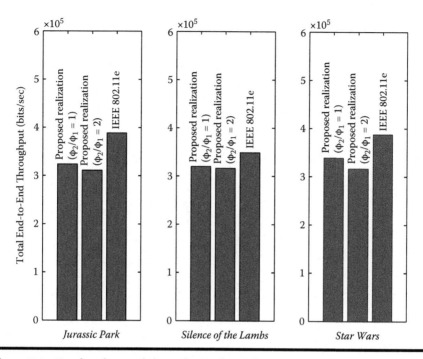

Figure 7.4   Total end-to-end throughput of two flows with different video traces.

7.4.2   Transmitting Video through Mobile Nodes

When building a mobility scenario, a routing protocol support is necessary. The Ad Hoc On-Demand Distance Vector (AODV) routing protocol, which is widely adopted in evaluations of ad hoc networks, is used in our simulations. There are 12 nodes that move within a 2,000 × 2,000 m² area. Initial locations of the nodes are random. Figure 7.5 shows an example of this network topology.

All of the nodes move according to the random waypoint model, with speeds defined by a uniform distribution function $U[0, y]$. Two flows, denoted $F1$ and $F2$, are deployed in such a network. We define that $F1$ initiated by node 0 destines at node 10, and $F2$ initiated by node 1 destines at node 11. $F1$ and $F2$ are both video traffics of *Jurassic Park*, but with different offsets of start time, 0 and 300 s, respectively. The differentiation ratio, ϕ_2/ϕ_1, is set as 2:1 between $F1$ and $F2$. We choose five speed scenarios for evaluation: $U[0, 20]$, $U[0, 40]$, $U[0, 60]$, $U[0, 80]$, and $U[0, 100]$ (m/s).

In Figure 7.6, good differentiation ratios between the average end-to-end delays of two flows are exhibited in all speed scenarios. This shows that PDMED is still able to achieve good proportional differentiation in end-to-end QoS between flows

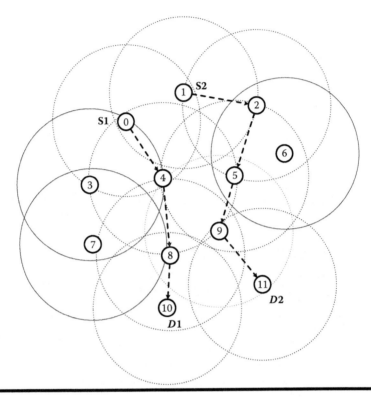

Figure 7.5   An example of network topology with mobile nodes.

even if nodes move randomly at various speeds. Although the hop counts of the flow change randomly due to random movement of nodes, PDMED can still accurately control the ratios of resource utilization between flows. The accuracy of proportional differentiation is shown in Figure 7.7. We also see in Figure 7.6 that the average end-to-end packet delay becomes larger when the moving speed increases from $U[0, 20]$ to $U[0, 40]$ (m/s), and then becomes smaller slowly after the moving speed increases over $U[0, 40]$ (m/s). The reason is that, when moving speeds of nodes are slow, the links between nodes are stable. Thus, rerouting seldom happens so that the end-to-end packet delay is low. When speed increases, link breakage happens more often. Hence, the end-to-end packet delay increases due to rerouting. However, when speed keeps increasing, although rerouting may happen more frequently, it is easier to reroute a path for a flow because nodes are faster in moving close to each other. Although rerouting action can be faster for finding a new path when nodes move at high speeds, it is impossible to trade off the packet delay due to waiting for a new path. Therefore, we see such a trend of the average end-to-end packet delay in Figure 7.6.

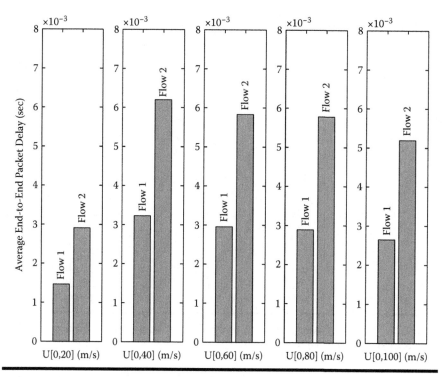

Figure 7.6 **Average end-to-end delay at different speeds using video trace of** *Jurassic Park.*

Besides the moving speed of nodes, the density of nodes in the network also affects the link stability. If there are fewer nodes moving in a network, it is more difficult to find a new path. Thus, we also repeat the above simulation in a 3,000 × 3,000 m² area. Because the moving area is increased 1.5 times, it is obvious that the number of hops of a flow is potentially increased. Thus, the average end-to-end packet delay should increase. In the following, we focus on comparing the proportional differentiation performances from two different area sizes. In Figure 7.7, we compare the achieved differentiation ratio, D_2/D_1, between the two scenarios. Here, D_2 and D_1 are the average end-to-end packet delays of flow 2 and flow 1. Obviously, when the moving area of nodes increases to 3,000 × 3,000 m², the accuracy of proportional differentiation performance is affected. Although there is still differentiation between two flows, the ratio is far from the target ratio, ϕ_2/ϕ_1. This is because the lower density of nodes in the network increases the difficulty of finding a new path, and the duration of the link breakage is also increased. Thus, link breakages happening can dynamically vary the end-to-end packet delay. The figure shows that PDMED is not capable of compensating for this variation completely. And

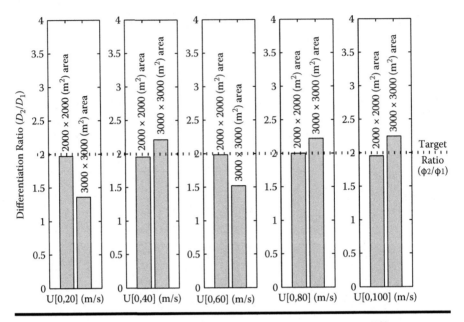

Figure 7.7 **Ratios of average end-to-end delay between flows at different speeds using video trace of** *Jurassic Park.*

this phenomenon also indicates that the effect of link breakage on proportional differentiation of PDMED is much larger than the dynamically changing hop count. To support proportional differentiation in such a network, with a long link breakage, another mechanism, such as QoS routing in our framework or a controller on node distribution in the network, is needed to cooperate with PDMED.

Because link breakages reduce the network resources, the total end-to-end throughput of two flows in the 3,000 × 3,000 m² scenario is reduced, compared to the 2,000 × 2,000 m² scenario, as shown in Figure 7.8. From the figure, we also see a trend of total throughput with increasing speeds. When the speeds are low, the total throughput of the network is highest because the wireless links between nodes are robust. With increasing speed, the total throughput of the network is reduced due to more occurrences of link breakages. The packets have to wait for a new route. When the speed keeps increasing, the total throughput becomes high again because of the faster establishment of a new route. However, after that, a further increase in speed leads to a decrease in throughput. This is because frequent rerouting reduces available network resources.

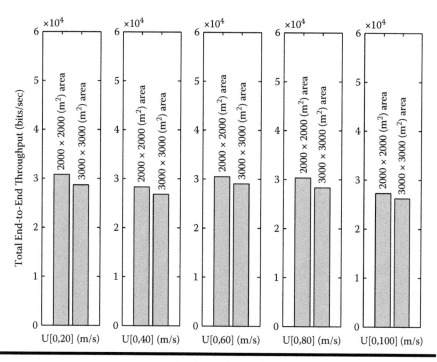

Figure 7.8 Total throughput at different speeds using video trace of *Jurassic Park.*

7.5 The Improved PDMED Scheme

Recall that PDMED dynamically adjusts the scheduling parameters based on the instantaneous end-to-end QoS performances of all flows in networks to achieve the proportional differentiation between flows. The variation of the end-to-end performance of one flow will bring variation on the end-to-end performances of its neighboring flows and even all other flows in the network. Thus, when the mobility of nodes and the time-varying channel affect the instantaneous end-to-end QoS value of one flow, the end-to-end QoS values of other flows are also affected soon after. And the more variations of the wireless channel that happen due to mobility and the time-varying channel, the larger the deviation of the instantaneous proportional differentiation ratio among flows from the target ratio. In PDMED, to compensate this effect and achieve proportional differentiation accurately, the performance of flows, such as delay or throughput, has been traded off. A support for PDMED that handles variations on packet transmissions due to mobility of nodes and the time-varying channel in the physical layer is quite necessary for increasing utilization of network resources, robustness, and accuracy of PDMED in the environment with mobile nodes and time-varying channel quality.

In the literature, there exist methods to handle node mobility and time-varying channel quality. Among them, some track and predict the mobility of nodes [28–30]. Others capture the time-varying channel quality based on cellular wireless networks [31–33]. Some give an estimation of the upper bound of the signal-to-noise ratio from the point of view of a whole ad hoc network [34]. None has captured the instantaneous effects of node mobility and time-varying channel on packet transmissions in CSMA/CA-based wireless multi-hop ad hoc networks.

We observed that, with the random waypoint mobility model, the received SINR for a packet exhibits self-similar characteristics. Based on the self-similarity of SINR, we suggested a forecasting method to predict the value of SINR series in one step ahead. Then, we proposed an improvement to PDMED, namely, PDMED+, for better network throughput while providing proportional differentiation in end-to-end packet delay.

7.5.1 Self-Similarity in SINR of Ad Hoc Networks

The widely adopted random waypoint mobility model appears to create realistic mobility patterns for the way people might move [35]. Consider the same network scenario as in Figure 7.5. There are 12 nodes whose movement is limited in an area of 3,000 × 3,000 m^2.

Because in the random waypoint model the nodes move in a straight line from a current site to the next site, no obstacle is considered between two sites. So we choose free space propagation model [36] to define the received signal strength as (8.10), which assumes that transmitter and receiver have a clear, unobstructed line-of-sight path:

$$P_r(d) = \frac{P_t G_t G_r \lambda^2}{(4\pi)^2 d^2 L} \tag{7.10}$$

where P_r is the received power, which is a function of the T–R (transmitter to receiver) separation; P_t is the transmitted power; G_t and G_r are the transmitter antenna gain and receiver antenna gain, respectively; d is the T–R separation distance in meters; L is the system loss factor not related to propagation ($L > = 1$); and λ is the wavelength in meters. For simplicity, so that we can focus on mobility and wireless channel characteristic, G_t, G_r, and L are set to 1.

Based on the definition of received signal strength in (8.10), the SINR of a transmission from node i to node j is given as follows:

$$SINR(i,j) = \frac{P_r(i,j)}{\sum_{k \notin B_t, B_r} P_r(k,j) + N_b} \tag{7.11}$$

where $P_r(i, j)$ is the received power of a packet from node i to node j. k denotes the nodes that are out of radio coverage of the transmitter of the target transmission. Also, B_r is the set of nodes located in the radio coverage of the receiver node of the target transmission, B_t is the set of nodes located in the radio coverage of the transmitter node of the target transmission, and N_b is background noise.

With the SINR definition, we conducted a simulation experiment to study the characteristics of SINR due to the mobility of nodes. We recorded the position of every node and calculated the distance and SINR between any two nodes with a sampling interval of 0.01 s for a period of 1,000 s. The SINR value is calculated for a transmission from a transmitter node and a receiver node.

We used variance-time plot methodology [37] to analyze the data. Specifically, a data series X with a length of N is divided into N/m blocks by a block size m. The estimation of variances of data in each block, $\left(\overline{V}^m_{X_k}\right)$, are calculated, $k = 1,2, \ldots,$ N/m. Using $\left(\overline{V}^m_{X_k}\right)$, we can obtain the variance of the whole data series as follows:

$$V^m_X = \frac{1}{N/m} \sum_{k=1}^{N/m} \left(\overline{V}^m_{X_k}\right)^2 - \left(\frac{1}{N/m} \sum_{k=1}^{N/m} \overline{V}^m_{X_k}\right)^2 \tag{7.12}$$

With the logarithm of the different block size, m, and the logarithm of the corresponding V^m_X, we can use least squares line fitting to calculate the estimated slope of the fitting line and the correlation coefficient, r, which is calculated by the following formula:

$$r = \frac{\sum\limits_{i=1}^{n} V^{m_i}_X m_i - n \overline{V}^m_X \overline{m}}{\sqrt{\left(\sum\limits_{i=1}^{n} m_i^2 - n\overline{m}^2\right)\left(\sum\limits_{i=1}^{n} \left(V^{m_i}_X\right)^2 - n\left(\overline{V}^m_X\right)^2\right)}} \tag{7.13}$$

where n is the number of different block sizes. We denote the block sizes from the smallest to the largest as $m_i = m_1, m_2, \ldots, m_n$. Here, $\overline{m}$ is the average of m_i. $\overline{V}^m_X$ is the average of $V\{M_i\}$ for every m_i. Also, r expresses how a perfect linear fit exists between the discrete points. In general, a reasonable fit requires $\overline{V}^{m_i}_X$.

Table 7.2 shows three numerical results: the slope of the fitting line by log-log correlogram, Hurst parameter, and correlation coefficient of least squares line fitting of any two nodes when their speeds are uniformly distributed with $U[0, 20]$

Table 7.2 Variance-time Plot Analysis on Data of Speed Scenario of $U[0, 20]$ (m/s)

n-n	β	H	r	n-n	β	H	r	n-n	β	H	r
0-1	-0.056	0.972	0.821	1-0	-0.074	0.963	0.841	2-0	-0.608	0.696	0.975
0-2	-0.629	0.686	0.974	1-2	-0.052	0.974	0.857	2-1	-0.040	0.980	0.890
0-3	-0.008	0.996	0.814	1-3	-0.008	0.996	0.835	2-3	-0.065	0.967	0.851
0-4	-0.014	0.993	0.901	1-4	-0.008	0.996	0.858	2-4	-0.043	0.979	0.881
0-5	-0.095	0.953	0.855	1-5	-0.009	0.995	0.858	2-5	-0.046	0.977	0.873
0-6	-0.015	0.992	0.749	1-6	-0.043	0.978	0.800	2-6	-0.045	0.977	0.864
0-7	-0.014	0.993	0.902	1-7	-0.020	0.990	0.853	2-7	-0.199	0.901	0.891
0-8	-0.026	0.987	0.866	1-8	-0.022	0.989	0.812	2-8	-0.031	0.985	0.915
0-9	-0.072	0.964	0.759	1-9	-0.008	0.996	0.847	2-9	-0.026	0.987	0.812
0-10	-0.319	0.841	0.841	1-10	-0.084	0.958	0.899	2-10	-0.042	0.979	0.885
0-11	-0.009	0.995	0.818	1-11	-0.017	0.991	0.864	2-11	-0.044	0.978	0.901
3-0	-0.031	0.985	0.779	4-0	-0.008	0.996	0.887	5-0	-0.131	0.935	0.845
3-1	-0.022	0.989	0.749	4-1	-0.004	0.998	0.845	5-1	-0.007	0.996	0.874
3-2	-0.039	0.980	0.752	4-2	-0.028	0.986	0.738	5-2	-0.022	0.989	0.782
3-4	-0.022	0.989	0.776	4-3	-0.006	0.997	0.877	5-3	-0.005	0.998	0.883
3-5	-0.025	0.988	0.792	4-5	-0.012	0.994	0.849	5-4	-0.012	0.994	0.842
3-6	-0.028	0.986	0.776	4-6	-0.063	0.969	0.808	5-6	-0.022	0.989	0.816
3-7	-0.034	0.983	0.772	4-7	-0.012	0.994	0.827	5-7	-0.046	0.977	0.822
3-8	-0.016	0.992	0.774	4-8	-0.012	0.994	0.806	5-8	-0.072	0.964	0.714
3-9	-0.033	0.984	0.744	4-9	-0.205	0.897	0.886	5-9	-0.044	0.978	0.838
3-10	-0.019	0.990	0.779	4-10	-0.012	0.994	0.801	5-10	-0.006	0.997	0.870
3-11	-0.018	0.991	0.748	4-11	-0.011	0.995	0.799	5-11	-0.032	0.984	0.778
6-0	-0.016	0.992	0.752	7-0	-0.046	0.977	0.775	8-0	-0.085	0.958	0.747

6-1	-0.038	0.981	0.877	7-1	-0.062	0.969	0.810	8-1	-0.018	0.991	0.795
6-2	-0.008	0.996	0.810	7-2	-0.185	0.907	0.868	8-2	-0.034	0.983	0.857
6-3	-0.014	0.993	0.826	7-3	-0.036	0.982	0.757	8-3	-0.006	0.997	0.837
6-4	-0.059	0.970	0.817	7-4	-0.034	0.983	0.768	8-4	-0.016	0.992	0.838
6-5	-0.015	0.992	0.830	7-5	-0.053	0.974	0.792	8-5	-0.008	0.996	0.846
6-7	-0.018	0.991	0.809	7-6	-0.037	0.982	0.800	8-6	-0.007	0.997	0.865
6-8	-0.016	0.992	0.879	7-8	-0.031	0.985	0.808	8-7	-0.008	0.996	0.871
6-9	-0.111	0.945	0.724	7-9	-0.041	0.979	0.800	8-9	-0.006	0.997	0.883
6-10	-0.350	0.825	0.848	7-10	-0.033	0.984	0.778	8-10	-0.008	0.996	0.822
6-11	-0.010	0.995	0.778	7-11	-0.044	0.978	0.806	8-11	-0.139	0.930	0.866
9-0	-0.072	0.964	0.756	10-0	-0.004	0.998	0.859	11-0	-0.005	0.998	0.847
9-1	-0.012	0.994	0.862	10-1	-0.014	0.993	0.876	11-1	-0.013	0.993	0.857
9-2	-0.024	0.988	0.906	10-2	-0.006	0.997	0.827	11-2	-0.013	0.994	0.838
9-3	-0.042	0.979	0.846	10-3	-0.065	0.968	0.750	11-3	-0.006	0.997	0.854
9-4	-0.225	0.888	0.886	10-4	-0.006	0.997	0.862	11-4	-0.030	0.985	0.783
9-5	-0.073	0.963	0.861	10-5	-0.004	0.998	0.830	11-5	-0.007	0.997	0.824
9-6	-0.015	0.992	0.795	10-6	-0.004	0.998	0.818	11-6	-0.009	0.996	0.806
9-7	-0.012	0.994	0.856	10-7	-0.012	0.994	0.812	11-7	-0.058	0.971	0.797
9-8	-0.106	0.947	0.797	10-8	-0.001	1.000	0.231	11-8	-0.140	0.930	0.863
9-10	-0.060	0.970	0.852	10-9	-0.002	0.999	0.760	11-9	-0.031	0.985	0.827
9-11	-0.027	0.986	0.865	10-11	-0.002	0.999	0.851	11-10	-0.009	0.996	0.860

(m/s). Similar numerical results are available for $U[0, 40]$, $U[0, 60]$, $U[0, 80]$, and $U[0, 100]$ (m/s). In the table, **n–n** denotes a pair of nodes (transmitter node to receiver node) by its ID number. Here, β denotes the slop of the log-log correlogram (the log of the sample variance against the log of the sample size), H denotes the Hurst parameter that is given by $H = 1 - \beta/2$, and **r** is the correlation coefficient for least squares line fitting.

From the analysis results, we can see that the SINR series between any two nodes in the network exhibit self-similarity because the Hurst parameters calculated are all between 0.5 and 1.0. This means that the SINR series between any two nodes have short- or long-range dependency characteristics. Therefore, through signal processing methods, we are able to find out the parameters of the characteristics of the SINR series and forecast the values of SINR ahead. We are also able to predict the value of SINR before transmitting a packet that provides information on the condition of the physical layer before putting the packet into the physical layer from the MAC layer. This discovery serves a good basis for designing an improvement to PDMED.

7.5.2 Prediction Method and Estimation of Prediction Error

With self-similarity, it is feasible to forecast the SINR value based on the history data in a SINR series. We use the fractionally integrated autoregressive moving average (F-ARIMA) process time-series to model our SINR series. There are a lot of prediction methods based on the F-ARIMA process in literature, such as [38–40]. We choose the method in [38] for prediction because of its simplicity.

The steps of this prediction method are as follows:

1. Estimate Hurst parameter, H, of the SINR series, which is denoted by $x(n)$. Then, compute the differential factor, $d = H - 0.5$.
2. Convert $x(n)$ from F-ARIMA(p, d, q) process to an ARMA(p, q) process, denoted by $w(n)$, as follows:

$$w(n) = \nabla^d \left(x(n) - \mu \right). \tag{7.14}$$

where μ is the expected value of $x(n)$, and

$$\nabla^d = \left(1 - B\right)^d = \sum_{k=0}^{\infty} \binom{d}{k} \left(-1\right)^k B^k. \tag{7.15}$$

Here, B is a lag operator such that $x(n-1) = Bx(n)$, and

$$\binom{d}{k} = \frac{\Gamma(d+1)}{\Gamma(k+1)\Gamma(d-k+1)} \tag{7.16}$$

where Γ represents the gamma function.

3. Estimate $\phi(B)$ and $\theta(B)$ of $w(n)$ using the Prony method [41]:

$$\phi(B)=1-\phi_1 B-\phi_2 B^2-\cdots-\phi_p B^p$$

(7.17)

$$\theta(B)=1-\theta_1 B-\theta_2 B^2-\cdots-\theta_q B^q$$

(7.18)

4. Convert F-ARIMA(p, d, q) process to F-ARIMA(0, d, 0) process as denoted by $y(n)$ through $y(n) = \theta(B)^{-1}\phi(B)x(n)$.

5. Predict one-step-ahead value of $y(n)$ by applying the following formula:

$$\hat{y}(n)=\sum_{j=1}^{k}\beta_{kj}y(n-j)$$

(7.19)

where

$$\beta_{kj}=-\binom{k}{j}\frac{\Gamma(j-d)\Gamma(k-d-j)}{\Gamma(-d)\Gamma(k-d+1)}$$

(7.20)

6. Compute the predicted value of F-ARIMA(p, d, q) process, i.e., $x(n)$, according to $\hat{y}(n)$ through $\hat{x}(n)=\theta(B)^{-1}\phi(B)\hat{y}(n)$.

With the method above, we take the SINR series from node 0 to node 1 of the speed scenario by $U[0,20](m/s)$ as an example, to do the prediction and show the performance of the prediction method. We predict the one-step-ahead value of SINR series by Matlab and compare the original SINR series (Figure 7.9) to the predicted values (Figure 7.10). The two figures show that the method is able to accurately predict the value of SINR series so as to track the trend of variation of SINR series.

From the figures, it is obvious that the difference values between predicted values and original ones, which may lead to a wrong decision when predicting whether a packet is transmitted successfully, are mostly small, i.e., within the range of $\pm0.5 \times 10^{-4}$. To capture the characteristics of the difference, we estimate its probability density function. The estimation method is that we separate the range of the difference values into continuous intervals with 1×10^{-6}, which is the accuracy degree of numeric data. Then, we count the number of difference values that fall into every interval from all of the data, and calculate the probability of an interval as the number of difference values in it divided by the total number of data. With

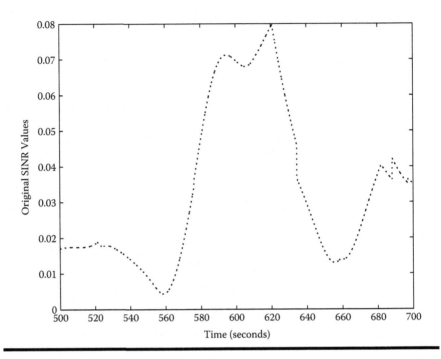

Figure 7.9   The actual SINR series.

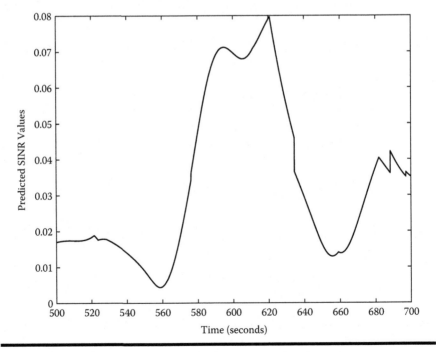

Figure 7.10   The predicted SINR series in one step ahead.

the probabilities of these continuous intervals, we can approximate the shape of the probability density function of the difference, which is shown in Figure 7.11. From the shape of Figure 7.11, we find that the shape of the probability density of all the values of the difference can be approximated by a modified normal distribution.

7.5.3 The Improved Scheme: PDMED+

With the predicted SINR on packet transmissions in the physical layer, now we are going to design an improved scheme of PDMED. For simplicity, we set a SINR threshold to decide whether a packet is received successfully.

Let us recall how PDMED works with mobile nodes moving randomly. In PDMED, when a transmitter node suffers bad wireless channel quality when transmitting a packet, no matter whether due to mobility or channel error, it still transmits the packet because it does not know the channel quality. Then, after waiting for a time-out period without ACK from the receiver node, the transmitter assumes that it failed to transmit the packet due to collisions. Then, the transmitter node will double its contention window size, generate a back-off duration, and start to back off to retransmit the packet. However, there is no collision happening, just a bad channel quality. It is wasteful for the transmitter node to back off again with a

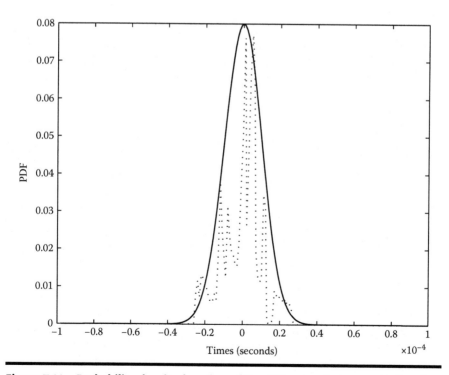

Figure 7.11   Probability density function of the difference between predicted and real SINR values.

doubled contention window. Getting this insight, we proposed a simple modification of PDMED: PDMED+.

In PDMED+, when a transmitter node is going to transmit a packet, it predicts the SINR that the packet will suffer in the point of view at the receiver node and decides whether the packet will be successfully received according to a threshold to SINR. If the node predicts that the transmission will fail, it defers the packet's transmission until the environment makes SINR of the packet increase over the threshold, i.e., the channel quality turns out to be good and the packet is able to be successfully transmitted. This method has two drawback situations. First, if the channel is bad for a longer time than the total duration of retrying seven times in IEEE 802.11, the delay of a packet is increased tremendously. However, in PDMED, which is based on the retransmission algorithm in IEEE 802.11, the packet will be dropped if it is not able to transmit successfully after seven retransmissions. Therefore, the delay of a packet is controlled with an upper limit. Second, when there are two nodes deferring themselves, waiting for a good channel, during their deference period one of their neighboring transmitters successfully transmits a packet. During the period of the transmission, the channel becomes good for both of the deferring nodes. After the transmission finishes, the two deferring nodes will begin to transmit their packets at the same time. At this time, collision definitely happens. The two nodes have to double their contention windows and back off for retransmission.

We proposed the following methods to handle the above-mentioned drawbacks. To overcome the first drawback, we also set a limit period for deferring a packet transmission because of bad channel quality. Borrowing the retransmission mechanism in IEEE 802.11, we set a similar long time for the total deferring period:

$$T = 2^7 \times W_{\min} \times \delta, \tag{7.21}$$

where δ is the slot time. After the time, the packet is dropped.

For the second drawback, when two deferring nodes predict good channel quality after hearing a transmission, we give differentiated periods for the two nodes before they transmit the deferred packets. We set the differentiated period as follows:

$$\sigma_i = U[0, r_i \times W_{\min}] \times \delta, \tag{7.22}$$

where $U[x, y]$ is a uniform distribution function that generates random integer numbers within the range $[x, y]$, σ_i is the time for differentiated period of flow i, r_i is the rank of flow i, W_{min} is the minimum contention window size of all flows, and δ is the slot time. This improved scheme, i.e., PDMED+, is illustrated in Figure 7.12.

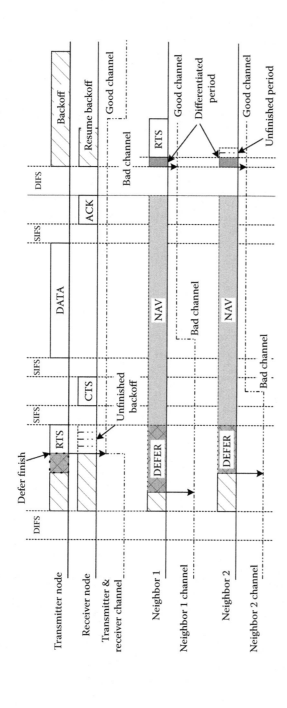

Figure 7.12   Illustration of PDMED+.

7.5.4 Performance Evaluation

We repeat the PDMED simulations for PDMED+ using a video trace of *Jurassic Park*. Figure 7.14 shows the total throughput of two flows. From the figure, it is obvious that PDMED+ is able to increase the total throughput of two flows, compared to PDMED. This confirms that compared to the retransmission method in PDMED, the way of deferring the packet until the channel becomes good in PDMED+ is to utilize the channel ability more efficiently. Although PDMED+ may increase a little bit of average end-to-end delay sometimes as show in Figure 7.13, it achieves a more accurate differentiated ratio between the two flows, as depicted in Figure 7.15. In the figure, the achieved ratio of PDMED+ is closer to the line-of-target ratio. The reason is that more packets are able to be transmitted by PDMED+ so as to increase the chances to adjust the differentiation ratio between two flows.

7.6   Summary

Wireless multi-hop ad hoc networks that are demanded by more and more applications with different QoS requirements suffer from time-varying and limited network resources. Thus, it is not easy to support QoS provisions to satisfy the

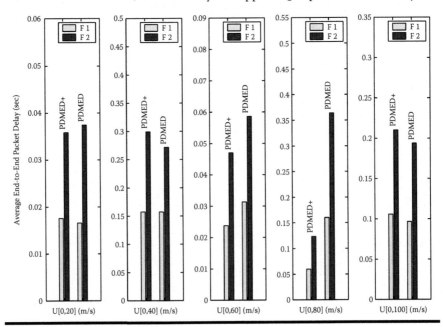

Figure 7.13   Average end-to-end delay at different speeds with video trace of *Jurassic Park*.

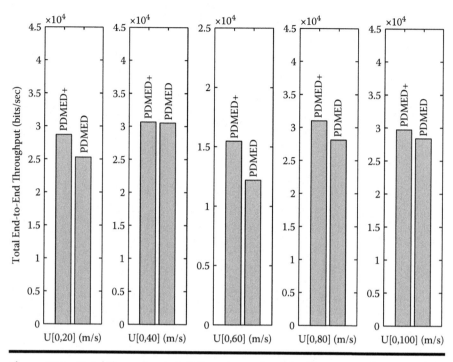

Figure 7.14 Total throughput at different speeds with video trace of *Jurassic Park*.

expectations of all users. An efficient network allocation between users according to their end-to-end QoS performances would be desirable. Due to the time-varying characteristic of network topology and wireless link capacity, several network components have to cooperate to achieve this kind of optimization.

This chapter first introduces a cross-layer framework to present a conception of providing proportional differentiation on end-to-end performances in wireless multi-hop ad hoc networks. Through four mechanisms in different layers and three monitors, the necessary information is exchanged between layers and adapts the functions of different network components so as to achieve proportional differentiation on end-to-end performance between users.

After proposing the framework, we introduced a realization called PDMED to provide a consistent and accurate proportional differentiation on the average end-to-end delay based in CSMA/CA-based wireless multi-hop ad hoc networks. Specifically, the distributed scheduler dynamically adjusts the back-off duration of a flow based on its instantaneous deviation from the maximum average end-to-end packet delay. QoS monitor functions via a feedback method and information sharing due to broadcasting wireless medium together with the store-and-forward multi-hop transmission. The destination nodes feed back the instantaneous average

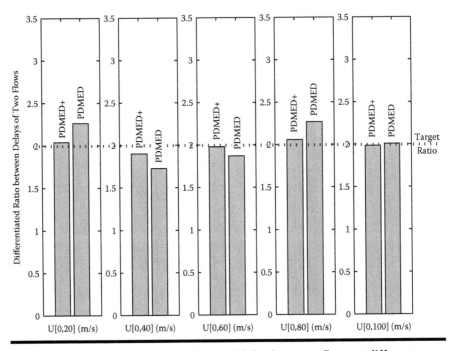

Figure 7.15   Ratios of average end-to-end delay between flows at different speeds with video trace of *Jurassic Park.*

end-to-end packet delay of a flow along the transmission path in backward. And neighbor nodes along the path monitor the feedback information.

Further, time-varying network topology and wireless link capacity due to node mobility reduce network resource utilization while providing QoS in wireless ad hoc networks. Given that nodes move according to the random waypoint mobility model, we observe that SINR between two nodes in a multi-hop ad hoc network exhibits self-similarity. Based on this observation, a channel monitor method is suggested to predict the SINR by one step ahead as an improvement to PDMED, namely, PDMED+. PDMED+ can increase the total throughput of the network while maintaining the proportional differentiation on average end-to-end delay. Through simulation evaluations, we indeed see an improvement on total through-put of the network. Although PDMED+ costs a little extra average end-to-end packet delay, it still maintains a good, consistent, and accurate proportional differentiation on average end-to-end packet delay.

References

[1] M. S. Corson. 1997. Issues in supporting quality of service in mobile ad hoc networks. In *Proceedings of the IFIP International Workshop on Quality of Service.*

[2] I. Chlamtac and A. Farago. 1994. Making transmission schedules immune to topology changes in multi-hop packet radio networks. *IEEE/ACM Transactions in Networking* 2:23–29.

[3] J. Ju and V. O. K. Li. 1998. An optimal topology-transparent scheduling method in multi-hop packet radio networks. *IEEE/ACM Transactions in Networking* 6:298–306.

[4] J. L. Sobrinho and A. S. Krishnakumar. 1999. Quality of service in ad hoc carrier sense multiple access wireless networks. *IEEE Journal on Selected Areas in Communications.* 17(8):1353–11368.

[5] J. Sheu, C. Liu, S. Wu, and Y. Tseng. 2004. A priority MAC protocol to support real-time traffic in ad hoc networks. *ACM Wireless Networks* 10:61–69.

[6] X. Yang and N. H. Vaidya. 2002. Priority scheduling in wireless ad hoc networks. In *Proceedings of the ACM International Symposium on Mobile Ad Hoc Networking and Computing,* 71–79.

[7] I. Aad and C. Castelluccia. 2001. Differentiation mechanisms for IEEE 802.11. In *Proceedings of IEEE Infocom.*

[8] J. Zhao, Z. Guo, Q. Zhang, and W. Zhu. 2002. Performance study of MAC for service differentiation in IEEE 802.11. *IEEE Globecom.*

[9] M. Benveniste, G. Chesson, M. Hoehen, A. Singla, H. Teunissen, and M. Wentink. 2001. EDCF proposed draft text. IEEE working document 802.11-01/131rl.

[10] J. Kim and C. Kim. 2004. Performance analysis and evaluation of IEEE 802.11e EDCF. *Wireless Communications and Mobile Computing.*

[11] B. Li and R. Battiti. 2003. Performance analysis of an enhanced IEEE 802.11 distributed coordination function supporting service differentiation. In *Proceedings of QofIS,* pp. 152–61.

[12] C. Dovrolis, D. Stiliadis, and P. Ramanathan. 2002. Proportional differentiated services: Delay differentiation and packet scheduling. *IEEE/ACM Transactions on Networking* 10:12–26.

[13] H. Luo and S. Lu. 2000. A topology-independent fair queueing model in ad hoc wireless networks. In *Proceedings of IEEE ICNP,* 325–35.

[14] H. Luo, S. Lu, and V. Bharghavan. 2000. A new model for packet scheduling in multihop wireless networks. In *Proceedings of ACM MOBICOM,* pp. 76–86.

[15] H. Luo, S. Lu, V. Bharghavan, J. Cheng, and G. Zhong. 2004. A packet scheduling approach to QoS support in multi-hop wireless networks. *Mobile Networks and Applications* 9:193–206.

[16] A. K. Somani and J. Zhou. 2003. Achieving fairness in distributed scheduling in wireless ad-hoc networks. In *Proceedings of the IEEE International Performance, Computing, and Communications Conference,* pp. 95–102.

[17] J. Chen and A. K. Somani. 2003. Fair scheduling in wireless ad hoc networks of location dependent channel errors. In *Proceedings of the IEEE International Performance, Computing, and Communications Conference,* 103–10.

[18] V. Kanodia, C. Li, A. Sabharwal, B. Sadeghi, and E. Knightly. 2002. Ordered packet scheduling in wireless ad hoc networks: Mechanisms and performance analysis. In *Proceedings of the ACM International Symposium on Mobile Ad Hoc Networking and Computing*, 58–70.

[19] C. R. Lin and M. Gerla. 1997. MACA/PR: An asynchronous multimedia multihop wireless network. In *Proceedings of IEEE INFOCOM '97* 1:118–125.

[20] V. Kanodia, C. Li, A. Sabharwal, B. Sadeghi, and E. Knightly. 2002. Distributed priority scheduling and medium access in ad hoc networks. *ACM Wireless Networks* 8:455–66.

[21] K. Wang and P. Ramanathan. 2003. End-to-end delay assurances in multi-hop wireless local area networks. In *Proceedings of IEEE Globecom 5:2962–2966*.

[22] K. Wang and P. Ramanathan. 2003. End-to-end throughput and delay assurances in multi-hop wireless hotspots. In *Proceedings of the ACM International Workshop on Wireless Mobile Applications and Services on WLAN Hotspots*, pp. 93–102.

[23] Y. Yang and R. Kravets. 2004. Distributed QoS guarantees for realtime traffic in ad hoc networks. In *Proceedings of the IEEE International Conference on Sensor and Ad Hoc Communications and Networks*, 118–127.

[24] D. Li and P.-Y. Kong. 2006. A scheme to provide proportionally differentiated end-to-end packet delay in wireless multi-hop ad hoc networks. In *Proceedings of IFIP Networking*, 1–12.

[25] B. G. Chun and M. Baker. 2002. Evaluation of packet scheduling algorithms in mobile ad hoc networks. *Mobile Computing and Communications Review* 1(2) ACMSIGMOBILE 6(3):36–49.

[26] OPNET (Optimum Network Performance). http://www.opnet.com.

[27] Video trace. http://www-tkn.ee.tu-berlin.de/research (accessed March 2005).

[28] D. Schafhuber and G. Matz. 2005. MMSE and adaptive prediction of time-varying channels for OFDM systems. *IEEE Transactions on Wireless Communications* 4(2):593–602.

[29] Y. Zhou, P. C. Yip, and H. Leung. 1997. On the efficient prediction of fractal signals. *IEEE Transactions on Signal Processing* 45:1865–68.

[30] P.-Y. Kong and K.-H. The. 2004. Performance of proactive earliest due date packet scheduling in wireless networks. *IEEE Transactions on Vehicular Technology* 53(4):1224–1234.

[31] T. Liu, P. Bahl, and I. Chlamtac. 1998. Mobility modeling, location tracking, and trajectory prediction in wireless ATM Networks. *Journal of Selected Areas in Communications* 16(8):922–36.

[32] X. Shen, J. W. Mark, and J. Ye. 2000. User mobility profile prediction: An adaptive fuzzy inference approach. *Wireless Networks* 363–74.

[33] S. Capkun, M. Hamdi, and J. Hubaux. 2001. GPS-free positioning in mobile ad hoc networks. In *Proceedings of the 34th Annual International Conference on System Science.*

[34] R. Hekmat and P. V. Mieghem. 2004. Interference in wireless multi-hop ad hoc networks and its effect on network capacity. *Wireless Networks* 10(4):389–399.

[35] T. Camp, J. Boleng, and V. Davies. 2002. A survey of mobility models for ad hoc network research. *Wireless Communication and Mobile Computing* 2:483–502.

[36] T. S. Rappaport. 2001. *Wireless communications: Principles and practice.* Englewood Cliffs, NJ: Prentice Hall.

[37] H. F. Zhang, Y. T. Shu, and O. Yang. 1997. Estimation of Hurst parameter by variance-time plots. In *Proceedings of IEEE PACRIM*, 883–86.

[38] N. Sadek, A. Khotanzad, and T. Chen. 2003. ATM Dynamic bandwidth allocation using F-ARIMA prediction model. In *Proceedings of ICCCN*, 359–63.

[39] Y. Shu, Z. Jin, L. Zhang, L. Wang, and O. W. W. Yang. 1999. Traffic prediction using FARIMA models. In *IEEE International Conference on Communication*, pp. 891–95.

[40] C. G. Dethe and D. G. Wakde. 2004. On the prediction of packet process in network traffic using FARIMA time-series model. *Journal of Indian Institute of Science* 84:31–39.

[41] Matlab. http://www.matlab.com.

Chapter 8

Topology-Transparent Scheduling Protocols for QoS-Robust Wireless Ad Hoc and Sensor Networks

Carlos H. Rentel and Thomas Kunz

Contents

219

8.1 Introduction

Wireless ad hoc network applications continue to emerge at a rapid pace. Advances in integrated circuit miniaturization, the availability of more efficient energy sources and energy-scavenging methods, and advances in microprocessor technology are taking us closer to the paradigm in which tiny, intelligent, and inexpensive sensor nodes can form a wireless distributed network that can be used to monitor and control the physical world with unparalleled resolution and speed. Scientific, environmental, life monitoring, building and home automation, machine-to-machine, traffic and road monitoring (e.g., WAVE IEEE 1609 standard suite), surveillance, military communications, real-time location, and disaster relief are just a few of the applications that are being enabled through wireless ad hoc and sensor networks [1–7].

Several challenges still remain to make wireless ad hoc networks a reality in some market segments. There is still growing interest, particularly among the industrial and military communities, to find more cost-effective, energy-efficient, and practical solutions to the problem of communication reliability. Designers and researchers are faced with the challenge to ensure nontrivial performance levels to attain a more widespread use of wireless ad hoc networks within the critical applications found in industrial and military environments.

This chapter introduces topology-transparent scheduling protocols for wireless ad hoc and sensor networks. Topology-transparent scheduling is an attractive scheduling approach that can potentially enhance the reliability of wireless ad hoc or sensor networks by ensuring free-colliding transmissions regardless of network topology variations. Assurance of performance has proved to be an elusive problem over traditional medium access mechanisms based on random strategies, such as Carrier Sense Multiple Access with Collision Avoidance (CSMA/CA) found in the increasingly popular IEEE 802.11 and IEEE 802.15.4 standards. One of the reasons for this is that detailed topology information is needed, which cannot always be available on time, or if available, can introduce a large amount of communication and processing overhead. Industrial and critical military applications are demanding more reliable wireless networks than what existing solutions can offer. For instance, the Instrumentation, Systems, and Automation Society (ISA), through its ISA100 committee, is currently working on a wireless standard for manufacturing, automation, and control applications referred to as SP100. The main issues addressed by SP100 are those of assurance of confidence and integrity of wireless information transfer. Additionally, the Zigbee Alliance is looking into incorporating new mechanisms to enhance the reliability of the original specification. We believe the assurance offered by topology-transparent scheduling protocols make them an attractive solution to the reliability problem. In this chapter we treat the QoS problem primarily as a way of providing assurance for a minimum service level requirement. However, we also briefly describe ways to simultaneously satisfy diverse QoS traffic requirements at the end of the chapter.

Topology-transparent scheduling protocols are a family of link layer scheduling techniques that are resilient to network topology variations and that need, at most, some slowly varying or static network topology information. The terms *static* and *slowly varying* must be understood within the context of a specific wireless ad hoc network application. Topology information may include the number of nodes in the network and the network node density, which in some applications may change slowly (i.e., hours, days). A topology-transparent scheduling protocol must also be robust to topology changes and have a good average throughput and delay performance. Topology-transparent scheduling protocols differ from their topology-dependent counterparts in that no topology update messages are exchanged among nodes in the network. Therefore, bandwidth is not utilized for a great deal of the overhead tasks found in topology-dependent approaches. The latter was the main motivator for the study and creation of topology-transparent scheduling protocols in networks with rapid topology dynamics, such as mobile ad hoc networks (MANETs). We note, however, that frequent topology changes can also be observed in networks in which the nodes are not moving, such as in the rich radio frequency environments encountered on factory and industrial floors, where interference and obstacles can cause frequent link failures and changes [8].

The term *QoS robust* implies that a *minimum* set of QoS guarantees, with respect to a set of performance constraints, are maintained regardless of topology variations. QoS robustness in the context of the following exposition must be understood only at the link layer level.

Topology-transparent scheduling protocols can be synchronous or asynchronous. Synchronous protocols utilize network synchronization to form time frames and time slots. Asynchronous ones do not rely on network synchronization, which usually forces the use of some random strategy, such as ALOHA, or CSMA/CA. Another classification of topology-transparent scheduling protocols can be made depending on the access being random or based on predefined sequences or codes. Existing random topology-transparent scheduling protocols lack the ability to provide QoS robustness because they can become unstable, and free-collision guarantees are not possible. Code-based topology-transparent protocols, on the other hand, use codes usually derived from a finite Galois field [9], which allows them to guarantee a nonzero performance level by ensuring free-colliding transmissions. Combinations of the previous classifications are possible, and for instance, slotted-

ALOHA and slotted-CSMA/CA are considered synchronous random scheduling protocols. Another topology-transparent protocol variation includes the space-based approach in [10], which exploits the physical location of the node to schedule the transmissions. Common to all the previous protocols is the fact that they achieve some level of communication performance without the need to exchange topology-related information. However, we will focus mostly on the class of protocols referred to as code-based topology-transparent scheduling protocols and the use of coding theory for their analysis.

8.2 Background and Existing Work

We describe next what we refer to as code-based scheduling and leave the description the rest of the remaining protocols to the existing literature (e.g., [10–12]).

The scheduling of channels using polynomials in a Galois field seems to have originated in earlier works ([13] and more notably [14]). Solomon [14] proposed an optimal solution for the hit problem in a star-topology frequency hopping multiple access system. A *hit* is defined as a collision between two or more transmitters at the base station's receiver caused by the use of the same time-frequency channel. Solomon's idea was to use maximum distance separable (MDS) sequences over a Galois field to minimize the hit probability. The sequences used were in fact the codewords of a block code referred to as Reed–Solomon (RS) code. In the theory of code-based scheduling protocols, any code can in principle be used [15, 16]. The latter opens a new paradigm in which scheduling and medium access can be studied based on well-established coding theory principles. A thorough treatment of coding and finite field theory is not the goal of this chapter. However, some fundamental definitions are given for the benefit of the reader.

> **Definition 1:** Let Ω be a finite set comprised of all the elements in $GF(q)$ (i.e., Galois field of order q), and let Ω be the set of all codewords of length n over Ω. Any nonempty subset $C \subset \Omega^n$ is referred to as a q-ary block code. If C has dimension k over Ω^n, then C has q^k codewords, and it is denoted as $C(n,q,k)$. k is sometimes referred to as the rank of the code, and a Galois field of order q (i.e., $GF(q)$) is a finite field of q elements, where q is a prime, or power of a prime number.

> **Definition 2:** Having two codewords c_x and c_y from a code C, the Hamming distance $d(c_x,c_y)$ is the number of positions in which c_x and c_y differ.
>
> The minimum distance of a code C is then $d(C) = \min_{c_x \neq c_y \in C} d(c_x,c_y)$. For illustration purposes, assume that each codeword in Ω^n is represented by the empty circles in a space, as seen in Figure 8.1, and C codewords are represented as filled circles. The distance between any two filled cir-

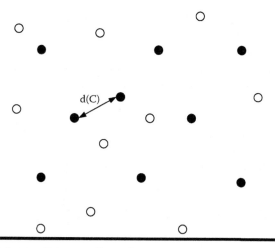

Figure 8.1 Minimum distance of a code.

cles represents the Hamming distance between those two codewords, and the minimum of all these possible distances is the minimum distance of the code. Solomon [14] was interested in codes with the largest possible minimum distance for a given dimension to minimize collisions. From the Singleton bound it can be deduced that any *linear* code satisfies $d(C) \leq n-k+1$ [17]. Therefore, the following definition applies.

Definition 3: An MDS linear code $C(n,q,k)$ has $d(C) = n-k+1$.
An example of a trivial MDS code is $C(n,q,k)$, which is the set of q n-tuples with all elements of each codeword different from one another. The latter will translate into the classical case of a different channel assigned to each user (e.g., channels 0, 1, ..., $q-1$ to users 0, 1, ..., $q-1$, respectively), which limits the system to a number of q connections at any given time (in bandwidth or time), and if no topology information is used, to a limited number of q users.

An approach referred to as topology-transparent scheduling for a multi-hop packet radio network was first proposed in [18] based on polynomials over a Galois field. Each radio node is assigned a different polynomial with degree k and coefficients in GF(q), which it uses to select a time slot in a frame. The difference between two polynomials also results in a polynomial of degree less than or equal to k; therefore, the number of roots of the difference between two polynomials will be bounded by k. This translates into the fact that the maximum number of common points between any two different polynomials will be bounded by k as well. Additionally, it is possible to guarantee a minimum of performance different than zero (i.e., assured number of free-colliding transmissions) as long as the scheduling is performed following the unique polynomial evaluation. The procedure depends on knowledge of the maximum number of interferers that a node could have and the total number of nodes in the network.

A well-known set of sequences with MDS properties can be constructed by a generalization of the method used in [18] and constitute the codewords of an RS code. This was first identified in [15]. The scheduling procedure designed in [18] corresponds to the codewords of a singly extended RS code. RS codes can be extended, augmented, truncated, and shortened. Therefore, the proposed procedure in [18] can be seen as a specific case of a more general scheduling approach based on RS codes. In fact, RS codes are a subset of the family of codes built from algebraic curves, as will become clearer in Section 8.3.

Other relevant works include the optimization of the procedure in [18] by the authors in [19], the use of Latin squares for multichannel TDMA [20], and the work in [21] identifying a generalization to the procedure in [18] based on combinatorial arrangements referred to as orthogonal arrays (OAs). We next describe in more detail some representative topology-transparent scheduling protocols.

8.2.1 Chlamtac–Faragó Topology-Transparent Algorithm

In the Chlamtac and Faragó [18] algorithm the network is viewed as an undirected graph $G(V, E)$ with N nodes, where V is the set of nodes ($|V| = N$) and E is the set of edges or links. The degree of a node v ($\deg(v)$) is the number of its neighbors, and the maximum degree D_{max} is the global maximum degree in the network (i.e., max $\deg(v)$).

Time is slotted, and a frame is comprised of L slots. A frame F is seen as a set of slots: $F = \{s_0, s_1, ..., s_{L-1}\}$. The scheduling assignment of node v is then given by a set $S_v \in F$, where S_v is the set of slots in which node v can transmit. The objectives of the algorithm are as follows:

1. For each node v, each neighbor u of v, and each neighbor $u \neq v$ of u, there should be at least one slot $s \notin S_v$ such that $s \notin S_u$ and $s \notin S_u$. This will allow node v to transmit without collisions at least once in every frame. This requirement is equivalent to stating that at least once in a frame a node will use a slot that is different from the ones used by its intended receiver and the neighbors of its intended receiver.

2. The slot assignment depends on global parameters N and D_{max} only.

3. The frame length L should be significantly smaller than N. The values of N and D_{max} could prove to be difficult to obtain in a highly dynamic wireless environment. However, it is assumed that an upper bound to the actual values is either known or enforced.

Let GF(q) be a Galois field of order q. Let $q = p^m$, p a prime, and $m \geq 1 \in Z^+$ is an arbitrary positive integer. Elements in GF(q) are labeled with the integers 0, 1,

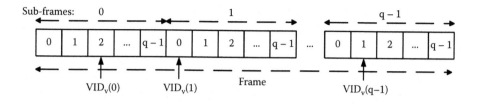

Figure 8.2   Frame structure of the Chlamtac–Faragó algorithm.

..., $q - 1$. Assign every node v in the network a unique, otherwise arbitrary, vector-identifier polynomial $VID_v[x]$ of degree k with coefficients in $GF(q)$. That is, $VID_v[x] = a_k x^k + a_{k-1} x^{k-1} + ... + a_0$, where $a_j \in GF(q)$, for $j = \{k, k\_1, ..., 0\}$. A frame has a size of $L = q^2$ slots. The authors in [18] devised an algorithm to determine the set of slots $S_v \in F$ for every node in the network in such a way as to get the smallest possible frame size while guaranteeing at least one free-colliding slot. Each unique polynomial is evaluated for every value between 0 and $q - 1$ and assigned to the node's slot assignment set S. Note that an alternative way to visualize the frame is as comprised of q subframes with q slots each, as shown in Figure 8.2. In this case, node v will evaluate its unique polynomial with the subframe value to obtain the slot number assigned to it in every subframe. Requirements 1 and 2 above are satisfied by constraining the values of q and k as follows [18]:

$$q^{k+1} \geq N \tag{8.1}$$

$$q \geq kD_{max} + 1 \tag{8.2}$$

8.2.2 Ju–Li Topology-Transparent Algorithm

Ju and Li [19] proposed an optimal topology-transparent scheduling approach departing from [18] that optimizes an expression representing the minimum throughput of any node in a multi-hop packet radio network. A way was found to compute values of q and k that maximize minimum throughput rather than focusing on the minimization of the frame size as in [18]. In particular, recall that kD_{max} is the maximum number of collisions a node can suffer in a frame, and the number of transmission attempts of a node in a frame is q. The minimum throughput is defined as

$$G_{min} = \frac{q - kD_{max}}{q^2} \tag{8.3}$$

where the size of the frame is $L = q^2$. G_{min} is the ratio of the difference between the number of transmission opportunities and the maximum number of collisions to the frame size. The maximal G_{min} for a given value of k is found following a classical optimization procedure over the field of real numbers (the reader is referred to [19] for details). The optimality of Ju–Li's algorithm was shown to be valid only within a given construction, and not in a more global sense, in [15] because, for instance, RS codes always exist that are longer than the given order of the code; therefore, frames of more than q subframes are always possible with the longer code keeping the MDS property (i.e., $d_{min} = n - k + 1$) (see Section 8.3.1).

8.2.3 Latin Squares TDMA Multichannel Topology-Transparent Algorithm

Ju and Li [20] proposed the used of Latin squares (ref. [4] in [20]) for a multichannel TDMA-based multi-hop packet radio network. In this approach every node is capable of half-duplex communication with multiple receivers and a single transmitter. Therefore, each node is capable of receiving multiple packets simultaneously over different channels, but it is only able to transmit a single packet in a given time slot. The half-duplex operation implies that the node cannot transmit and receive at the same time. The number of nodes N and the maximum degree D_{max} are assumed known as in previous algorithms. Figure 8.3(a) shows the arrangement of time slots and channels in this approach.

> **Definition 4:** A Latin square of order q is a $q \times q$ array comprised of q symbols in such a way that each symbol appears once in each row and once in each column. (Note that the popular Sudoku game is a special case of a Latin square.)

> **Definition 5:** A pair of Latin squares $\mathbf{A} = (a_{ij})$ and $\mathbf{B} = (b_{ij})$, where a_{ij} and b_{ij} represent the elements of $\mathbf{A}$ and $\mathbf{B}$ in the ith row and j^{th} column, respectively, are orthogonal if $(a_{ij}, b_{ij}) \neq (a_{kl}, b_{kl})$, $\forall (i, j) \neq (k, l)$, and $i, j, k, l \in \{1, \cdots, q\}$.

An example of two Latin squares is

$$\mathbf{A} = \begin{bmatrix} 0 & 1 & 2 \\ 1 & 2 & 0 \\ 2 & 0 & 1 \end{bmatrix}, \quad \mathbf{B} = \begin{bmatrix} 0 & 1 & 2 \\ 2 & 0 & 1 \\ 1 & 2 & 0 \end{bmatrix}$$

There are a total of $r \leq q - 1$ orthogonal Latin squares of order q. In the Latin squares TDMA multichannel algorithm each node is assigned one of the symbols in any of the r orthogonal Latin squares, then each node selects the time slot and

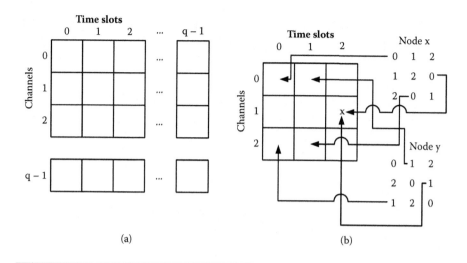

Figure 8.3 (a) General Latin square representation of channels and time slots. (b) Example of time slot and channel assignment.

frequency channel that corresponds to the column and row where that symbol is respectively located. For example, node x may be assigned symbol 0 in **A** above, and node y symbol 1 in **B**. Then node x uses time slot 0 on channel 0, time slot 1 on channel 2, and time slot 2 on channel 1. Node y uses time slot 0 on channel 2, time slot 1 on channel 0, and time slot 2 on channel 1. Therefore, the nodes will collide on a common receiver, as shown in Figure 8.3b, if both have packets to transmit by the time they reach time slot 2. Note that different nodes can be assigned a different symbol from any Latin square of the orthogonal family; therefore, different symbols in the same Latin square, or same symbols in different Latin squares are possible. A frame is comprised of the $q \times q$ array of time slots and channels, and the frame length is defined as the number of time slots (i.e., q).

It can be proved that two nodes that have been assigned different symbols in the same Latin square will never collide, and if two nodes are assigned symbols from different Latin squares, there is at most one collision [20]. Furthermore, in a network where a node is surrounded by D neighbors, there are at most D collisions and a minimum number of collisions given by $\max(D + 1 - p, 0)$.

The inequality $q \cdot r \geq N$ must hold to make sure there are sufficient unique symbols to assign to every node. When the number of channels M is less than the number of slots q, a Latin square becomes a Latin rectangle $M \times q$, which can also be used to assign channels and time slots in this method.

Results in [20] show average packet delays for different Poisson arrival rates per time slot. Robustness is demonstrated, for instance, by showing that delay does not change substantially even in a case where the system is designed for D_{max} in a network with $D = 33$. Guarantees, however, can only be fulfilled if D_{max} is known.

8.2.4 Orthogonal Array Topology-Transparent Algorithm

The authors in [21] generalized the work in [18] and [19] via orthogonal arrays (OAs).

> **Definition 6:** A $z \times q^t$ array with entries from the set $\Omega = \{0,1,...,q-1\}$ is an OA with q levels and strength t, with $0 \le t \le z$, if every $t \times q^t$ subarray of **A** contains each t-tuple exactly once as a column. Such an array is denoted as $OA(t,z,q)$.

For instance, the $OA(2,4,4)$ is given by

$$OA(2,4,4) = \begin{bmatrix} 0 & 0 & 0 & 0 & 1 & 1 & 1 & 1 & 2 & 2 & 2 & 2 & 3 & 3 & 3 & 3 \\ 0 & 1 & 2 & 3 & 0 & 1 & 2 & 3 & 0 & 1 & 2 & 3 & 0 & 1 & 2 & 3 \\ 0 & 1 & 2 & 3 & 1 & 0 & 3 & 2 & 2 & 3 & 0 & 1 & 3 & 2 & 1 & 0 \\ 0 & 1 & 3 & 2 & 3 & 2 & 0 & 1 & 2 & 3 & 1 & 0 & 1 & 0 & 2 & 3 \end{bmatrix}$$

$OA(2,4,4)$ has $q^t = 4^2$ columns, $z = 4$ columns, and every 2×16 subarray contains different 2-tuples as columns. A given column of the array intersects any other column in less than t positions.

In the OA topology-transparent algorithm a different column of an OA is assigned to a node in the network. Therefore, if a node has at most D_{max} neighboring nodes, the maximum number of collisions between those neighbors and the given node is $D_{max}(t-1)$ (i.e., the minimum number of free-colliding time slots is Z-$D_{max}(t-1)$). It is possible to guarantee a minimum throughput and delay performance if an OA is used such that $z - D_{max}(t-1) > 0$.

It turns out that there exist OAs that outperform the proposed constructions in [18], and that perform better than the maximally optimum algorithm proposed in [19]. The reason is that [18] and [19] restricted their attention to cases for which the frame size could only be q^2 time slots long. There are, however, some OAs with $z > q$ for which better results are obtained.

The same authors in [22] showed the use of another combinatorial structure referred to as the Steiner system, which outperforms OAs in delay performance. The Steiner system is also an array of symbols, and the time slot assignment is done in the same manner as in the OA case. Steiner systems admit shorter schedules and, therefore, have better delay performance. However, the existence of Steiner systems is noted to be unsettled at present.

8.3 Code-Based Topology-Transparent Scheduling

The previous topology-transparent scheduling algorithms are based on specific polynomial constructions and some form of a combinatorial array (i.e., Latin squares,

Latin rectangles, orthogonal arrays, or Steiner systems). OAs generalize the work in [18–20]. However, a different generalization and different results are possible if coding theory principles are used.

This section presents a generalization to code-based topology-transparent scheduling protocols based on the use of q-ary codes with minimum Hamming distance greater than zero. The authors in [15], [16], and [23] explored a different perspective in which codes traditionally used for error control coding were used for the scheduling of transmissions. In particular, it is found that RS codes [17] generalize and improve over the approaches in [18] and [19] as well, and that Hermitian codes have better performance than RS codes in some cases (see Section 8.3.1).

Code-based scheduling rests on the principle that the time intervals a node uses to transmit its information can be mapped to a sequence of numbers in a Galois field GF(q). Assume that time is divided in n subframes $0,1,\ldots,n-1$, and each subframe is divided in q time slots $0,1,\ldots,q-1$, as illustrated in Figure 8.2. Then every node uses a time slot to transmit within each subframe of the frame.

For instance, assuming GF($q = 7$), $n = 3$, and N nodes within the set [{1,2,\ldots,i, \ldots$N$}. The ith node could select the time slots as dictated by the set {0,5,6}. That is, the ith node will transmit in the time slots 0, 5, and 6 of subframes 0, 1, and 2, respectively. Under this perspective, each sequence of time slots can be seen as codewords in a vector space of dimension n over a finite field of dimension q. This perspective opens an opportunity for synergy between the fields of coding theory and scheduling of transmissions in a wireless network.

Note that, intuitively, a design goal could be to have the codewords as separate as possible in order for the different nodes to use as many different time slots as possible, and avoid collisions at a common receiver, where separation among codewords is measured in Hamming distance. However, this is not necessarily true because two or more nodes can use the same codewords if they are sufficiently apart from one another. Additionally, as found in the upcoming discussions, minimum relative distance is a more suitable metric.

Assume a code $C(n, k, q)$ of length n and dimension k in GF(q). $C(n, k, q)$ has q^k codewords of length n in an alphabet comprised of the elements of GF(q). In principle, the codewords of any such code can be utilized as scheduling patterns for the nodes of a wireless ad hoc network. For the sake of brevity, we refer here to a TDMA scheme. A node i accesses the medium assuming a time slotted structure like the one shown in Figure 8.4, and using the unique code C_i assigned to it. Time is divided in frames of size qn time slots, or n subframes with q time slots each. Note that the use of a $C(n, k, q)$ code generalizes over the frame size q^2 used in [18] and [19].

We define the lower-bound throughput of a generalized code-based scheduling algorithm as the ratio of the minimum number of free-colliding slots to the frame size, that is,

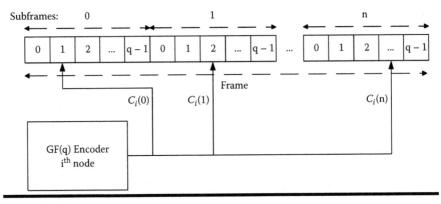

Figure 8.4 Code-based scheduling generalization time slot assignment.

$$G_{min} = \frac{n - (n - d_{min})I_{max}}{nq} \qquad (8.4)$$

where I_{max} is the estimated maximum number of interferers a node can have at any given time, d_{min} is the minimum Hamming distance of the code, q is the order of the finite field in which the code is defined, and n is the length of the code. The maximum degree D_{max}, used in previous works (e.g., [18, 19]) and obtained by viewing the network as an undirected graph, has been changed for the more realistic parameter I_{max}; in this way nodes more than two hops away could be considered when modeling the interferers of a given node by lumping their contribution in I_{max}. Note that the destruction of a packet in a given receiver does not depend only on first- and second-hop neighbor transmissions because the transmission and interference ranges are usually different in real wireless networks; additionally, the combined effect of numerous interferers at large distances may be sufficient to destroy a packet even when those interferers are not at range. I_{max} is an upper-bound value that needs to be known a priori, or that could be enforced with techniques such as power control or topology control mechanisms. However, this is a topic beyond the scope of this chapter.

The numerator in equation (8.4) is the minimum possible number of successful slots in a frame of nq slots given the minimum distance d_{min} of the code, where n is the number of opportunities a node accesses the medium, and $(n - d_{min})I_{max}$ is the maximum number of collisions expected in the worst case when there are a maximum number of interferers (I_{max}). Note that a random encoder, that is, one that generates random codewords in GF(q), is excluded from our analysis because $d_{min} = 0$ for such code. We are interested in codes in GF(q) with $d_{min} > 0$ to be able to guarantee a minimum throughput greater than zero (i.e., $G_{min} > 0$). These codes can in principle be linear or nonlinear. However, we focus on linear codes. G_{min}

is a lower bound because the codewords of any nontrivial code will be separated by Hamming distances larger than or equal to d_{min}. The following constraints are imposed on the parameters of a code to guarantee a positive value of the lower bound throughput,

$$q^k \geq N \tag{8.5a}$$

$$n \geq (n - d_{min})I_{max} + 1 \qquad n \leq \frac{1 - d_{min}I_{max}}{1 - I_{max}} \tag{8.5b}$$

$$q \geq I_{max} + 1 \tag{8.5c}$$

Inequality (8.5a) guarantees a unique codeword for every node in the network, and inequality (8.5b) ensures that a node will have at least some successful transmissions in a frame. That is, a node should be assigned a number of opportunities to transmit in a frame greater than the maximum number of collisions possible if one wishes to have a minimum of performance greater than zero. Inequality (8.5c) is necessary to guarantee a minimum throughput greater than zero because a node transmits once in every subframe of size q slots. This implies that the size of the subframes must be larger than the maximum number of interferers a given node may have in a given locality (i.e., $q > I_{max}$); otherwise, it is not possible to claim that the node will always have a chance to transmit free of collisions in an interval of time where there are more contending nodes than slots available.

One important goal is to find the parameters of the specific code for which equation (8.4) is maximized constrained to inequalities (8.5a), (8.5b), and (8.5c). First, let us rearrange equation (8.4) as follows:

$$G_{min} = \frac{1}{q} - (1 - \lambda)\frac{I_{max}}{q} \tag{8.6}$$

where $\lambda = d_{min}/n$ is typically referred to as the minimum relative distance of a q-ary code. Note that constructions that produce codes with larger λ ratio will have larger G_{min}. In other words, larger lower-bound throughput guarantees are possible for codes with more relative minimum distance for the same values of q and I_{max}. Because $\lambda \leq 1$, the maximum possible G_{min}, given q and I_{max}, is $G_{min}^o = 1/q$ with $\lambda^o = 1$. However, not all codes with $\lambda = 1$ have enough codewords to support networks of arbitrary sizes. A code with $\lambda = 1$ has its minimum distance equal to its length, which implies that every codeword is different in every digit with respect to the rest. The performance of different scheduling codes can therefore be exactly compared through their minimum relative distance for the same values of q and I_{max}.

Theorem 8.1

A *q*-ary code in which *all* of its codewords are $d = n$ from one another can only support a network of size $N \leq q$, and its maximum throughput is equal to $1/N$ (proof in [24]).

Note that whenever $N = q$, or $N = I_{max+}$, the maximum throughput performance of code-based scheduling will be the same as the one attainable by a round-robin scheduling procedure (i.e., $1/N$). Therefore, code-based scheduling proves to be as effective as a simple round-robin scheduling protocol in a single-hop star-topology network (e.g., a group of terminals communicating with a base station in a cellular radio network). It is important to realize that theorem 8.1 is valid only for codes with $d = n$ across all codewords in the code. It does not say anything about codes with $\lambda = 1$ and some of its codewords satisfying $d \leq n$.

To have a minimum throughput greater than zero ($G_{min} > 0$), we have, from equation (8.6), that λ must satisfy

$$\lambda > 1 - \frac{1}{I_{max}}.$$

Therefore, for a code to be effective in terms of providing a nonzero minimum throughput guarantee, its minimum relative distance must lie within the following bounds:

$$1 - \frac{1}{I_{max}} < \lambda \leq 1.$$

The next theorem proves a tighter upper bound on G_{min} achievable with any linear code in GF(*q*) for a given *q*, *k*, and I_{max}.

Theorem 8.2

For any linear code in GF(*q*),

$$G_{min} \leq \frac{1}{q} - \left(1 - \frac{1}{\sum_{i=0}^{k-1} \frac{1}{q^i}}\right)^{\frac{I_{max}}{q}}$$

(proof in [24]).

Theorem 8.2 is an upper bound on the minimum performance level achievable by any linear code in GF(q). This result is possible because we see topology-transparent scheduling from a coding theory perspective. Next, we present the performance comparison of some codes with good minimum relative distance properties and comparisons with a popular contention-based approach.

8.3.1 Reed–Solomon and Hermitian Codes

We noted in Section 8.3 that code constructions with larger d_{min}/n ratios for a given q and I_{max} will have larger G_{min} guarantees. In other words, larger G_{min} are possible for codes with larger *relative minimum distance*. Hermitian (Her) codes are an example of long codes (i.e., longer than RS codes) that possess a good d_{min}/n ratio. Hermitian codes are constructed using algebraic geometry (AG) principles [25, 26]. In particular, Hermitian codes are derived from a Hermitian curve in a finite field. In fact, RS codes can be seen as AG codes over a straight line in a finite field and therefore are a specific case of the more general set of AG codes. AG codes derived from many algebraic curves can be constructed, including elliptic and hyperelliptic curves. However, Hermitian curves have more points per given order; this is one of the reasons that, for some code parameters, the Hermitian codes possess a larger relative minimum distance than the RS codes of the same order. First, however, we prove that a doubly extended RS code offers a larger minimum throughput guarantee than a nonextended or singly extended RS code and compute the optimal value of q that maximizes equation (8.6).

An RS code can always be doubly extended without losing its maximum distance separable (MDS) property (the method of constructing this extension can be found, for instance, in [17]). A triple extension of an RS code is also possible without losing the MDS property, but only when the code has the following parameters: $(n,k,q) = (2^m + 2, 3, 2^m)$ or $(2^m + 2, 2, 2^m - 1, 4)$ [17]. Note that the construction given in [18], and used in [19], is the same construction used for RS codes. However, this is first realized in [23], and the following analysis is possible due to that realization. The minimum distance of an RS code is given by $d_{min} = n - k + 1$. Substituting the latter in equation (8.6) yields

$$G_{min}^{RS} = \frac{n - (k-1)I_{max}}{nq} \tag{8.7}$$

Nonextended (*ne*), singly extended (*se*), and doubly extended (*de*) RS codes have $n = q - 1$, q, and $q + 1$, respectively. Substituting the latter values of n in equation (8.7) we obtain the minimum throughput for each code version,

$$G_{\min}^{neRS} = \frac{q-1-(k-1)I_{\max}}{(q-1)q}$$

$$G_{\min}^{seRS} = \frac{q-(k-1)I_{\max}}{q^2}$$

$$G_{\min}^{deRS} = \frac{q+1-(k-1)I_{\max}}{(q+1)q} \tag{8.8}$$

We take the ratios of the minimum throughputs in equation (8.8) and expand the factors to get

$$\frac{G_{\min}^{seRS}}{G_{\min}^{deRS}} = \frac{q^2 + (1-KI_{\max})q - KI_{\max}}{q^2 + (1-KI_{\max})q} < 1 \qquad G_{\min}^{seRS} < G_{\min}^{deRS}$$

$$\frac{G_{\min}^{neRS}}{G_{\min}^{seRS}} = \frac{q^2 - (1+KI_{\max})q}{q^2 - (1+KI_{\max})q + KI_{\max}} < 1 \qquad G_{\min}^{neRS} < G_{\min}^{seRS} \tag{8.9}$$

where $K = k - 1$. Therefore, $G_{\min}^{deRS} > G_{\min}^{seRS} > G_{\min}^{neRS}$. The latter implies the fact that the results found in [19] are not optimal in a more global sense because larger G_{min} are possible with the same values of k and q. Note that k in this and the following sections represents the rank of the code as it is frequently used in the coding theory literature. This is different from the k used in [19], which represents the maximum degree of the polynomial used to construct the codes. The relationship between both is $k_{[19]} = k - 1$, where $k_{[19]}$ is the k used in [19].

The value of q that maximizes $G_{\min}^{deRS}$ is

$$q^* = KI_{\max} - 1 + \sqrt{KI_{\max}(KI_{\max} - 1)} \tag{8.10}$$

which after substitution in $G_{\min}^{deRS}$ results in

$$\max\{G_{DE\,\min}\} =$$

$$\begin{cases} \dfrac{\sqrt{KI_{\max}(KI_{\max}-1)}}{\left(KI_{\max}+\sqrt{KI_{\max}(KI_{\max}-1)}\right)\left(KI_{\max}-1+\sqrt{KI_{\max}(KI_{\max}-1)}\right)} & \text{for } q \geq KI_{\max} \\[4mm] \dfrac{N^{1/(K+2)}+1-KI_{\max}}{\left(N^{1/(K+2)}+1\right)N^{1/(K+2)}} & \text{otherwise } (q = N^{1/(K+2)}) \end{cases} \tag{8.11}$$

Figure 8.5 plots $\{G_{DE\,\min}\}$ versus the Ju–Li algorithm for the case when $q \geq KI_{\max}$ and assuming $D_{max} = I_{max}$ to make the comparison meaningful. Note that a doubly extended RS code outperforms the Ju–Li algorithm, which is actually the performance of an optimal singly extended RS code. However, the difference becomes negligible for $KI_{max} > 10$. That is, as the node density increases in the network, the performance difference between both codes becomes negligible.

Next, we compare the RS and Hermitian code constructions in terms of minimum throughput and maximum and minimum delay. Hermitian codes and their construction are described briefly in [15]. For the following comparisons we find

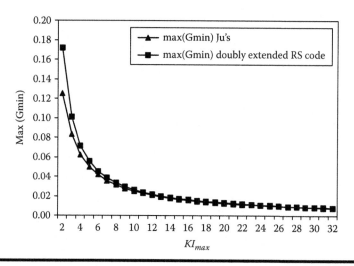

Figure 8.5 *max{G_min}* **for Ju–Li's algorithm and for a doubly extended Reed–Solomon code.**

the Hermitian and RS codes that maximize the minimum throughput in equation (8.4) subject to inequalities (8.5a), (8.5b), and (8.5c). Note that the only orders allowed in a Hermitian code are of the form q^2, where q is the power of some prime.

Minimum and maximum delay (i.e., DT_{min}, DT_{max}) are defined in [19]. For instance, for a doubly extended RS code DT_{min}, and DT_{max} take the following form for any value of q:

$$DT_{max} = (q+1)q / (q+1-(k-1)I_{max})$$
$$DT_{min} = q \tag{8.12}$$

Figures 8.6 and 8.7 show max $\{G_{min}\}$ and DT_{min}, DT_{max} for $N = 100$ nodes. Note that both constructions have roughly the same performance, except at lower densities, in which the doubly extended RS code (DE-RS) has a better performance in terms of minimum throughput than the Hermitian code.

In classic TDMA each node has a unique slot assigned to it out of the N slots in a frame. Therefore, the throughput of a node is guaranteed to be $1/N$, and the delay is equal to N slots; this is shown in Figures 8.6 and 8.7 as well. Note that the performance of code-based scheduling falls below classic TDMA as the number of possible interferers increases beyond a certain threshold. In general, the performance of code-based approaches will degrade with respect to classic TDMA as the node density increases (assuming that node density will increase the number of interferers a node's receiver can have). In the limit, when all the nodes are one hop from

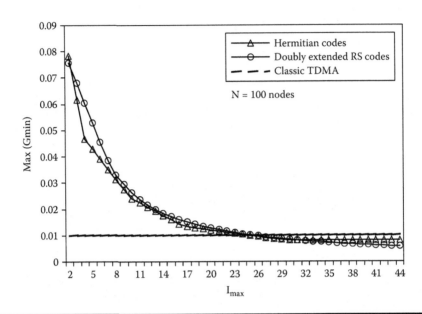

Figure 8.6 *max{G_{min}}* **using DE-RS and Hermitian codes in a network of 100 nodes.**

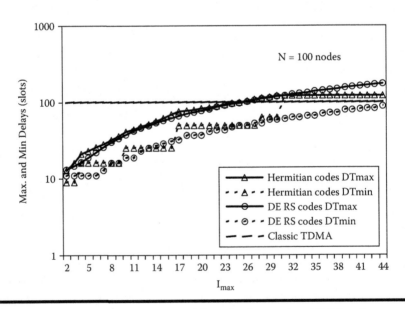

Figure 8.7 **Maximum and minimum delays (in slots) using DE-RS and Hermitian codes.**

one another (e.g., an access point or base station serving some nodes, or a single-hop wireless local area network), the classic TDMA approach will have a performance better than or equal to that of any code-based approach, as theorem 8.1 proved.

Figures 8.8 through 8.11 show the max $\{G_{min}\}$, DT_{min} and DT_{max} for $N = 500$ and 10,000 nodes. As the number of nodes increases, the Hermitian code construction offers higher minimum throughput guarantees than the RS code, particularly at a lower number of maximum interferers. As the number of nodes increases, the I_{max} threshold below which Hermitian codes offer better performance than RS codes also increases. The minimum and maximum delays of Hermitian codes tend to be less than the corresponding ones for RS codes as the number of nodes increases. The previous characteristic makes Hermitian codes attractive for large sensor networks. However, the maximum number of interferers of a given node must be controlled or accurately known, if some performance advantage is desired. In any case, the control of I_{max} is advantageous for both constructions.

The reason why Hermitian codes possess a larger minimum throughput guarantee for certain I_{max} values can be explained as follows. Table 8.1 shows the parameters of some Hermitian and RS codes as a reference. Let us start by noting that the minimum Hamming distance for codes of rank 1 are the same as the RS codes of the same order; however, as the rank of the code increases beyond 2, the Hermitian codes show smaller n/d_{min} ratios, as observed in Table 8.1, and this is a crucial factor for the better performance of the Hermitian codes in certain regions. Take, for instance, the RS and Hermitian codes of order $q = 16$ and rank $k = 3$. A doubly extended RS code of these characteristics will have $n = 17$, and $d_{min} = 17 - 3 + 1 = 15$.

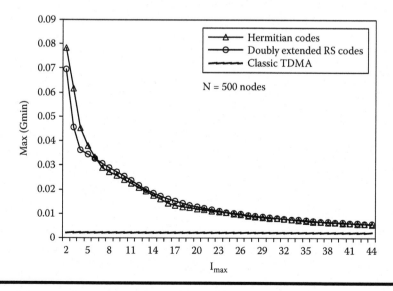

Figure 8.8 *max{G_{min}}* **using DE-RS and Hermitian codes in a network of 500 nodes.**

Table 8.1 Parameters of Some Hermitian and Doubly Extended RS Codes

q^2	n (Her)	k	d_{min} (Her)	n/d_{min} (Her)	d_{min} (RS)	n/d_{min} (RS)
4	8	1	8	1.00	5	1.00
4	8	2	6	1.33	4	1.25
4	8	3	5	1.60	3	1.67
4	8	4	4	2.00	2	2.50
9	27	1	27	1.00	10	1.00
9	27	2	24	1.13	9	1.11
9	27	3	23	1.17	8	1.25
9	27	4	21	1.29	7	1.43
16	64	1	64	1.00	17	1.00
16	64	2	60	1.07	16	1.06
16	64	3	59	1.08	15	1.13
16	64	4	56	1.14	14	1.21

Note: The length n of the DE-RS code is q^2+1 to make comparisons fair.

The Hermitian code has $d_{min} = 59$ with $n = 64$. Note that the Hermitian code is not MDS with these parameters. However, the ratio n/d_{min} is smaller for the Hermitian code, and this makes G_{min} larger. The d_{min}/n ratio is known in error control coding terminology as the relative minimum distance of a code. When the number of nodes in the network is considerably increased, the rank of the codes will need to be increased to satisfy inequality (8.5a) while maintaining a value of q considerably smaller than N. This is the behavior observed in the codes that maximize G_{min} in Figures 8.6 through 8.11 when the number of nodes is increased.

An important factor in topology-transparent scheduling protocols, first discussed in [19], is how robust the scheduling approach is to errors in the estimation of the number of nodes in the network (N) and the maximum number of interferers (I_{max}). References [15] and [16] describe in more detail the robustness of RS and Hermitian codes. In general, these codes are shown to be relatively robust to variations in the two design parameters above.

The advantage of the Hermitian codes (i.e., possessing a smaller n/d_{min} ratio than the RS codes when the rank of the code is higher) could also be utilized, for instance, to increase the number of codewords available when the number of nodes increases without the need to increase the order of the code. Changing the order (q) of the code translates into a different frame size, which implies a need to redistribute all the codewords in the entire network. If, however, a higher-rank code with an unchanged order is used, the original codewords assigned to the old nodes can still be used because they form a subset of the new code set, and therefore, only additional codeword assignments will be needed for the newly arriving nodes. The latter

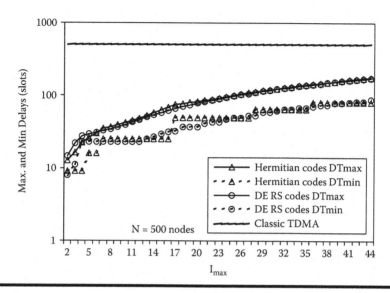

Figure 8.9 **Maximum and minimum delays (in slots) using DE-RS and Hermitian codes.**

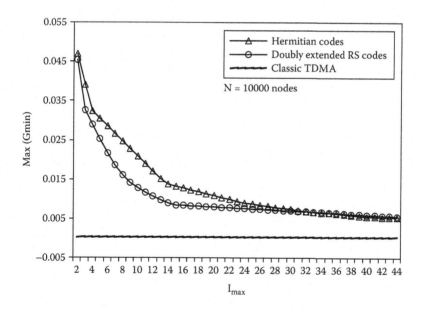

Figure 8.10 $max\{G_{min}\}$ **using DE-RS and Hermitian codes in a network of 10,000 nodes.**

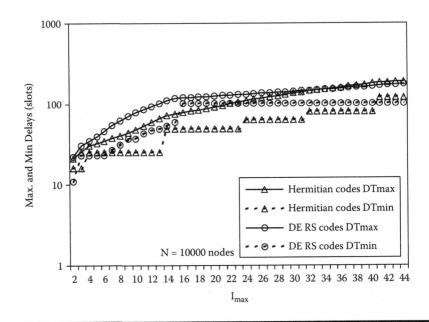

Figure 8.11 Maximum and minimum delays (in slots) using DE-RS and Hermitian codes.

is possible only in linear codes and represents a considerable practical advantage. Another potential advantage of Hermitian codes is that higher-rank codes with good performance could be used to generate an extremely large set of codewords, which in turn means that a node would select randomly a codeword from this large set rather than having a centralized manager that assigns codes to every node in the network. The possibility of two nodes picking the same codeword would potentially be negligible assuming the number of codewords is much larger than the number of nodes. Therefore, this would represent a reduction in the overhead created by the need to assign unique codewords to each node in the network. Finally, note that Hermitian codes show larger improvement over RS codes for large networks. Large networks of 500 or more nodes are quite possible in practice. Commercial wireless lighting control networks, for example, can be comprised of thousands of nodes connecting the light fixtures of a large common area to a central control/monitoring workstation (e.g., lights in a large hangar or store).

The next section explores the differences between a contention-based topology-transparent scheduling protocol and a code-based scheduling approach based on the RS code construction.

8.3.2 Comparative Evaluation of Code-Based and Contention-Based Scheduling

The first analytical comparison between a contention-based and a code-based scheduling protocol was presented in [27]. A code-based scheduling protocol has the important advantage of being able to guarantee a minimum throughput and packet delay performance. A contention-based scheduling approach is not able to claim such guarantees due to its random nature. However, contention-based approaches are attractive due to their simplicity. In this section, we analytically compare the average performance of slotted-ALOHA with that of OAs.

An expected throughput of a code-based approach based on OAs was derived in [28] for OAs of strengths 2 and 3 (i.e., the strength of an OA is analogous to the rank of a code [25]). The expected throughput of a given node with i neighbors is defined as [28]

$$\bar{G}_i^c = \sum_{w=0}^{n}(n-w)\frac{\binom{n}{w}C_i^w}{\binom{q^k-1}{i}}\frac{1}{nq}$$

(8.13)

Equation (8.13) is valid for a frame with n subframes and q slots per subframe; therefore, we have a frame of nq slots. w represents the number of slots in which two codewords coincide in some specific positions out of the n possible positions (i.e., n is the length of the codeword). Given a codeword W, equation (8.13) finds all the possible codewords that *collide* or coincide with W in w positions, where w takes values between 0 (i.e., no collisions at all, or Hamming distance of n) and n (i.e., all elements of the codeword collide, or Hamming distance of 0). C_i^w is the number of different ways in which i codewords coincide in w specific positions with the given codeword W (the union of all coincidences is taken). A method is given in [28] to compute $\bar{G}_i^c$ efficiently.

The slotted-ALOHA protocol is a relatively simple contention-based scheduling protocol that requires network synchronization (e.g., [29]). Slotted-ALOHA has the advantage of being relatively simple to implement compared to a code-based approach; therefore, it is interesting to compare the average performances of these two methods.

Assuming that we have knowledge of the number of neighbors of a given node x, the probability of successful transmission of x in slotted-ALOHA is

$$\bar{G}_i^s = p(1-p)^i$$

(8.14)

where p is the transmission probability of a node and i is the number of neighbors of the intended receiver. The optimum value of p can be found by differentiating equation (8.14), equating the result to zero and solving for p as follows:

$$\frac{\partial \bar{G}_i^s}{\partial p} = (1-p)^i - i(1-p)^{i-1} p = 0 \quad p^o = \frac{1}{i+1}$$

(8.15)

Substituting p^o in equation (8.14) yields

$$\bar{G}_i^s = \frac{1}{i+1}\left(1-\frac{1}{i+1}\right)^i$$

(8.16)

Equation (8.16) is also the average best throughput of slotted-ALOHA assuming knowledge of i. Figure 8.12 shows $\bar{G}_i^s$ and $\bar{G}_i^c$ for a number of neighbors between 1 and 40. The $\bar{G}_i^c$ curves shown are for different subframe sizes (q values) between 3 and 27 and for strength 2 OAs with $n = q + 1$. The OA curves with smaller values of q decay more rapidly as the number of neighbors of the given node increases; however, they have a higher expected throughput with fewer number of neighbors. As can be observed, the expected throughput of slotted-ALOHA is always larger than or equal to the expected throughput of OAs for all the number of neighbors considered. The same result is obtained for RS codes.

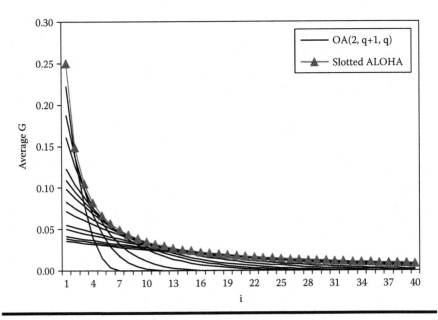

Figure 8.12   Expected throughput of OAs of strength two and slotted-ALOHA.

8.3.3 Code Selection in Code-Based Scheduling Protocols

The previous result does not necessarily mean that code-based scheduling approaches are always worse than or equal to the average performance of the classic slotted-ALOHA protocol. The reason is that $\bar{G}_i^c$ defines average throughput across all possible column (or codeword) combinations. However, as shown next, better average performance can be obtained if codewords are selected. We compare the use of SE-RS (singly extended RS) codes when codeword selection is performed against slotted-ALOHA. The codewords are selected following the procedure proposed in [15].

Table 8.2 shows the SE-RS code parameters that maximize G_{min} constrained to inequalities (8.5a), (8.5b), and (8.5c) for different values of I_{max} and when 20 nodes are active in the network. Figure 8.13 shows a comparison between the SE-RS codes in Table 8.2 (i.e., $n = q$) using the previous codeword selection algorithm against slotted-ALOHA. We can see that SE-RS codes can outperform slotted-ALOHA when the codewords are selected.

The minimum throughput of the codes used in Figure 8.13 is shown in Figure 8.14. Note that the minimum throughput shown is the actual value of minimum throughput. The actual value of minimum throughput is larger than or equal to the lower-bound G_{min}. The actual and lower-bound minimum throughputs are shown in Figure 8.14.

The performance improvement achieved by using codeword selection is more pronounced as the order of the code is increased to cope with larger neighborhood

Table 8.2   RS Codes That Maximize G_{min} with N = 20 Nodes

I_{max}	q	k
2	5	2
3	7	2
4	8	2
5	11	2
6	13	2
7	13	2
8	16	2
9	19	2
10	19	2
11	19	2
12	19	2
13	19	2
14	19	2

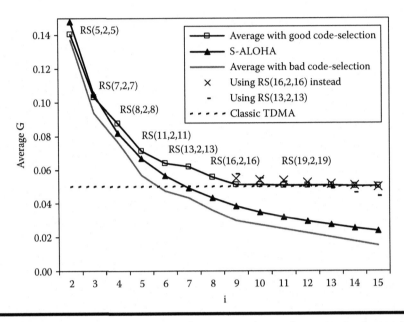

Figure 8.13   Average throughput performance using codeword selection and SE-RS codes.

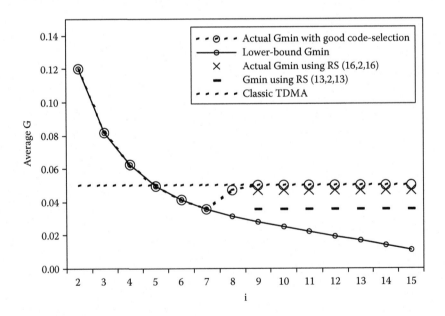

Figure 8.14 G_{min} using codeword selection for codes.

sizes. Figures 8.13 and 8.14 also show that codes that maximize equation (8.4) are not necessarily the best in terms of average throughput. However, using other code parameters affects the minimum throughput guarantee by decreasing it considerably in some cases. The final choice of code parameters, however, will depend on the particular application and design requirements.

8.4 Conclusion and Future Research

This chapter has presented an overview of topology-transparent scheduling protocols, particularly those referred to as code-based scheduling protocols. Relevant existing works have been presented along with an initial look at the use of coding theory in the area of wireless transmission scheduling. We present next some possible future research directions.

8.4.1 Multicode-Based Topology-Transparent Scheduling

From the total number of codewords possible in a code, each node could have been assigned a set of codewords $\mathbf{C}^x : x \in \{1,2,...,i,...,N\}$. For instance, the ith node could have $\mathbf{C}^i = [C_1^i = \{0,5,6\}; C_2^i = \{3,4,0\}; C_3^i = \{2,1,3\}]$, where the superscript identifies the node and the subscript the codeword. A node could allocate these codewords to its traffic streams based on QoS requirements, as shown in Figure 8.15a, where two codewords are assigned to traffic stream 1 and one codeword to traffic stream 2, effectively giving stream 1 more bandwidth.

Every node could make use of the link bandwidth differently. depending on the number of codes assigned to it. Figure 8.15(b) shows a network graph with $D_{max} = 3$. However, the ith node uses three codewords and the jth node only one codeword; therefore, the i–j link will be utilized by four codes, $C_1^i, C_2^i, C_3^i, C_1^j$, in a time length equal to a frame. Note that from the jth node's perspective potential collisions could come from the kth node in the k–i *physical* link caused by overlapping between C_1^p and C_1^j, the pth node in the p–i *physical* link caused by overlapping between C_1^p and C_1^j, and the ith node caused by overlapping between C_1^i, C_2^i, C_3^i, and C_1^j. Therefore, the jth node sees five *effective* links created by the five codewords of the neighboring nodes. In fact, the effective maximum degree of the network in Figure 8.15b is five. The node with more codes will have more performance guarantee, in this case at the expense of the other nodes, giving the option to provide different QoS services to some users. A complete analysis of this case, along with the case in which all the nodes have an equal number of codes (more than one) is an open question.

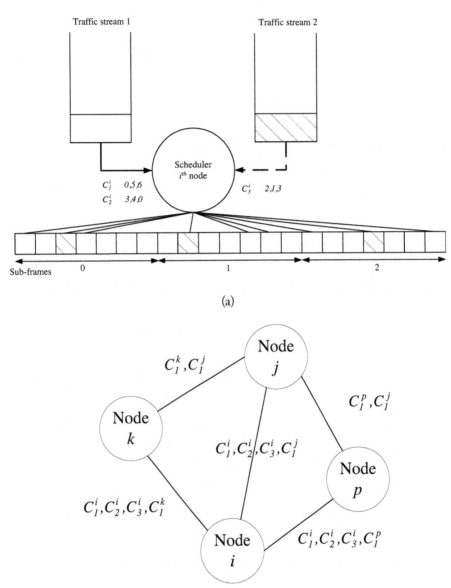

Figure 8.15 (a) Example of a multitraffic code assignment. (b) Example of a network with a multicode assignment.

8.4.2 Implementation

Known code-based scheduling protocols rely on: (1) knowing the maximum number of nodes, (2) knowing the maximum number of interferers a node can have, and (3) ensuring that every node has a unique code. Knowing or ensuring these may be difficult in some applications. Therefore, techniques to estimate or enforce them are necessary.

8.4.3 Exploiting Regional Topology Information

It may be beneficial to find techniques that make a node select its codewords based on what codewords its neighbors are using. A node could determine where code conflicts are and avoid them. This may prove to be useful if done cleverly and with minimum overhead, as it contradicts a global topology-transparent property.

8.4.4 Multichannel Code-Based Scheduling Algorithms

It is possible to extend the work on Latin squares in [20] by using different code constructions for the rows and columns that represent the channels and time slots (see Figure 8.4a). Different combinations are possible, including random codewords (i.e., slotted-ALOHA) in one dimension and RS, or other codes, in the other dimension. What is important is that other codes may prove to have better performance than the optimal results in [20], which are particular to the Latin square constructions. This takes us to the last, and probably most important, future research task.

8.4.5 The Best Codes

Finally, the quest for optimum codes that can guarantee the best lower-bound throughput and minimum delay is still an open problem. The approaches found so far in the literature are based on existing combinatorial constructions or codes with some desirable characteristics. A different perspective is needed in which we depart from the problem and arrive at an optimum code construction. Part of the challenge is that this is a multidisciplinary problem that involves, at least, networking and coding (or combinatorial) theory analysis. A first step may involve finding the best codes for a given static network topology and then studying variable network topologies. Additionally, a code may be, for instance, better at offering performance guarantees, but may not be the best in terms of average performance; therefore, different objective functions are possible. A first step was given in [15] and [16] by identifying minimum relative distance as an important code parameter, as well as in this chapter via theorems 8.1 and 8.2.

References

[1] Arampatzis, Th., Lygeros, J., and Manesis, S. 2005. A survey of applications of wireless sensors and wireless sensor networks. In *Proceedings of the 2005 IEEE International Symposium on Intelligent Control*, Mediterranean Conference on Control and Automation, 719.

[2] Herscovici, N., et al. 2006. Wireless sensor motes for small satellite applications. *IEEE Antennas and Propagation Magazine* 48:175.

[3] Egan, D. 2005. The emergence of ZigBee in building automation and industrial control. *Computing and Control Engineering Journal* 16:14.

[4] Kawasaki, K., et al. 2006. Wireless ad-hoc network-based tourist information delivery system. In *Proceedings of the 20th International Conference on Advanced Information Networking and Applications*, 2:138.

[5] Obana, S. S., Kadowaki, N., and Davis, P. 2006. Breakthroughs in large-scale ad hoc wireless networking and application for vehicles. In *Proceedings of the 7th International Conference on Mobile Data Management*, 88.

[6] Szewczyk, R., et al. 2004. Habitat monitoring with sensor networks. *Communications of the ACM* 47:34.

[7] Elgan, H., et al. 2005. Towards commercial mobile ad hoc network applications: A radio dispatch system. In *Proceedings of the 6th ACM International Symposium on Mobile Ad Hoc Networking and Computing (MobiHoc '05)*, 355.

[8] Andreas, W., et al. 2002. Measurements of a wireless link in an industrial environment using an IEEE 802.11-compliant physical layer. *IEEE Transactions on Industrial Electronics* 49:1265.

[9] McEliece, R. J. 1987. *Finite fields for computer scientists and engineers*. Dordrecht, The Netherlands: Kluwer Academic.

[10] Amouris, K. U.S Patent 7,08211, July 2006.

[11] Raphael, R., and Sidi, M. 1990. *Multiple access protocols: Performance and analysis*. New York: Springer.

[12] Bertsekas, D., and Gallager, R. 1992. *Data networks*. Upper Saddle River, NJ: Prentice Hall.

[13] Eniarsson, G. 1980. Address assignment for a time-frequency-coded, spread-spectrum system. *Bell Systems Technical Journal* 59:1241.

[14] Solomon, G. 1973. Optimal frequency hopping sequences for multiple-access. In *Proceedings of the Symposium of Spread Spectrum Communications*, 1:33.

[15] Rentel, C. H., and Kunz, T. 2005. Reed-Solomon and Hermitian code-based scheduling protocols for wireless ad hoc networks. In *Proceedings of the 4th International Conference on Ad Hoc and Wireless Networks ADHOCNOW 2005*, 221.

[16] Rentel, C. H., and Kunz, T. 2005. MAC coding for QoS guarantees in multi-hop mobile wireless networks. In *Proceedings of the 1st ACM International Workshop on QoS and Security for Wireless and Mobile Networks*, Montreal, 39.

[17] MacWilliams, F. J., and Sloane, N. J. A. 1977. *The theory of error-correcting codes*. Amsterdam: North Holland.

[18] Chlamtac, I., and Faragó, A. 1994. Making transmission schedules immune to topology changes in multi-hop packet radio networks. *IEEE/ACM Transactions on Networking* 2:23.

[19] Ju, J.-H., and Li, V. O. K. 1998. An optimal topology-transparent scheduling method in multi-hop packet radio networks. *IEEE/ACM Transactions on Networking* 6:298.

[20] Ju, J.-H., and Li, V. O. K. 1999. TDMA scheduling design of multihop packet radio networks based on Latin squares. *IEEE Journal on Selected Areas in Communications* 17:1345.

[21] Syrotiuk, V. R., Colbourn, C. J., and Ling, A. C. H. 2003. Topology-transparent scheduling for MANETs using orthogonal arrays. In *Proceedings of the International Conference on Mobile Computing and Networking*, 43.

[22] Syrotiuk, V. R., Colbourn, C. J., and Ling, A. C. H. 2003. Steiner systems for topology-transparent access control in MANETs. In *Proceedings of the International Conference on Wireless and Ad Hoc Networks*, 247.

[23] Rentel, C. H. 2006. Network time synchronization and code-based scheduling for wireless ad hoc networks. PhD thesis, Carleton University, Ottawa, Canada.

[24] Rentel, C. H., and Kunz, T. 2007. Unpublished manuscript.

[25] Pretzel, O. 1998. *Codes and algebraic curves.* New York: Oxford Science Publications.

[26] Berlekamp, E. R. 1973. Goppa codes. *IEEE Transactions on Information Theory* 19:590.

[27] Rentel, C. H., and Kunz, T. 2005. On the average-throughput performance of code-based scheduling protocols for wireless ad hoc networks. Poster presented at the ACM MobiHoc Conference, Urbana-Champaign.

[28] Colbourn, C. J., Ling, A. C. H., and Syrotiuk, V. R. 2004. Cover-free families and topology-transparent scheduling for MANETs. *Designs Codes and Cryptography* 32:65.

[29] Rentel, C. H., and Kunz, T. 2005. A clock-sampling mutual network synchronization algorithm for wireless ad hoc networks. In *Proceedings of the IEEE Wireless Communications and Networking Conference*, New Orleans, 638.

Chapter 9

Guaranteeing QoS in Wireless Sensor Networks

*José Fernán Martínez Ortega, Ana B. García,
Iván Corredor, Lourdes López,
Vicente Hernández, and Antonio da Silva*

Contents

9.1 Introduction

We have recently been witness to a great evolution in the area of wireless sensor networks (WSNs), mainly the improvement in the hardware of the sensor nodes (miniaturization of the pieces, increase in ROM and RAM memory, greater energy resources, etc). This, together with the application possibilities that this type of network offers us, has brought about a renewed interest in WSN. One definition of WSN could be as follows: a network of small devices of limited resources, equipped with a CPU, sensors, and transceivers that are embedded in a physical setting where they operate in an unattended manner. A lot of research has been carried out on the architecture and design of protocols, energy saving, and location; however, very little of the research has centered on the optimization of the transmission of data through WSN, that is, the quality of service (QoS) provided to the data traveling through these networks. The objective of this chapter is to summarize all of the research carried out in this area, especially that known as the communications protocol stack. To finish, we propose an application scenario in which we put into practice all of the knowledge compiled on QoS presented throughout this chapter.

9.1.1 The Scope of WSN Applications with QoS Requirements

WSNs have a wide range of applications. It is envisaged that in the short to medium term this technology is going to facilitate the appearance of new areas of application, as well as renewing already existing ones to make them more efficient. Some of the applications that will make a great impact on the market will be those whose objectives require a certain level of QoS, mainly in real-time. As will be seen in Section 9.3, QoS in a WSN can be defined by many parameters, among them the delay in detecting an event, the reliability of the reception of the data sent by the sensor nodes to the sink nodes, and the quality of the resolution of the information obtained. Below is a summary of the WSN areas of application with demanding QoS requirements that, according to the application, can be characterized by different parameters, although in the majority of cases the demands for real-time and reliability are common factors.

One of the types of application for WSNs that is frequently mentioned is emergencies in disaster situations. Typical scenarios are the detection and localization of fires in natural habitats (measuring the temperature and the existence of NOx and COx gases) or controlling the escape of hazardous products from chemical plants. Applications for the prediction of natural disasters are also related. In this case, the potentially large size of the extension in the deployment of the nodes usually requires less expensive nodes. In both types of applications, parameters such as delimited delay, the reliability of the reception of the information, the precision (e.g., geographical) with which the event is described, and even the low rate of false alarms are part of the QoS.

Another area of WSN application is the control of intelligent buildings in a more accurate way than traditional methods, because it allows better monitoring of the environmental parameters to adjust the environmental systems, thus providing greater efficiency in its use and a greater degree of comfort to the inhabitants. A WSN can also be used to monitor the levels of mechanical stress to which the structure of a building may be exposed. This type of application, very useful in areas of higher seismic activity, is also in real-time and with a high critical nature.

In the management of large facilities, WSN can also have a wide range of possible applications: from the control of access to restricted areas, only allowing access to authorized people (with the help, for example, of radio frequency identification (RFID) labels), to the localization and tracking of objects in a military scenario. These applications combine a series of rather demanding requisites as a large number of sensor nodes are needed, which must also collaborate among themselves (for example, in tracking applications), and must be able to work for a long period of time using only energy supplied by their batteries.

In the area of medicine and health care WSN applications open up new possibilities such as observing the vital signs of postoperative and intensive care patients, observation of long-term patients (for example, old people), and even the automatic administration of medicines according to the state of the patient. This is an area in which there are not only technical QoS requirements but also ethical factors, and even controversy, always present.

9.1.2 The Differences between Classic Ad Hoc Networks and WSN

An ad hoc network is established for a specific reason that rapidly covers a communication need. In these networks, autoconfiguration is a very important aspect (the network must be able to work autonomously, without management or any manual configuration whatsoever).

Within the ad hoc network there is a subset known as mobile ad hoc networks (MANETs), normally associated with multi-hop wireless communications; also, as the name itself suggests, the mobility of the nodes is a typical characteristic. The two main challenges of MANETs are the reorganization of the network when the nodes are moved and the correct handling of the wireless communications.

These problems are shared between the MANETs and the WSN. However, there are several differences between these two technologies that make the research efforts diverge, and the solutions found for MANETs are of no use in WSN. The most significant differences are set out below.

- **Applications and equipment:** MANETs are usually used for voice communications between distant users or to access a remote infrastructure, such as a Web server. Therefore, the equipment that make up these networks must be powerful enough to support the requirements of these applications (for example, laptop computers or PDAs), which are fare less restrictive than the sensor nodes of a WSN.
- **Interactions with the environment:** Because WSNs have to interact with the environment, their traffic characteristics will be very varied. On occasion, low data transmission rates can be seen for long periods of time, which alternate with small periods (seconds or minutes) of high activity, brought about by the appearance of events (a phenomenon known in real-time systems as a shower *of* events or an alarm storm). MANETs, for their part, are designed to support conventional applications (Web, voice, etc.) with their own, often predictable, traffic characteristics.
- **Scale:** WSNs have to be prepared to contain large quantities (thousands or even hundreds of thousands) of ad hoc network nodes, requiring different and more scalable solutions. A usual practice in WSNs (but not in MANETs) is to do direction profiles, without identifier-based addressing schemes have direction outlines based on identifiers (similar to IP), something that increases the scalability but brings about new challenges.
- **Energy:** The impact of energy considerations in the architecture of a network is much more profound in WSNs than in MANETs because the nodes of a WSN have greater energy restrictions and the replacement of the batteries might not be feasible.

9.1.3 Structure of the Rest of the Chapter

Section 9.2 contains a summary of the traditional QoS mechanisms for communication networks. This section concludes with a subsection that will explain which of these mechanisms could be taken advantage of in WSNs. In Section 9.3, a detailed analysis is carried out of the WSN characteristics that could determine or influence the mechanisms so as to guarantee QoS in applications and services. Section 9.4 will evaluate and analyze the proposed protocols and mechanisms to guarantee QoS in a WSN, making a distinction between the different layers of communication (link, network, and transport) in which they can be applied. In Section 9.5, an application scenario is selected and described to demonstrate how to choose the QoS mechanisms according to the performance requirements imposed by the services deployed and configured in a WSN. The validity of the decisions taken on the QoS mechanisms selected will be verified by means of quantitative simulation results. Section 9.6 summarizes the most relevant aspects of the chapter, as well as any open questions that might encourage future work and contributions in the area of QoS in WSNs.

9.2   QoS Fundamentals Applicable to WSNs

9.2.1 Basic QoS Mechanisms

To provide QoS in a packet-switching network, not just one mechanism is required but a set of methods that in general will have to combine several for every particular case. The most significant mechanisms are summarized below:

- **Overprovisioning:** Consists of providing a sufficient amount of resources so that their occupation is always low and the quality good. It is obviously not a very good technique for environments low in resources.
- **Buffer storage:** The packets can be temporarily stored in a buffer on the receptor side before going on to a higher level. The objective is to attenuate the jitter (delay variation), somewhat fundamental in audio or video traffic. Using this technique, the reliability or bandwidth is not affected and the average delay increases.
- **Traffic shaping and traffic policing:** Both refer to the traffic parameters that approximately describe the temporary pattern of the generation of packets of a source. In the first case, this temporary pattern can be altered (by delaying some packets) to adapt it to the parameters agreed upon in the network. In the second (traffic policing), the network determines which packets do not comply with the parameters and can take action against this excessive traffic so as not to prejudice the quality perceived by other clients and not cause congestion.
- **Proportional routing:** This is a proposed routing technique to provide a greater QoS, consistent in dividing the traffic for a determined destination by means of different routes. The only feasible way of dividing the traffic through multiple routes is usually to use information available locally in the node. A simple method is to divide the traffic into equal parts or in proportion to the capacity of the output links, although there are other, more refined algorithms, such as that presented in [1].
- **Packet scheduling:** The order in which the waiting packets are served in one or more buffers of an output link can be a determinant in providing a differentiated QoS to different traffic flows. One of the first algorithms was fair queuing [2], in which there is an exclusive buffer for each flow, serving all of them as a round-robin. In [3] an improvement in this algorithm is proposed, where the round-robin is carried out byte by byte, instead of packet by packet, to avoid assigning a wider bandwidth to flows with greater packets.

Using these algorithms, the same priority is given to all of the flows, but on occasion, it is better to give a greater bandwidth to some flows with respect to others (e.g., to flows with real-time requirements). *Weighted fair queuing* can be used for this, in which a flow can be provided with more than one byte per pulse. It has been proven that weighted fair queuing can provide the latency guarantees necessary for real-time traffic and multimedia.

9.2.2 Traditional QoS Models

- **Internet Engineering Task Force (IETF): IntServ and DiffServ models.**
  The IntServ model complements that of the Internet to provide quality-of-service capacities differentiated to different flows. In a DiffServ environment, the traffic emitters have to signal the characteristics of the flows they generate in such a way that the interested receptors in turn signal the opportune reserves. This signaling is carried out through the Resource Reservation Protocol (RSVP) and makes the routers keep the state of reserves made from the different flows. IntServ has a problem of scalability, especially in a network core with a lot of active flows. In this respect, the current trend for quality of service in IP networks is DiffServ, which proposes a quality-of-service model based on a limited set of types of traffic in which the packets are classified. These packets are marked as they enter the network, thus determining the treatment that they receive as they pass through the nodes.

- **ATM Forum model.** For the ATM networks, the ATM Forum [4] defined different types of ATM connections (service categories), according to the quality-of-service (QoS) parameters that the user can request and the traffic parameters requested by the network. The types of connection that provide stronger QoS guarantees are constant bit rate (CBR), real-time variable bit rate (rt-VBR), and non-real-time VBR (nrt-VBR). The first is for a constant rate of traffic, while the other two specify variable traffic parameters. All of them guarantee a certain rate of loss, and the first two (CBR and rt-VBR) also guarantee delay parameters and delay variations (thus they are suitable for real-time traffic).

- **Service model for wireless networks.** In accordance with the nature of wireless networks and the guaranteed QoS offered, the network services can be classified into three categories: assured statistical QoS service, adaptive service, and best effort (only the latter does not guarantee QoS). The adaptive services provide mechanisms to adapt traffic flows during periods of QoS and handoff fluctuations [5], and have shown themselves to be capable of effectively mitigating the fluctuations in the availability of resources in wireless networks [6]. In the case of a guaranteed statistical QoS service, the QoS provided to the user is specified by means of a triplet: $\{r_s, D_{max}, \varepsilon\}$, where r_s

is the source rate, D_{max} is a delay value, and ε is the maximum probability of experiencing a delay greater than D_{max}. This QoS triplet is essential in the design of provision mechanisms of statistical QoS [7].

9.2.3 QoS Mechanisms Applicable to Wireless Sensor Networks

As we will see in the following section, WSNs have characteristics and functions never before seen in other types of telematic network. This fact inevitably leads to having to go back to the drawing board with the design, as there are very few conventional QoS mechanisms that can be directly applied to WSN. However, when modeling the QoS of a WSN it is possible to take advantage of certain algorithms and techniques used for other wireless or cable networks. The adaptation process of these algorithms and techniques must be carried out with great care, never losing sight of the enormous restriction on resources (energy, computational, memory, etc.) that are currently present in WSN nodes, as well as other singularities of the WSN, as described in the following section. For example, it is necessary to avoid planning algorithms that require intensive CPU use, as in *weighted fair queuing*, for which it makes little sense to apply to a WSN because of the type of traffic generated by the sensors. Neither is it recommended to use routing techniques that send all of the packets in one data flow through a single route, prior to reserving resources in each of the nodes of the said route (see the IntServ model of IETF), because the traffic in a WSN is usually in bursts; therefore, many of the memory and CPU resources may be wasted. Although there is also continuous traffic, this technique will continue without being viable, owing to the topology of the WSN having a high dynamism (the nodes may fail, or could even suffer mobility), which is why it runs the risk of the route becoming unusable at a given moment. In general, the protocols that require the storage of the state of the network, partial or complete, are not viable, or are based on an addressing scheme similar to IP.

The solutions that can avoid, to a certain extent, the problems arising from the aforementioned algorithms and techniques are those based on the classification of traffic coming from different flows, similar to DiffServ. The advantages of using this type of mechanism on WSN are evident. First, it avoids the traffic congestion generated by the procedures for the establishment of sessions (resource reserve), control, and session freedom. Second, the algorithms used are *light*. This will have beneficial consequences: it takes up less memory space in the nodes, they do not need so may CPU cycles, and therefore, energy in the batteries is saved. The disadvantage of using this type of mechanism is that it requires a coherent configuration in the network nodes in such a way that the treatment of each node is in accordance with the level of QoS necessary for each flow.

9.3 Challenges to Guarantee QoS in Wireless Sensor Networks

9.3.1 Challenges in Wireless Sensor Networks

This section describes some of the challenges that must be faced when designing a system to provide QoS in a WSN.

- **Great restriction in the resources in the nodes:** Resource restriction involves energy (the main worry), bandwidth, memory, buffer size, computational capacity, and radio power transmission. As a result of these heavy restrictions, in any QoS mechanism for WSN, a special requirement is imposed: *simplicity.* Intensive-use CPU algorithms, protocols with excessive signaling, and maintenance of the state of the sensor network are not feasible.

- **Unbalanced traffic:** In many WSN applications, the traffic generally flows from a large number of sensor nodes to a single sink node or a small set of these nodes. This characteristic must be taken into account when designing the QoS mechanisms.

- **Data redundancy:** WSNs are characterized by their high information redundancy, which is received from the sensors, mainly due to the geographical proximity of some nodes. Although data redundancy can be taken advantage of to increase reliability and robustness, considerable waste of energy occurs when handling this data. The fusion and data aggregation mechanisms are possible solutions to maintaining the robustness yet reducing the redundancy. However, these mechanisms also introduce latency and complicate the QoS design in WSNs.

- **Network dynamics:** The topology of the WSN has a high level of dynamism resulting from certain factors, such as node failure, failure in the wireless links, node mobility, or transition of the state of the nodes owing to the use of a power manager or energy efficiency mechanisms. This network dynamism increases, to a large extent, the complexity of the QoS support.

- **Energy balance:** To increase the lifetime of the network, the load must be uniformly distributed between the nodes in such a way that the energy of a single node or group of nodes does not run out too soon. QoS support must take this factor into account.

- **Scalability:** A generic WSN focuses on containing hundreds or thousands of sensors distributed around a determined area. Therefore, QoS support for WSN must be able to adapt to a large number of nodes, that is, the QoS must not degrade rapidly as the number of nodes or their density increases.

- **Multiple sinks:** A WSN itself could have multiple sink nodes, with each one imposing different requirements. For example, a sink can ask a group of sensors, situated in the northeast of a field of sensors, to send it information every minute on temperature, while another sink might only be interested in exceptional events such as high temperature in the southeast area. WSNs must be able to support different degrees of QoS associated with different sinks.

■ **Multiple types of traffic:** Some applications can require different types of sensors to monitor different environment parameters, for example, the temperature, pressure, and humidity of the surrounding area, detecting movement from acoustic signals, or the capture of single or video images of moving objects. These special sensors may share the same set of nodes or not. In any case, the readings generated could be subject to different QoS requirements, follow different data-sending models to the sink and require different routing mechanisms.

■ **Critical packets:** The contents of the packets or high-level descriptions reflect the critical nature of the real physical phenomenon, and these degrees of critical nature or priority finally increase the quality of the applications. The QoS mechanisms may require a differentiation in the importance of the packets and the establishment of a priority structure.

In conclusion, QoS support for WSNs must take into account at least a subset of the previously described characteristics when specifying the mechanisms to be applied.

9.3.2 Architecture of the System and Design Issues

9.3.2.1 Sensor Network Scenarios

■ **Types of sources and sinks:** A source is any entity within the network that can provide information. This is normally a sensor node, but it could also be an actuator node that can provide feedback on an operation. On the other hand, a sink is the entity where the information is required. There are three basic options for the sinks (see Figure 9.1): it could belong to the sensor network as such, and simply be another sensor or actuator node, or it could be an entity external to this network.

For the second case, the sink could be a device, such as, for example, a PDA, used to interact with the sensor network; it could simply be a gateway to another large network such as the Internet, where the request for information could really come from some "distant" node, connected indirectly to the sensor network. The latter option is the most common.

■ **Single-hop networks versus multi-hop:** The transmission power necessary for a radio link is proportional to the square of the distance or even more if there are obstacles in the way. Therefore, the viable distance between the sink and the receptor is limited, and the multi-hop routing could end up consuming less energy than a direct communication (see Figure 9.2).

This is especially important in WSNs, which occasionally are deployed over wide areas or operating in environments with strong attenuations (for

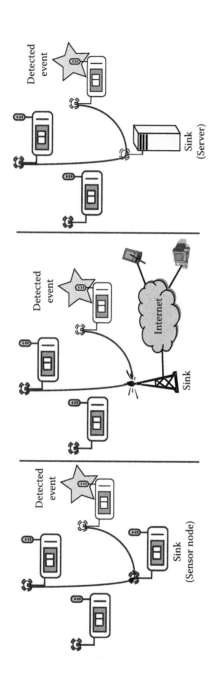

Figure 9.1 Three types of sink in a single-hop network architecture.

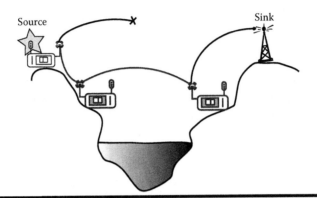

Figure 9.2   When a direct communication is impossible owing to distance or obstacles, the solution will be a multi-hop communication.

example, within buildings) or in the presence of obstacles. In a WSN with multi-hop functioning the nodes themselves could act as forwarding nodes. Depending on the application, the probability of having an intermediate sensor node in the correct place could be very high. The disadvantage of using this strategy in multi-hop routing is that it introduces an overload in the management of the topology and in the medium access control.

■ **Multiple sources and sinks:** In many cases there are multiple sources and sinks in the network (see Figure 9.3). In the more complex case, multiple sources can send information to multiple sinks, where all or part of the information can reach all or some of the sinks.

■ **Types of mobility:** One of the main virtues of a wireless communication is its capacity to support participating mobiles. In WSN, mobility can appear in three ways:

– Sensor node mobility. The wireless sensor nodes themselves could be mobile. The level of mobility will depend on the application. For example, in a cattle observation application (where the nodes are attached to the animals) mobility is quite common.

– Sink mobility: The sink nodes of a WSN could also be mobile. This type of mobility is usually associated with devices not related to the WSN. One example would be a human user requesting information through a PDA while moving inside an intelligent building.

– Event mobility: In event detection applications, and more specifically tracking applications, the reason for wanting to track the events or objects could be their mobility. In these scenarios, it is usually important for the event to be covered by a sufficient number of sensors all of the time.

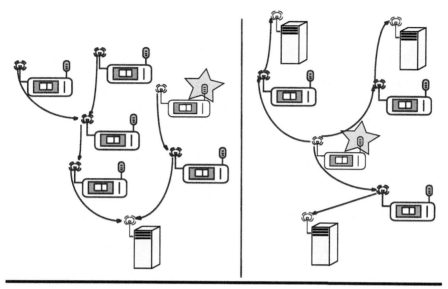

Figure 9.3 Multiple sources and multiple sinks.

Providing QoS, when some kind of described mobility is present, becomes a very complicated task that especially affects the routing protocols. Mobility in WSNs is a field that has not been researched very much yet.

9.3.2.2 Node Deployment

Another factor to bear in mind is the topological distribution of the nodes. It will depend on the application and will condition the choice of routing protocol considerably. The distribution may be deterministic or self-organized. In a deterministic deployment, the sensor nodes are placed manually and the packets are routed by following predetermined routes. In self-organized WSNs, the sensor nodes are randomly situated (for example, throwing them out of an airplane over the area to be covered), creating an ad hoc infrastructure.

9.3.2.3 Sending Models toward the Sink

Depending on the application of the sensor network, the data-sending models toward the sink can be continuous, event-driven, query-driven, or hybrid.

■ **Continuous:** In the continuous model, the sensors send the data continuously to the sinks at a pre-established transmission rate. Some of the applica-

Table 9.1 Comparison between the Different Data-Sending Models

	Delivery model to sink		
	Event-driven	*Query-driven*	*Continuous*
End-to-end	No	No	No
Interactivity	Yes	Yes	No
Delay tolerance	No	Depends on application	Yes
Criticity	Yes	Yes	Yes

tions that use this sending module are those of voice and image in real-time or those that periodically need to store measurement data from the sensors.

■ **Event-driven:** The majority of the event-driven applications are interactive, nontolerant to delays (in real-time), have critical objectives, and are multi-point to point. This means that the events that the sensors are expecting to see are very important so that the application can obtain satisfactory results. Applications of this type, when they detect an event, carry out the opportune actions as soon as possible and in as reliable a way as possible.

■ **Query-driven:** The majority of the query-driven applications are interactive, tolerant to delays, have objective criteria, and are multipoint to point. This model for sending data is similar to the event-driven model with the exception that in this model the data is pulled by the sink, while in the event-driven model the data is pushed to the sink.

■ **Hybrid:** In many applications there are different data-sending models in the same network. In these cases providing QoS is complicated, because mechanisms are required that are adjusted to all types of traffic.

Table 9.1 shows a comparison between the different data models.

9.3.2.4 Quality of Service (QoS)

QoS in WSNs can be defined from two perspectives, the same as conventional networks: from the network perspective (low-level QoS) and from the applications/users perspective (high-level QoS). Low-level QoS, that is, that observable by the network devices, is defined by a series of service attributes such as bandwidth, delay, delay fluctuation (jitter), and rate of packet loss. The bandwidth in a WSN is not in general very important, because the proportion of packets sent is an insufficient metric; the most important thing is the quality and quantity of the information that we can extract from the WSN. With regard to the high-level QoS attributes, WSNs depend a great deal on the application. Some of the generic possibilities are as follows:

- **Probability of event detection/notification:** In applications with critical objectives, such as the detection of fires, the nonnotification of the event could have disastrous repercussions. The probability of detecting an event depends on the efforts invested in establishing network structures that allow the notification of the event (for example, routing tables) during the execution time of the network.
- **Error in the classification of events:** If the event must be not only detected but also classified, the classification error must be minimal.
- **Delay in the detection of events:** Establishes the maximum time limit permissible between the detection of an event in a sensor node and its notification to any or all of the interested sink nodes.
- **Exactness of the tracking:** The tracking applications must not lose the object being tracked, the information on the position must be as close to real-time as possible, and the error rate must be small. Other aspects regarding the exactness of the tracking include, for example, the resolution of the movement information.

9.3.2.5 Energy Efficiency

Many studies on the energy resources of WSNs have reached the same conclusion on the importance of energy efficiency as an optimization objective. The term *energy efficiency* in the area of WSNs can encompass many aspects. The most common are described below:

- **Energy necessary to inform of an event:** In the notification of events, the sending of more information than is strictly necessary must be avoided; if not, it is a waste of energy in the nodes. Mechanisms of fusion and data aggregation can be used to avoid it.
- **Energy-delay balance:** Some applications require data to be sent "urgently" coming from events considered to be of major importance and for which an increase in energy consumption can be justified in order for the data to reach its destination quickly. The design of this type of application must establish the levels of delay and appropriate energy to inform the user of the event.
- **Lifetime of the network:** The period during which the network is operative or, in other words, the time period during which it is able to carry out its tasks (commencing with a determined amount of energy in each node). The concept of WSN lifetime can be understood in three ways: the time elapsed up to the death of the first node, the time elapsed until 50 percent are left without energy, or the time elapsed until two or more parts of the network cease to be connected due to the death of nodes.

9.3.2.6 Robustness

Robustness is related to QoS and, to a certain degree, the scalability requirements. WSNs also have an appropriate degree of robustness. They must be capable of remaining active in spite of some nodes failing due to the batteries running out, or because there has been a severe change in the environment, or because an obstacle is interfering with the radio link between the nodes (these failures must be resolved through, for example, the search for alternative routes). In practice, it is very difficult to carry out a precise evaluation of the robustness, as WSNs are mainly dependent on the failure models for the nodes as well as the communication links.

9.4 Solutions to Guarantee QoS in a WSN: Protocols and Mechanisms

The network and link layers of the protocol stack have been a focus of attention for many researchers looking to increase the energy efficiency of WSNs. Routing and Medium Access Control (MAC) algorithms have come about from this research, which will probably form part of WSN standards in the near future. However, very little research has taken place on what is known as QoS support. Although there has been some research into routing with QoS for ad hoc mobile networks, no exhaustive research has yet taken place in the context of WSN.

In this section we will analyze the mechanisms and protocols to provide a suitable level of QoS in WSNs. Research carried out has centered mainly on the analysis of the state of the art of the protocols and mechanisms at the level of link, network, and transport.

9.4.1 Mechanisms and Protocols at the Link Layer

9.4.1.1 Justification for the Design of WSN-Specific Link Layer Protocols and Mechanisms

As with all networks with shared medium, the Medium Access Control (MAC) in WSNs is an essential technique to get the network to operate successfully. A fundamental task in MAC protocols is to avoid or manage the collisions produced by the simultaneous transmissions of two or more nodes. There are many MAC protocols that have been developed for wireless voice and data networks. Typical examples are time division multiple access (TDMA), code division multiple access (CDMA), and those based on contention, such as the IEEE 802.11 standard. MAC protocols in WSNs are usually based on contention (derived from the CSMA algorithm) or the division of time (derived from the TDMA algorithm), depending on the application that is being given support. Owing to the special characteristics of WSNs, especially to the restrictions of the nodes and the changing nature of the topologic,

Table 9.2 Comparative Table of MAC Protocols in Wireless Sensor Networks

	Data aggregation/ fusion	*Scalability*	*Priority mechanisms*	*Energy-aware*	*Contention-based*
B-MAC	No	High	No	Yes	Yes
Z-MAC	No	High	Yes	Yes	Hybrid
Watteney, 2005	No	Low	Yes	No	Yes
MAC 802.15.4 with i-GAME	No	Medium	Yes	Yes	No

the classic MAC protocols, such as the IEEE 802.11 or Bluetooth standards, are not applicable to WSNs, thus making new designs necessary. Based on the peculiarities of WSNs, we could determine that a good MAC design for a WSN must take into account the following points:

- **Energy efficiency:** Because there are generally great difficulties in changing or charging the batteries of the sensor nodes in a WSN, energy efficiency is fundamental in prolonging the life of a network.
- **Scalability, density of the nodes, and topology:** The topology of a WSN can go through dynamic changes, which is why the MAC protocol must be able to adapt to them very quickly.
- **Latency, delay, bandwidth, etc.:** Attributes of vital importance in a WSN with applications in real-time that the MAC protocol must consider.

A selection of MAC protocols are set out in the following paragraphs designed specifically for WSNs, with the capacity to provide QoS in different environments.

Table 9.2 summarizes a comparison between these protocols.

9.4.1.2 B-MAC

B-MAC [8] stands out for its simplicity of design and implementation, which has an immediate effect on the memory size and power saving. B-MAC does not implement any specific QoS mechanism; however, this fact is compensated by its good design. Some parts of this design are addressed to improving the efficiency by avoiding collisions, efficiency in the channel occupation at low and high data rates, tolerance to changeable environments, or scalability for a large number of nodes. Although B-MAC was designed for monitoring applications, it is possible to take advantage of this approach in other applications, such as target tracking, localization, triggered events, and multi-hop routing. B-MAC also has a high degree of configurability. If we bear all of these characteristics in mind, we will be able to affirm that B-MAC is a good alternative for applications based on event-driven data delivery models with minimum delay requirements.

9.4.1.3 Z-MAC

Z-MAC (Zebra MAC) [9] is a hybrid scheme that combines the advantages of CSMA and TDMA while isolating their weaknesses. Z-MAC is characterized by an initial functioning period in which wide time-slot scheduling is carried out. To achieve this task, Z-MAC uses DRAND, a very efficient distributed scheduling algorithm. Although the initial assignment of slots incurs in high overheads, this is eventually amortized by a long network operation period and compensated with improvements in power saving and throughput. Z-MAC implements a contention control by avoiding congestion situations. Thus, under low contention it has CSMA-like behavior and under high contention TDMA-like behavior. This approach is also sufficiently robust for dynamic topology changes. These two characteristics are very important for applications with delay or reliability requirements.

9.4.1.4 MAC Protocol for Hard Real-Time for Linear Networks

In [10] the authors propose a hard real-time MAC protocol for a low-cost network (e.g., only one frequency) with identical, randomly deployed sensors without global clock synchronization. This protocol was designed for linear topologies, with the sink in one extreme receiving all events, and thus it is free from routing considerations. There are two alternating operation modes: *protected* and *unprotected*. When the network is in unprotected mode, the transmission speed is close to optimal but collisions may occur. However, if it is in protected mode, the transmission speed is slower but the frames are transmitted reliably because the network will be collision free. This characteristic can be interesting for real-time applications with critical requirements.

9.4.1.5 i-GAME Mechanism for the Improvement of the 802.15.4 Standard

The MAC protocol included in the IEEE 802.15.4 standard implements a mechanism called guaranteed time-slot (GTS). GTS tries to assign an additional time slot for applications with delay requirements. However, this mechanism is less efficient in WSNs with a large number of nodes. To correct this deficiency, [11] proposes the so-called implicit GTS allocation mechanism (i-GAME). The main idea of i-GAME is to share the same GTS between multiple nodes, instead of it being exclusively dedicated to a single node. The assignment of GTS resources is based on an admission control algorithm. This algorithm admits a request if its requirements do not exceed the available resources.

9.4.2 Mechanisms and Protocols at the Network Layer

9.4.2.1 Justification for the Design of WSN-Specific Network Layer Protocols and Mechanisms

Routing in sensor networks is quite complicated, mainly as a result of the following characteristics:

- WSNs usually lack an overall direction scheme similar to that used by IP.
- The majority of applications for WSNs require a flow of data originating in multiple regions to be directed to a determined sink.
- Many WSNs are routing networks based on *data-centric* information, where the data is sent or solicited based on certain attributes.
- The traffic of generated data has a significant redundancy, because the neighboring sensors that detect the same event will generate similar data.
- As with any decision on WSN design, the great limitations of the nodes with regard to transmission power, energy, processing, and memory capacity require special care.
- In the presence of mobile nodes, frequent and unexpected changes in the topology must be taken into account.
- It is important to know the positions of the sensor nodes at all times because the collection of the data is normally based on localization (localization systems based on GPS are not viable due to the consumption of energy).

The routing techniques and protocols recently developed for WSNs are set out in the following sections and pay a great deal of attention to those based on QoS. The protocols studied have been classified according to the structure of the network and its functioning (routing criteria). However, it is possible that many of the protocols could be included in more than one category. Table 9.3 summarizes a comparison between these protocols.

9.4.2.2 Routing in Flat Networks

In flat networks, all of the nodes in a WSN have the same role and collaborate together with the aim of capturing events. Due to the large number of nodes, it is not viable to assign an overall identifier to each node. Data-centric routing is more

Table 9.3 Comparative Table of Routing Protocols in Wireless Sensor Networks

	Network topology	Data delivery model	Data aggregation/fusion	Traffic guarantees	Several traffic classes	Networks dynamics	Resources reservation	Scalability
Directed Diffusion	Flat	Query-driven and event-driven	Yes	Reliability	No	Limited	Yes	Medium
SPIN	Flat	Query-driven and event-driven	Yes (by means of data negotiation)	No	No	Limited	No	Low
TEEN and APTEEN	Hierarchical	Query-driven, event-driven, and continuous	Yes	Certain guarantees of real-time	Yes	Fixed sink	No	High
SAR	Flat	Query-driven and event-driven	Yes	Real-time and Reliability	Yes	No	Yes	Low
SPEED	Flat	Query-driven and event-driven	No	Soft real-time	No	No	No	Low
1MMSPEED	Flat	Event-driven and continuous	No	Reliability and real-time	Yes	Limited	No	High
Akkaya, 2003	Hierarchical	Event-driven and continuous	No	Reliability and real-time	Yes	Fixed sink	No	Low

suitable, where the sink sends requests to certain regions and waits for the data to arrive at the sensors located in the selected regions. In protocols based on information, such as SPIN and Directed Diffusion [12], energy is saved through the negotiation and elimination of redundancy.

- **SPIN:** A family of adaptive protocols called Sensor Protocols for Information via Negotiation (SPIN) are proposed in [13] and [14]. These protocols do not implement any concrete QoS system; on the contrary, they are based on a data negotiation mechanism. SPIN uses it to eliminate redundant data by means of metadata exchange, assigning a high-level name to describe the data that sensor modes have collected and carrying out metadata negotiations before any data has been transmitted (similar to typical aggregation systems). However, this mechanism has an advantage over other systems: it avoids redundant data transmissions for later processing. Thus, the network increases its lifetime and the available bandwidth. Additionally, nodes are free from load processing, which supposes the data aggregation.
- **Directed diffusion:** This is a data-centric and application-aware paradigm because all data generated by sensor nodes is named by attribute-value pairs. Unlike traditional end-to-end routing, it tries to find routes from multiple sources to a single destination that allows redundant data aggregation [15]. The Directed Diffusion paradigm consists of aggregating different data coming from several sources by deleting redundancy. This feature reduces the number of transmissions drastically. It has two main consequences: first, the network saves energy, thus extending its lifetime, and second, it has a higher bandwidth in the links close to the sink node, which could be decisive for real-time applications with QoS requirements. In addition, Directed Diffusion is based on a query-driven model, where the sink node requests data through broadcasting *interest*. The request can originate from humans or systems and is defined as a pair-value, which describes a task that has been carried out by the network. The interests are disseminated through the network and set up gradients to create data that will satisfy queries about the requesting node. When the events appear, they start to flow toward the originators of interests along multiple paths providing data transmission reliability in the network. Another Directed Diffusion characteristic is caching network data (generally attribute-value pair interests), increasing coordination efficiency, robustness, and scalability between sensor nodes; this is the essence of the Directed Diffusion paradigm.
- **Routing based on energy saving:** A QoS-aware protocol for real-time traffic generated by a WSN consisting of image sensors is proposed in [16]. It implements a priority system that divides the traffic flows into two types: *best effort* and *real-time*. All nodes use two queues, one for each type of traffic. Thus, different kinds of service can be provided to these types of traffic. On the other hand, a routing mechanism based on multipath that uses an extended

version of Dijkstra's algorithm is implemented, which can provide reliability in the data transmissions. The source node chooses a route to achieve the end-to-end requirements and then forwards the packet to the next hop neighbor on the route. Each intermediate node classifies the received packet into real-time or best-effort types. The scheduling algorithm avoids the best-effort traffic, thus reducing resources to real-time traffic. The main disadvantage of this protocol is that it supports only one real-time traffic priority. This characteristic can be appropriate for a network with a single application; however, a network with multiple applications could have several types of real-time traffic with different priorities.

9.4.2.3 Routing in Hierarchical Networks

Hierarchical routing or that based on cluster, originally proposed for cable networks, allows scalability and communication efficiency to be improved. In a WSN defined with a hierarchical architecture, the nodes with the highest energy charges (cluster heads) can be used to process and send information, while the nodes with lower energy reserves can be used for the development of capturing tasks in the proximity of the object, which increases the scalability, lifetime, and energy efficiency of the entire network. It is usual for the cluster heads to implement data aggregation and fusion algorithms to reduce the number of messages transmitted.

- **TEEN and APTEEN:** TEEN and APTEEN, proposed in [17] and [18], have been defined for time-critical applications. These are designed to work even though an abrupt change happens in the attribute values that are being measured by the sensors. APTEEN (Adaptive TEEN) is a modification of TEEN that additionally considers the case of periodic transmissions of measurements toward a sink node. It implements a very complex query system that allows three types of queries (historical, one-time, and persistent) to be achieved. These queries are carried out by an external user through the sink node. The historical and persistent queries do not need QoS requirements. However, one-time queries become critical data with respect to time. In this case, the end user should be aware of his or her geographical position with minimum delay. To achieve minimum delay, the system carries out a special time-slot management assigned to each node by a TDMA schedule. Furthermore, APTEEN carries out the important task of data aggregation, which is equivalent to having free bandwidth and energy saving.

9.4.2.4 Protocols Based on QoS

The network layer in applications with requirements in real-time in WSNs is very important, mainly because: (1) it is responsible for facilitating routes to guarantee the two-point union and provide energy efficiency and stability, and (2) it serves as an intermediary between MAC and the application in the exchange of performance parameters. Owing to the intensive use of resources inherent in the applications in real-time and the low availability of resources of the WSN, the work of a routing protocol is very difficult. As has already been commented on, the nature of the topology of a WSN varies very much over time, which makes it difficult to give assured guarantees in *hard* real-time. However, the guarantees in *soft* real-time or soft QoS [19] can be covered by the network protocols.

Although the nature of the topology changes, the routing protocols must provide routes that are robust and stable during the transmission of a data flow. A multipath routing could be of great interest in increasing the number of possible routes and increase the robustness of the throughput of the transmissions. However, multipath routing could result in an increase in the delay caused by both the queuing and processing in the intermediate nodes. The more hops to the destination, the greater the delay. In this scenario, a routing protocol has to achieve a balance between the number of hops and the required delay, to provide the minimum guarantees. This is commonly known as the *energy-latency trade-off*.

- **SAR:** Sequential assignment routing (SAR), proposed in [20], was one of the first protocols for WSNs that has considered QoS issues in routing decisions. SAR carries out routing decisions based on three factors: energy resources, QoS planned for each path, and the type of traffic to which the packet belongs (types of traffic are implemented by means of a priority mechanism). SAR uses two systems for resolving reliability problems, which consist of a multipath approach and a localized path restoration (this path restoration is done by means of communications between neighboring nodes). The multipath tree is defined by avoiding nodes with low energy or QoS guarantees, taking into account that the root tree is located in the source node and ends in the set of sink nodes. Finally, SAR will create a multipath table whose main objective is to obtain energy efficiency and fault tolerance. Although this ensures fault tolerance and easy recovery, the protocol entails certain overheads when table and node states have to be maintained (refreshed). This problem increases especially when there are a huge number of nodes.
- **SPEED:** This is another QoS routing protocol for WSNs that provides light real-time end-to-end guarantees [21]. The QoS mechanism used by SPEED is based on estimation procedures. The application in a node then estimates the required speed for a certain delay, taking into account its distance to the sink node. The network layer will admit the packet depending on the required speed. Moreover, SPEED will be able to recover if the network becomes congested. The routing module in SPEED is called stateless geographic nondeterministic forwarding (SNFG). This module implements a distributed database where a node can be selected to reach the speed requirement.

■ **MMSPEED:** Multi-Path and Multi-SPEED Routing Protocol [22] is a novel packet delivery mechanism for the provision of QoS. Its main goal is to provide QoS differentiation in two quality domains: *timeliness* and *reliability*. Therefore, traffic flow can be carried out with a combination of service options based on reliability and timeliness requirements. The method used by MMSPEED to obtain reliability is the typical multipath routing, with a number of paths that depend on the required degree of reliability for the traffic flows. On the other hand, the method used by MMSPEED to obtain timeliness is a dynamic system that guarantees the packet delivery speed. MMSPEED uses localized geographic forwarding by using only local node neighbor information. The local decisions imply a problem of inaccuracy, which is resolved through dynamic compensation. Thus, traffic flow requirements can be fulfilled with a high probability. With this mechanism, the intermediate nodes are able to increase the transmission packet speed to higher levels, if they estimate that the packet cannot fulfill its delay deadline associated with the current speed, but it could be met at higher speeds.

With the aim of providing functionality to the QoS mechanisms implemented by MMSPEED, a MAC protocol with a prioritization mechanism should be established. Thus, the MMSPEED specification recommends the use of 802.11e (with several add-ons) at the MAC layer with its inherent prioritization mechanism based on differentiated interframe spacing (DIFS). Each speed value is mapped onto a MAC layer priority class.

The MMSPEED protocol solves many QoS issues related to real-time traffic in WSNs. However, many other aspects, such as network layer aggregation or handling the energy-delay trade-off, still need to be dealt with profoundly to guarantee good performance in a deployed WSN.

9.4.3 Mechanisms and Protocols at the Transport Level

9.4.3.1 The Unsuitability of the Traditional Transport Protocols

The transport protocols currently used in conventional networks (Transmission Control Protocol [TCP] and User Datagram Protocol [UDP]) cannot be implemented directly in WSNs. For example, UDP does not provide reliability in sending, something that is often necessary in many applications for WSNs, nor does it offer flow control and congestion that could bring about the loss of a packet as well as the unnecessary consumption of energy. On the other hand, TCP also has its disadvantages: overload associated with the establishment of the connection, degradation in the performance of wireless systems due to the loss of packets, delayed response in congestion mitigation, etc.

9.4.3.2 Characteristics and Design of Protocols at the Transport Layer in WSNs

The protocol at the transport layer must implement the following functions: ordered transmission, control of flow and congestion, recovery of losses, and possible QoS guarantees. In WSNs, several new factors, such as the nature of the ascending traffic and the bandwidth limits of the wireless links, could cause congestion. The congestion alters the normal exchange of data and could bring about the loss of packets. The wireless channels also bring about a packet loss factor because of the bit error rate, which not only affects reliability but also is a waste of energy. As a result, the two main problems that have to be faced in the transport protocol in WSNs are the congestion and the loss of packets. In the design of a transport protocol for a WSN, the following must be taken into account:

- **Optimization factors:** The transport protocols for WSNs must provide reliability and extreme-to-extreme QoS using as little energy as possible. The performance of the transport protocols in WSNs could be evaluated using metrics such as energy efficiency, reliability, QoS (for example, rate of packet loss, latency in packet loss, among others), and balance in the assigning of bandwidth.
- **Control of congestion:** There are two main causes for congestion in WSN: The first could be due to the package arrival rate exceeding packet service rate. This is more probable in the nodes close to the sink because these normally transport more combined ascending traffic. The second cause is related to performance aspects in the link, such as contention, interference, and bit error rate. The congestion in WSN has a direct impact on energy efficiency and the QoS of the application. It could cause an overflow in the buffer, which would cause long queuing delays and large packet losses.
- **Recovery of losses:** In wireless environments both congestion and bit error rate are the main causes of packet loss, which can cause the loss of reliability to a great degree and QoS, as well as reducing energy efficiency. Other factors that could bring about the loss of packets are failures in the nodes, erroneous or out-of-phase routing information, and energy resources running out. To solve this problem, the sending rate of the source could be increased or a loss recovery based on retransmission could be introduced.

9.4.3.3 Transport Protocols

The transport protocols for WSNs can be classified into three categories: protocols for congestion control, protocols for reliability, and hybrid protocols for reliability and congestion control:

- **Protocols for congestion control:** There are several protocols for congestion control adapted to the ascending nature of the traffic for WSN. These protocols are differentiated for the mechanisms that are used by them for the detection of congestion, notification of the congestion, or rate adjustment. Among them, Fusion [23] and CODA detect the congestion, basing it on the distance of the intermediate nodes, while the Control and Fairness (CCF) [24] deduces the congestion on the basis of packet service time. Siphon [26] uses the same system as CODA to deduce the congestion, but based on the precision of the data perceived by the application in the sink. CODA uses notification of the explicit congestion, while others [23–25] use implicit congestion notification. CODA adjusts the sending rate in a way similar to that of Additive Increase Multiplicative Decrease (AIMD), while CCF uses a rate adjustment algorithm. However, in Siphon there is no rate adjustment; when there is congestion it redirects traffic to virtual sinks, which as well as having a primary, low-power radio range, has another wider radio range used as a "short cut" or siphon to mitigate the congestion.

- **Protocols for reliability:** The reliability protocols can be classified into two groups: those that analyze the reliability of the ascending flow and those that center on the reliability of the descending flow. In the ascending direction, ESRT handles the precision of the flow of events and only guarantees the *reliability of the event* by means of adjusting the source rate. On the other hand, Reliable Multi-Segment Transport (RMST) [27] and Reliable Bursty Convergecast (RBC) [28] provide *packet reliability* through the recovery of losses extreme to extreme. In the descending direction the traffic is multicast (point to multipoint). To provide reliability to descending traffic in WSNs, notification and loss detection mechanisms can be used, including a sequence number at the head of the packets. The continuity of the sequence numbers can be used to detect packet loss. This technique can be applied extreme to extreme (as TCP does), or hop to hop. In the extreme-to-extreme focus, the destination and source are responsible for the detection of the loss and its notification. In the hop to hop, it is the intermediate notes that detect and notify the loss of packets, saving energy consumption, for example. GARUDA and PSFQ use loss detection and notification based on NACK, and local retransmission for the recovery of losses.

- **Protocols for reliability and congestion control:** Sensor Transmission Control Protocol (STCP) is a generic transport protocol for ascending flow extreme to extreme. It provides both congestion control and reliability, assigning more responsibility to the sink. The intermediate nodes detect the congestion based on the length of the queues and the notifications coming from the sink through the establishment of a bit at the head of the packets. This is an assisted network congestion control, extreme to extreme. STCP provides variable reliability controlled by the application. For example, it makes extreme-to-extreme retransmissions based on NACK for applications that generate continuous flows and extreme-to-extreme retransmissions based on ACK for event-driven applications.

9.5   Case Study

In this section we will apply the study we have presented on the protocols for QoS in WSNs to a forest surveillance scenario. We will begin by extracting the QoS-related requirements that our real-time forest surveillance application has, which will allow us to select the network and MAC protocols later that best suit these requirements. However, it is possible that these protocols will not carry out all the necessary requirements. In that case, we will also propose which add-on features must be introduced into each protocol. We will also create a simulation model from the designed application, which we will submit to several simulation tests. In the conclusions section we will discuss the shortcomings that, in our opinion, the protocols studied have and which could be corrected in future research.

9.5.1   Description

The application will focus on both forest fire detection and event tracking in a natural environment (natural reserve) of great ecological importance. The main objective of the application will be the early detection of forest fires to avoid ecological disasters. Likewise, the application will have secondary objectives, such as the detection and tracking of intruders within protected spaces to avoid illegal actions such as poaching, bonfires, etc. In summary, the application will be used for forest surveillance, including the detection of dangerous activities and determining the conditions that increase the risk of fires; the detection and location of fires; fire monitoring and assistance in fire extinction; and detection and tracking of intruders who accede to restricted areas.

In our forest fire detection application, sensor nodes collect measurement data, such as relative humidity, temperature, infrared radiation, and COx and NOx gases (these factors are necessary for the detection of fires and determining the degree of danger in a forest fire). Other components of the WSN that will support our application are laptops and PDAs (as support to firemen and safety watchmen), a server, and a database. All WSN services will be accessible to remote users through Web services. Figure 9.4 illustrates the proposed scenario.

The sensors, with which every node will be equipped, will be able to measure the following parameters: infrared radiation, humidity, and gases such as NOx and COx. All sensors will be used to determine the risk of fire at a given moment. The infrared radiation sensor will also be used for the detection and tracking of intruders in restricted areas. Specifically, the application will have the following characteristics:

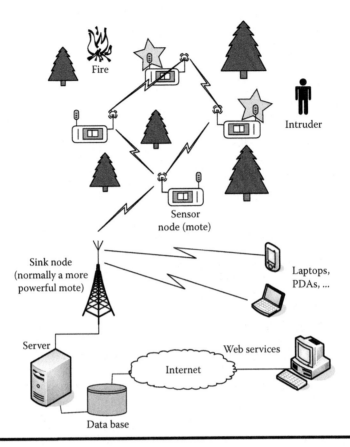

Figure 9.4   Forest surveillance application scenario.

1. *Topology and network dynamics*: The WSN topology is a design parameter that should be taken into account when guaranteeing QoS. The selected topology for the WSN will be flat. Therefore, every node will have the same hierarchy in the WSN as well as the same hardware components. The hierarchy will not be necessary in the proposed network because it will use a localized geographic routing. The use of this routing has several advantages with regard to guaranteeing QoS.

2. *Geographical information*: It will be necessary for the sensor nodes to obtain their geographical information (coordinates) to locate the events within the extension of the natural reserve. The methods are usually used with the aim of getting this information, which is based on GPS [29] or distributed location services [30]. For WSNs, a GPS-based approach will become too expensive, in which case our WSN will implement a distributed location service. This

method introduces certain overheads during the initial phase of the WSN, which could impede the guaranteeing of QoS.

3. *Real-time requirements*: Fire monitoring or target tracking reflects the physical status of dynamically changing environments, such as temperatures or positions of moving targets in forest areas. This sensory data is valid only for a limited time; hence, it needs to be delivered within a time deadline.

4. *Unbalanced mixture traffic*: Another characteristic of our application, which will considerably affect the QoS decisions, will be its *reactive-proactive* hybrid behavior. The reactive behavior will come from the fire or intruder detection and will generate traffic to the sink node according to the event-driven delivery model. This traffic type is generated nonperiodically through the detection of critical events at unpredictable points in time. The proactive behavior will come from the monitoring of the environmental status and tracking targets and will generate traffic to the sink node according to the continuous delivery model. In accordance with this mixture of periodic and nonperiodic traffic types, the selected QoS mechanisms for the WSN will be designed for an unbalanced mixture of QoS-constrained traffic.

5. *Data redundancy*: The high redundancy in the sensor data is a common characteristic in most WSNs. The redundancy could improve several QoS requirements, such as reliability and the robustness of data delivery. However, it uses a lot of unnecessary energy. To solve this problem, we will use data fusion or data aggregation to maintain robustness while decreasing redundancy in the data, but these mechanisms require a lot of computational activity in at least several nodes (usually cluster heads). Therefore, these mechanisms also introduce delay and complicate QoS design in WSNs. We prefer to exclude these mechanisms, as our application is based on two critical objectives, and the real-time requirements will prevail over the energy requirements. An alternative to data aggregation and fusion is metadata negotiation, which is able to eliminate redundancy without introducing excessive delay in data delivery.

6. *Energy efficiency*: An important challenge of this application will be energy efficiency. The large number of sensor nodes involved in the WSN and the need to operate over a long period of time (from six months to one year) will require careful management of the energy resources. However, to implement the QoS mechanism to support critical real-time traffic and at the same time save energy is not a trivial task. The key is to distribute the energy load among all sensor nodes so that the energy at a single sensor node or a small set of sensor nodes will not be drained too quickly. Nowadays, achieving this energy distribution without compromising the QoS requirements is very difficult because the mechanisms and protocols (detailed in previous sections) do not consider both possibilities at the same time.

7. *Sensor data priority*: Not all sensing data is equal; hence, the data has different levels of importance. For example, the data generated in a fire detection event will have more importance than that generated in the monitoring to

determine the conditions that increase the risk of fire. The QoS mechanisms will determine the data delivery priorities for the different data types existing in the WSN.

As a result, QoS support for the network will take into account almost all of the aforementioned characteristics in the application specification. The next section describes how to extract network and MAC layers from QoS-related requirements of the protocol stack according to the application characteristics analyzed.

9.5.2 QoS Modeling

9.5.2.1 Network Layer

To guarantee network layer QoS for diverse traffic types is a challenging problem as WSN characteristics such as dynamic topology change as a result of node failure, addition or mobility, a large scale with thousands of densely placed nodes, periodical and nonperiodical traffic generated by sensors with different priorities and real-time requirements, as well as possible data redundancy produced by correlated sensor nodes.

The traditional network layer methods based on the end-to-end path discovery, resources reservation along the discovered path, and path recovery in case of topological changes will not be suitable for our WSN: Initially, the time wasted in the path discovery is not acceptable for urgent aperiodic (event-driven) packets. In addition, it is not convenient to reserve resources for the unpredictable aperiodic packets. Even for periodic continuous flows, these methods are not practical in dynamic WSN because service disruption during the path recovery increases the data delivery delay, which is not acceptable in our mission-critical application. Finally, the end-to-end path-based approaches are not scalable because of the excessive overheads related to path discovery and recovery in large-scale sensor networks. As an alternative to the inefficient reservation-based approaches, the network layer will include an end-to-end QoS provisioning method based on local decisions at each intermediate node without path discovery and maintenance.

To solve dynamic topology changes, the network layer will implement the aforementioned localized geographic routing. This type of routing will mainly provide adaptability to dynamic topology changes because the nodes will not require the global topology information to be obtained. Consequently, no control packet will be generated in significant amounts with topology changes due to node addition, failure, or mobility. The WSN's nodes will able to take a localized packet routing decision without a global network state update or a priori path setup, which will increase the network scalability and decrease the control traffic. On the other hand, this routing scheme is suitable for both critical nonperiodic and periodic packets as a result of no path setup and recovery latency.

Another characteristic that should be considered by the network layer is the traffic priorities. In our WSN, the traffic priority will be characterized by two domains: reliability and timeless. The network layer will implement complex mechanisms to achieve this objective. For example, it could implement a priority queue system with the purpose of differentiating the traffic with different end-to-end deadlines. On the other hand, the mechanisms that will be implemented for providing reliability to the data transmissions could exploit the inherent multiple redundant paths to the final destination in a dense WSN to guarantee the required end-to-end level of reliability (end-to-end reaching probability) of a packet. Finally, the network layer will not implement a mechanism for eliminating data redundancy such as data aggregation, for two major reasons: First, in-network processing is not recommended to guarantee end-to-end deadlines due to the level of delay that is introduced by the high computational activity of these mechanisms. Second, the network topology will be flat (all nodes will have the same capacity). Hence, there will be no nodes capable of completing the process without using too much energy and time. Alternatively, the network protocol will implement a method for dealing with redundant data by means of exchanging metadata (inside the so-called data negotiation) [14]. This eliminates the inefficiencies that data aggregation mechanisms present as a result of flooding and the later processing of information. For instance, if a tracking event is detected and a data negotiation mechanism is used, the location information is transmitted once and no more data is transmitted until the target moves.

9.5.2.2 MAC Layer

Not all of the aforementioned QoS requirements could be provided by the network layer. In this way, our WSN protocol stack will have a MAC protocol capable of performing medium access control according to packet deadlines, the measurement of the average delay to individual neighbors, and the measurement of the rate of loss to individual neighbors. In addition, it could be necessary to have the capacity to deliver the packet to multiple neighbors reliably.

Along with the aforementioned functionalities, the MAC layer must implement mechanisms where each one of the deadlines assigned by the network layer is associated with a transmission priority level. Thus, medium access prioritization will be achieved through the MAC layer. Likewise, the MAC protocol will be able to measure the average delay to individual neighbors with the purpose of forwarding the packet according to its deadline.

However, packet forwarding will be carried out not only under deadline criteria but also under reliability criteria. For this reason, the MAC protocol will measure the rate of loss to individual neighbors.

The localized geographic routing used by the network layer will require the transmission of control packets with the position data of neighbors situated at least to one or two hops. For the transmission of these control packets, it will be necessary for the MAC layer to have the capacity of reliable multicast packet delivery.

9.5.3 The Selection of QoS Mechanisms

9.5.3.1 Selected Network Protocol

From the network layer point of view, MMSPEED will be the protocol used by the application. For this particular case, MMSPEED implements localized geographic routing, which is fundamental for the network layer of our protocol stack. These mechanisms increase the network's self-adaptability to dynamic changes. In addition, this protocol is suited to both periodic (real-time) and aperiodic traffic because of the routing local decisions (no path setup and failure recovery). MMSPEED also implements a multispeed mechanism for assigning diverse deadlines to the packets with different delay requirements. This mechanism is appropriate to support multiple traffic types (continuous, event-driven, etc.). Furthermore, it has a dynamic speed compensation mechanism, capable of correcting small inaccuracies produced in initial routing decisions immediately.

Routing decisions in MMSPEED are also carried out on the basis of the reliability level required by the packet. To route within the reliability domain, MMSPEED has an advanced method to provide reliability in data transmissions, which consists of using the frame loss rate of the MAC layer for estimating the level of reliability of each link. However, MMSPEED has a drawback, as it does not handle data redundancy. It was previously mentioned that the best methods for eliminating data redundancy in our application are those based on metadata exchange. Thus, we are looking at adding a metadata negotiation mechanism to MMSPEED.

9.5.3.2 Selected MAC Protocol

To select a MAC protocol that complements the MMSPEED protocol is not a trivial decision. MMSPEED specification proposes an extension of 802.11e for supporting all the mechanisms implemented by the network layer. The most important is the priorities mechanism. However, this MAC protocol is not specific to WSNs. We propose the Z-MAC protocol-like alternative to 802.11e. Although this protocol needs several add-on features to be completely compatible with MMSPEED, it is an excellent basis because it implements a priority mechanism appropriate to this case study. The add-on features are mainly related to the aforementioned hybrid nature of Z-MAC. This nature forces the priority mechanism to work in a different way, depending on its level of contention (low level, CSMA, or high level, TDMA). In addition, it is necessary for the Z-MAC to associate each MMSPEED speed layer with a priority type in the MAC layer. On the other hand, Z-MAC has a highly efficient contention method that can avoid any unnecessary back-off delay in the packet transmissions. Another characteristic that distinguishes the Z-MAC is its adaptability to changes in topology.

9.5.4 Validation Results

Before setting out the network as well as implementing the selected protocols in the nodes, it is appropriate to create a simulation model of the network as close to reality as possible. This model must undergo several tests that simulate all events and contingencies that could come about in a real environment, and likewise the hardware and functioning of the nodes on the sensor that execute the protocols and application. By carrying out this methodology, we can find out to a high degree of accuracy how the WSN will behave once set out, allowing us to make any corrections a priori. To simulate our WSN, J-SIM [31] has been selected. A MMSPEED adaptation for the simulator is available [22]. However, there is currently no version of the MAC protocol selected (Z-MAC) for J-SIM, which is why the communication protocol stack will be completed with the MAC IEEE 802.11e protocol (which gives QoS support to MMSPEED). Carrying out a rigorous QoS study in all of the layers of communication protocol stacks is outside the scope of this chapter, which is why we are centering our attention on the level of the network, defining a configuration of the QoS parameters of the protocol MMSPEED and describing its behavior. The scenario model of the application will be characterized according to that detailed in Table 9.4.

The deployment of the nodes will be distributed in four areas (*east, west, north,* and *south*), completely surrounding the mountain in the center of the terrain, just as set out in Figure 9.5. There will only be one sink node that will be situated at coordinates (0, 0).

Once the WSN is set out, it is necessary to plan a series of events that in one way or another put into effect the different WSN mechanisms (detection of events, communication of alarms, etc.). In accordance with the objectives of our appli-

Table 9.4   Simulation Parameters

Size terrain	600×600 m
Terrain morphology	A mountain of 400×400 m, centered in the terrain
Sensor node number	176 nodes (sink included)
Nodes deployment	Uniform
Radio range	80 m
Initial energy charge	1000 J
Bandwidth	200 Kbps
Payload	32 bytes

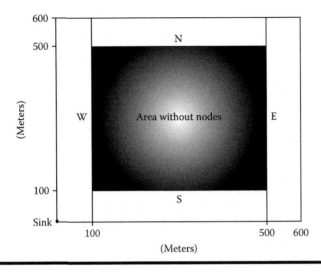

Figure 9.5 **Distribution of the nodes in the WSN: in the** *east* **and** *west* **sectors, 48 nodes are deployed; in the** *north* **and** *south* **sectors, 40 nodes are deployed. The separation between the nodes will be 40 m.**

cation (Section 9.5.1), two types of events have been planned: the appearance of people and the seat of the fire in the area of the natural reserve. The detection and notification of the events have to be almost instantaneous, which is why the traffic generated will be in real-time with some delay requirements for arrival at the sink (deadlines) and the reliability in getting the notifications to arrive at the sink within a minimum time period. This type of traffic will be governed by an event-driven model, that is, the traffic will be spontaneous and in bursts. A series of proactive environmental monitorings have been planned (temperature and gases). The monitorings will be carried out by a subset of nodes belonging to each of the tour sections; they will be carried out every hour and will last for five minutes. The monitorings will be programmed in such a way that at any given moment any sector can be monitored, thus avoiding congestion in the nodes close to the sink. The traffic generated by the monitorings will have a certain tolerance to delays and some average reliability requirements. This type of traffic will be governed by a continuous deterministic sending model: it is known a priori when it is going to appear and in what proportion; what cannot be predicted is the state that the WSN will be found in when the monitorings begin (how many nodes will be active, whether precise moment events are detected that will introduce additional traffic to the WSN).

According to these traffic characteristics, the parameters of the protocols are configured to guarantee QoS. In our WSN there will be two types of traffic: the first with strong real-time requirements, governed by an event-driven model, and a second, more tolerant to delays, governed by a continuous sending model. These

Table 9.5 Reliability and Delay Parameters for the MMSPEED Protocol

High-priority traffic (events)		Low-priority traffic (monitorings)	
Reaching sink probability	Max. delay (in seconds)	Reaching sink probability	Max. delay (in seconds)
0.4	0.5	0.2	4

two types of traffic must be treated differently by the communication protocols, giving greater QoS to the traffic coming from events detection than generated by monitoring. Thus, MMSPEED will be configured to handle two traffic classes, which will provide different levels of QoS as set out in Table 9.5. The most relevant results obtained from the simulation are described below.

9.5.4.1 Delays

The traffic differentiation implemented by MMSPEED works correctly and always prioritizes traffic coming from the events coming from monitorings. Figure 9.6 details the delays registered by the packets of traffic of both high and low priority during a simulation period, and the differentiation of MMSPEED traffic can be seen.

When the two traffic classes coincide in time, MMSPEED manages to maintain the level demanded by QoS for the high-priority traffic, which must arrive at the sink (without exceeding the established time limit—0.5 s). The jitter (delay fluctuation) is not too high, which will improve the data quality in real-time obtained by the application, especially if this data has been generated by the tracking of a

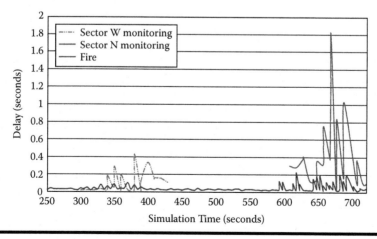

Figure 9.6   MMSPEED delays.

person inside the area controlled by the WSN. The low-priority traffic will manage to maintain acceptable levels of delay, although the jitters are somewhat high. This fact will not lead to a decline in the data quality obtained from the monitorings because they are not real-time data (they are generally stored in the database for later consultation).

9.5.4.2 Lost and Discarded Packets

It has been proved that in MMSPEED there are two situations that increase the loss or discarding of packets: when congestion appears in the nodes close to the sink (loss of packets) and when there is any inconsistency in the routing information in the intermediate nodes (discarding packets). On the other hand, traffic congestion in the nodes close to the sink has been impeded through the planning of the monitorings. Nevertheless, the possibility of congestion is not zero, because the WSN is not exposed to variable traffic and is unpredictable in appearance and proportion, coming from the detection of events.

In MMSPEED, when a flow of packets appears after a certain time without exchanged data, there is a high probability of having routing information inconsistencies. It is then probable that in one of the intermediate nodes, situated in one of the multiple routes used by the MMSPEED reliability mechanism, there is obsolete local information from the neighboring nodes to those that have to forward the packets. These, when detecting information, notify the nodes that the packet has passed through previously; this eventuality uses a back-pressure mechanism. After that, it discards the packet and updates its routing information. During this process, the network goes through a period of instability of variable duration (from tenths of seconds to seconds). However, the discarding of packets does not mean an effective loss of event notifications, as there is always a route whose nodes have information on the correct routing, and consequently can route the packets toward the sink.

9.5.4.3 Energy Consumed by the Nodes Close to the Sink

During the first 12 hours of simulated time, the consumption of energy of the eight nodes close to the sink has been registered (Table 9.6).

The average consumption of these sensor nodes is 3.5225 joules/hour. In the simulation environment their batteries would have run out after 12 days of simulated time. To transfer this data to the real world, we must bear in mind that the batteries that the nodes usually use can produce 3.6 watts/hour (in the best of cases), which is equivalent to 12,960 joules. Then the total energy for each node is 12,960*2 = 25,920 joules. Therefore, in a real environment, our WSNs would have a lifetime of 306 days (ten months) until the batteries run out. However, the useful

Table 9.6 Energy Consumed by MMSPEED

Node ID	Node coordinates	Total consumed energy (percentage)	Total consumed energy (Joules)
1	(40, 0)	41,388%	41.388
2	(80, 0)	40,743%	40.743
3	(0, 40)	40,742%	40.742
4	(40, 40)	39,594%	39.594
5	(80, 40)	42,666%	42.666
6	(0, 80)	42,835%	42.835
7	(40, 80)	42,995%	42.995
8	(80, 80)	47,199%	47.199

lifetime would be around nine months, because not all of the nodes consume their batteries at the same rate.

From the perspective of energy consumption, MMSPEED is not very effective owing to the excessive exchange of information and control as well as the lack of a mechanism for data aggregation.

9.6 Conclusions and Open Issues

In this chapter we have carried out an analysis and classification of mechanisms and protocols to provide and guarantee quality of service (QoS) in a type of wireless network with great future prospects: wireless sensor networks (WSNs). The WSN nodes have resource restrictions that make traditional QoS mechanisms for conventional networks (e.g., MANETs) not directly applicable, thus opening up a wide range of research opportunities in WSNs.

QoS is an issue that has not sparked much attention until now because of the difficulty in achieving compatibility in the mechanisms for the optimization of performance (in terms of latency, throughput, or jitter) with energy efficiency in the sensor nodes. Until now, different mechanisms and protocols have been proposed that have managed to solve some of the problems in providing QoS in WSNs. At the network level, the solutions that have given the best results are those based on traffic differentiation and multipath, together with the geographical routing localization techniques. At the MAC level, several equally valid techniques can be found, according to the situation. The MAC techniques that are best adapted to the dynamic nature of the WSNs are those that combine the TDMA and CSMA algorithms, according to the QoS requirements of the application and the level of contention in the wireless networks. Finally, it is worth mentioning that a solution for QoS in WSNs is being speculated and going through the stage for adoption of the protocol stack based on the components and cross-layer optimization [32], which assumes a considerable improvement with respect to the traditional mono-lithical protocol stack.

References

[1] Nelakuditi, S., and Zhang, Z.-L. 2002. A localized adaptive proportioning approach to QoS routing. IEEE Commun. Mag. 40:66.

[2] Nagle, J. 1987. On packet switches with infinite storage. *IEEE Trans.* Commun. 35:433.

[3] Demers, A., Keshav, S., and Shenker, S. 1990. Analysis and simulation of a fair queue-ing algorithm. *Internetwork Res. Exp.* 1:3.

[4] ATM Forum Technical Committee. 1996. Traffic management specification. Version 4.0.

[5] Wu, D., Hou, Y. T., and Zhang, Y. Q. 2001. Scalable video coding and transport over broadband wireless networks. *Proc. IEEE* 89(1).

[6] Balachandran, A., Campbell, A. T., and Kounavis, M. E. 1997. Active filters: Deliv-ering scalable media to mobile devices. In *Proceedings of the Seventh International Workshop on Network and Operating System Support for Digital Audio and Video* (NOSSDAV'97), St. Louis, MO.

[7] Wu, D. 2003. Providing quality of service guarantees in wireless networks. PhD dissertation, Department of Electrical and Computer Engineering, Carnegie Mellon University. http://www.wu.ece.ufl.edu/mypapers/Thesis.pdf.

[8] Polastre, J., Hill, J., and Culler, D. 2004. Versatile low power media access for wire-less sensor networks. In *Proceedings of the 2nd International Conference on Embedded Networked Sensor Systems* (SenSys '04), New York, 95.

[9] Rhee, I., Warrier, A., Aia, M., and Min, J. 2005. Z-MAC: A hybrid MAC for wireless sensor networks. Technical report, Department of Computer Science, North Caro-lina State University.

[10] Watteyne, T., and Augé-Blum, I. 2005. Proposition of a hard real-time MAC pro-tocol for wireless sensor networks. In *Proceedings of the International Symposium on Modeling, Analysis and Simulation of Computer and Telecommunication Systems* (MAS-COTS), Atlanta, GA, 532.

[11] Koubâa, A., Alves, M., and Tovar, E. 2006. i-GAME: An implicit GTS allocation mechanism in IEEE 802.15.4. In *Proceedings of the Euromicro Conference on Real-Time Systems* (ECRTS 2006).

[12] Ye, F., et al. 2002. A two-tier data dissemination model for large-scale wireless sensor networks. In *Proceedings of Mobicom'02*, Atlanta, GA.

[13] Heinzelman, W., Kulik, J., and Balakrishnan, H. 1999. Adaptive protocols for infor-mation dissemination in wireless sensor networks. In *Proceedings of the 5th ACM/ IEEE Mobicom Conference* (MobiCom '99), Seattle, 174.

[14] Kulik, J., Heinzelman, W. R., and Balakrishnan, H. 2002. Negotiation-based pro-tocols for disseminating information in wireless sensor networks. *Wireless Networks* 8:169.

[15] Intanagonwiwat, C., et al. 2000. Directed Diffusion: A scalable and robust communication paradigm for sensor networks. In *Proceedings of MobiCom'00*, Boston.

[16] Akkaya, K., and Younis, M. 2003. An energy-aware QoS routing protocol for wireless sensor network. In *Proceedings of the Workshops of the 23rd International Conference on Distributed Computing Systems*, 710.

[17] Manjeshwar, A., and Agarwal, D. P. 2001. TEEN: A routing protocol for enhanced efficiency in wireless sensor networks. In *Proceedings of the 1st International Workshop on Parallel and Distributed Computing Issues in Wireless Networks and Mobile Computing*.

[18] Manjeshwar, A., and Agarwal, D. P. 2002. APTEEN: A hybrid protocol for efficient routing and comprehensive information retrieval in wireless sensor networks. In *Proceedings of the Parallel and Distributed Processing Symposium* (IPDPS 2002), 195.

[19] Veres, A., Campbell, A., Barry, M., and Li-Hsiang, S. 2001. Supporting service differentiation in wireless packet networks using distributed control. IEEE *J. Select. Areas Commun.* 19:2081.

[20] Sohrabi, K., et al. 2000. Protocols for self-organization of a wireless sensor network. *IEEE Personal Commun.* 7:16.

[21] He, T., Stankovic, J., Chenyang, L., and Abdelzaher, T. 2003. SPEED: A stateless protocol for real-time communication in sensor networks. In *Proceedings of the 23rd International Conference on Distributed Computing Systems*, 46.

[22] Felemban, E., Lee, C. G., and Ekici, E. 2006. MMSPEED: Multipath multi-SPEED protocol for QoS guarantee of reliability and timeliness in wireless sensor networks. *IEEE Trans. Mobile Comput.* 5:738.

[23] Hull, B., Jamieson, K., and Balakrishnan, H. 2004. Mitigating congestion in wireless sensor networks. In *Proceedings of ACM SenSys '04*, Baltimore.

[24] Ee, C. T., and Bajcsy, R. 2004. Congestion control and fairness for many-to-one routing in sensor networks. In *Proceedings of ACM SenSys '04*, Baltimore.

[25] Wan, C. Y., Eisenman, S. B., and Campbell, A. T. 2003. CODA: Congestion detection and avoidance in sensor networks. In *Proceedings of ACM SenSys '03*, Los Angeles.

[26] Wan, C. Y., et al. 2005. Siphon: Overload traffic management using multi-radio virtual sinks in sensor networks. In *Proceedings of ACM SenSys '05*, San Diego.

[27] Stann, F., and Heidemann, J. 2003. RMST: Reliable data transport in sensor networks. In *Proceedings of IEEE SNPA '03*, Anchorage, AK.

[28] Zhang, H., et al. 2005. Reliable bursty convergecast in wireless sensor networks. In *Proceedings of ACM Mobihoc '05*, Urbana-Champain, IL.

[29] Karp, B., and Kung, H. 2000. Greedy perimeter stateless routing for wireless networks. In *Proceedings of the IEEE/ACM International Conference on Mobile Computing and Networking*, 243.

[30] He, T., Huang, C., Blum, B., Stankovic, J., and Abdelzaher, T. 2003. Range-free localization schemes for large scale sensor networks. In *Proceedings of Mobicom Conference*.

[31] Sobeih, A., Hou, J. C., and Kung, Lu-Chuan Kung, J. 2006. Sim: A simulation and emulation environment for wireless sensor networks. *IEEE Wireless Commun.* 13:104.

[32] Hill, J., and Culler, D. 2002. MICA: A wireless platform for deeply embedded networks. *IEEE Micro* 22:12.

Chapter 10

Congestion Control for Multicast Transmission over UMTS

Antonios Alexiou, Christos Bouras,
and Andreas Papazois

Contents

10.1 Introduction

Third-generation (3G) mobile cellular networks promise the provision of advanced services along with high data rates. In the meantime, the requirement for real-time multimedia data transmission that addresses user groups is increasing. Services like videoconferencing or distance learning are demanding features that load the network nodes and consume a large portion of the throughput provided by the network [1]. The Universal Mobile Telecommunications System (UMTS) constitutes the most prevalent standard of the 3G cellular networks. Despite the high capacity that UMTS networks provide, the expected demand will certainly overcome the available resources. This is the reason why multicast transmission is one of the major goals for UMTS and 3G networks in general.

Multicast is an efficient method for data transmission to multiple destinations. Its advantage is that the sender's data is transmitted only once over links that are shared along the paths to a targeted set of destinations. Data duplication is restricted only in nodes where the paths diverge to different subnetworks [2]. Multicast routing has been adopted by the Internet for more than ten years. Internet Protocol (IP) multicast is a bandwidth-conserving technology that reduces traffic by simultaneously delivering a single stream of data over a multicast tree. On the other hand, the wireless communication medium has itself a broadcast nature that is suitable for the adoption of multicast routing over cellular networks. The 3G Partnership Project (3GPP) is a global body dedicated to developing 3G specifications. In the beginning of the current decade, 3GPP recognized the need for the support of multicast routing in UMTS networks. As a result, the standardization of the Multimedia Broadcast/Multicast Service (MBMS) framework started in 2002 [3].

Congestion control is a policy that regulates the source transmission rate according to the network congestion. In IP multicast, the User Datagram Protocol (UDP) is used for the transport layer. This protocol does not implement any congestion control. Instead, the Transmission Control Protocol (TCP) regulates its transmission rate according to network congestion. This means that the coexistence of multicast traffic and TCP traffic may lead to unfair use of network resources. To prevent this situation, the deployment of multicast congestion control is indispensable. This kind of congestion control is well known as TCP friendliness [4].

The adoption of multicast congestion control in cellular networks poses an additional set of challenges related to the existence of wireless links and mobile terminals. In the first place, all the algorithms for congestion control treat the packet loss as a manifestation of network congestion. This assumption may not apply to networks with wireless links, in which packet loss is often induced by noise, wireless link error, or reasons other than network congestion. As a consequence, the network reaction should not be a drastic reduction of the sender's transmission rate [5]. Second, due to the fact that the physical radio resources (frequencies and code sequences) are limited, radio resource management (RRM) is a key process. It administers with high flexibility and efficiency the scarce radio resources while at the same time keeping service constraints. RRM performs congestion control over the radio links, and its strategy should be considered during the design of the congestion control mechanism [6]. Last but not least, the mobile terminals' computing power cannot afford complicated statistics and traffic measurements. Consequently, the holding of such operations on mobile equipment should be avoided.

In this chapter, we present a novel mechanism for multicast congestion control over UMTS networks. The proposed mechanism is based on the well-known TCP-Friendly Multicast Congestion Control (TFMCC) scheme. TFMCC is an equation-based multicast congestion control mechanism that extends the TCP-Friendly Rate Control (TFRC) [7] protocol from the unicast to the multicast domain in the Internet. It belongs to the class of single-rate congestion control schemes. Such schemes inevitably do not offer multiple transmission rates as layered schemes do. However, they are much simpler so as to meet a prime objective for UMTS multicast services, that is, scalability to applications with thousands of receivers [4].

In our proposed mechanism, the TFMCC scheme is partly modified and extended to support the particularities of the UMTS Terrestrial Radio Access Network (UTRAN). The major problem of the applicability of TFMCC over UMTS is the current limiting receiver (CLR) problem. The CLR problem is caused when the wireless channel quality is temporarily degraded. Minor modifications in the UMTS architecture are required by our proposed scheme. New functionalities are introduced in two nodes of the UMTS network to deal with the CLR problem. These impacts concern the user equipment (UE) and node B. The additional functionalities allow each UE to identify the reason of a packet loss. The UE can conclude whether a packet loss has been caused by wireless channel degradation or network congestion. Additionally, another aspect of the proposed mechanism is the handling of the permanent degradation of the wireless link.

This chapter is structured as follows: Section 10.2 provides an overview of the work related to the scientific domain. In Section 10.3, we briefly present concepts like the TFMCC algorithm, the UMTS networks, and the MBMS service. Moreover, we describe the problem of the applicability of congestion control over the wireless access networks. Section 10.4 is dedicated to the proposed congestion control mechanism. Section 10.5 describes the simulation experiments. And finally, some concluding remarks and planned next steps are given in Sections 10.6 and 10.7, respectively.

10.2 Related Work

The multicast congestion control problem in fixed networks is still a domain of active research, and a lot of solutions have been proposed. We use two distinct properties to classify the existing approaches [8]:

- The rates delivered to the receivers in a session. Existing approaches generally fall into three categories: single rate [4], multirate [9], and layered [10].
- The place where adaptation is performed. It is either at the end systems (end-to-end service) [4, 10, 11] or at the intermediate network nodes (active service) [12].

A technical problem of major importance in multicast congestion control is scalability. When the source receives a negative feedback of congestion notification inside the network, it regulates its transmission rate. To avoid a feedback implosion, the majority of the researchers, like the authors of [13] and [14], suggest that the receiver of the worst congestion level should be selected as the representative. In this approach, only the representative transmits feedback information for congestion control and the number of feedbacks is limited. Another advantage of the use of a single receiver is that the excessive restriction of transmission rate is avoided when the sender receives multiple negative feedbacks that originate from different receivers.

In contrast to the multicast congestion control problem in fixed networks, no specific solutions and algorithms have been proposed for the variation of this problem in cellular networks. Despite radio network congestion being a widely recognized and identified problem, few relevant studies have been published. The most strongly related publication is [15]. However, this publication refers to the extended class of wireless access networks (including WLANs), and it is not well aligned with 3GPP specifications for the UMTS cellular networks.

In [15], the authors investigate the wireless-caused representative selection fluctuation problem in wireless multicast congestion control. This problem is caused by the frequent degradation of the wireless channel and the subsequent bursty packet loss. This situation causes frequent change of the representative. The sender adjusts its transmission rate to the tentative worst receiver, which brings severe performance degradation to wireless multicast. In this chapter, two possible solutions are proposed, an end-to-end approach and an active approach. Finally, through performance evaluation in various situations, it is concluded that the end-to-end approach is sensitive for its inferring error. On the other hand, the active service leads to significant performance improvement.

10.3 Overview of the Domain

In this section we describe in brief some basic concepts of the examined scientific domain. The TFMCC algorithm and the UMTS system along with its MBMS service are presented. Finally, the CLR selection problem is analyzed.

10.3.1 TFMCC Mechanism

TFMCC is a well-known equation-based multicast congestion control mechanism that extends the TFRC protocol [7] from the unicast to the multicast domain. It constitutes a congestion control scheme that not only aims to reduce packet loss and improve bandwidth utilization, but also is fair toward competing TCP flows, i.e., is TCP friendly. TFMCC belongs to the class of single-rate congestion control schemes and applies at the end systems (end-to-end service). Such schemes inevitably do not offer multiple transmission rates as layered schemes do. However, they are much simpler so as to offer scalability to applications with thousands of receivers [4].

TFMCC uses a control equation derived from a model of TCP's long-term throughput to directly control the sender's transmission rate. The loss event rate and the round-trip time (RTT) are the parameters that define this target throughput. Each receiver calculates its target throughput and considers it the acceptable sending rate from the sender to itself.

TFMCC uses a feedback scheme that allows the receiver calculating the slowest transmission rate to always reach the sender. This scheme is based on the concept of the current limiting receiver (CLR). The CLR is the receiver that the sender believes currently has the lowest expected throughput of the multicast group. Moreover, the TFMCC design ensures that the sender gets feedback from the receivers experiencing the worst network conditions without being overwhelmed by feedback (feedback implosion is suppressed).

For full details of TFMCC, we refer the reader to [4].

10.3.2 UMTS Architecture

From the physical point of view, the UMTS network architecture is organized in two domains. This basic split considers the user equipment (UE) and the public land mobile network (PLMN). The UE is used by the subscribers to access the UMTS services, while the PLMN is a network established by an operator to provide mobile telecommunications services to the public. The PLMN is further divided into two land-based infrastructures: the UMTS Terrestrial Radio Access Network (UTRAN) and the core network (CN) (Figure 10.1). The UTRAN handles all radio-related functionalities. The CN is responsible for maintaining subscriber data and for switching voice and data connections.

The UTRAN consists of two kinds of nodes: the first is the radio network controller (RNC) and the second is node B. Node B constitutes the base station and provides radio coverage to one or more cells (Figure 10.1). Node B is connected to the UE via the Uu interface and to the RNC via the Iub interface. The Uu is a radio interface based on the wideband code division multiple access (WCDMA) technology. A single RNC with all the nodes B connected to it is called radio network subsystem (RNS).

The CN is logically divided into the circuit-switched (CS) domain and the packet-switched (PS) domain. All of the voice-related traffic is handled by the CS domain, while the PS domain handles the packet transfer. The entities of the CS portion of the CN will not be described because the purpose of this chapter is to focus on multicast. The PS domain is more relevant, and therefore, in the remainder of this chapter, more attention will be devoted to the PS functionality. The PS domain of the CN consists of two kinds of general packet radio service (GPRS) support nodes (GSNs): gateway GSN (GGSN) and serving GSN (SGSN) (Figure 10.1). The SGSN is the centerpiece of the PS domain. It provides routing functionality, manages a group of RNSs, and interacts with the home location register (HLR), which is a database permanently storing subscribers' data. The SGSN is connected to GGSN via the Gn interface and to RNCs via the Iu interface. GGSN provides the interconnection between the UMTS network and external packet data networks (PDNs) like the Internet [16].

Before a UE can exchange data with an external PDN, it must first establish a virtual connection with that PDN. Once the UE is known to the network, the

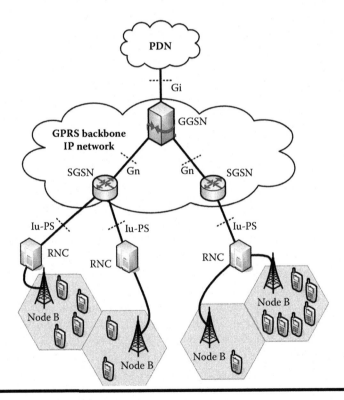

Figure 10.1   UMTS architecture.

packet transfer is based on the Packet Data Protocol (PDP). An instance of a PDP type is called PDP context. When a PDP context needs to be established, a PDP context activation procedure takes place. If this procedure is successful, it leads to the creation of two GPRS Tunneling Protocol (GTP) sessions dedicated to the sub-scriber: one between the GGSN and the SGSN over the Gn interface and another between the SGSN and the RNC over the Iu interface. The IP packets destined for an application using a particular PDP context are routed using the assigned GTP tunnels to the appropriate RNC. The RNC recovers the GTP-tunneled packet and transmits it to the UE [1].

Data transmission in the UTRAN is based on transport channels. The trans-port channels define the characteristics of the data transfer according to the service requirements. Despite the fact that there are several types of transport channels specified for UMTS, we will focus on the two most important types: the dedicated channel (DCH) and the forward access channel (FACH). The DCH carries infor-mation exchanged between a specific UE and the upper network levels. It exists in both the downlink and uplink directions. Instead, the FACH exists only in the downlink direction. FACH is a common channel, and consequently, a single FACH can carry information for more than one UE in a cell. The existence of mul-tiple types of transport channels in combination with the capability of switching between the different types allows higher flexibility and a more efficient use of the scarce radio resources while at the same time keeping service constraints [6].

10.3.3 MBMS Service

As we mentioned above, the 3GPP is currently standardizing the MBMS service [17, 18]. Actually, the MBMS is defined as an IP Datacast (IPDC) type of service, which can be offered via existing GSM and UMTS cellular networks. As the term *Multimedia Broadcast/Multicast Service* implies, two types of service mode exist in MBMS service: the broadcast and the multicast. In the broadcast mode, data is delivered to all the receivers roaming in a specific area. On the other hand, in the multicast mode the receivers have to declare their interest for the data reception. The service then decides whether the user may receive data. During the rest of our analysis we will focus on the multicast mode. The multicast mode is the most com-plicated and also covers all the aspects of the broadcast mode.

The basic MBMS architecture is almost the same as the existing UMTS archi-tecture in the PS domain. Figure 10.2 illustrates the basic MBMS architecture for UMTS. The most significant modification of the UMTS architecture is the addi-tion of a new node called Broadcast Multicast–Service Center (BM-SC).

The BM-SC is a data source unique to MBMS. In this node the MBMS data is scheduled and interfaces are provided for interaction with the content provider. The BM-SC may authorize and charge the content provider. At this point, it must

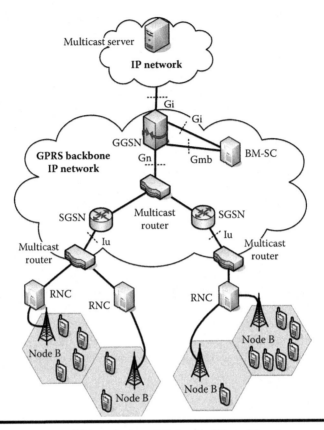

Figure 10.2 MBMS architecture for UMTS using IP multicast.

be clarified that the data source may not originate from an external PDN, but from within the UMTS network.

To reduce the implementation costs, the MBMS has been designed to introduce only minor changes to existing radio and core network architectures. For simplicity reasons, in our analysis, we will consider the functionality of the BM-SC incorporated in the GGSN.

The reception of an MBMS multicast service is enabled by certain procedures. These are subscription, service announcement, joining, session start, MBMS notification, data transfer, session stop, and leaving [17, 18]. Figure 10.3 presents the sequence of the MBMS multicast service phases.

10.3.4 CLR Selection Problem

In wireless communication systems like UMTS, the packet loss may not mean network congestion. The quality of a wireless link may be degraded due to signal

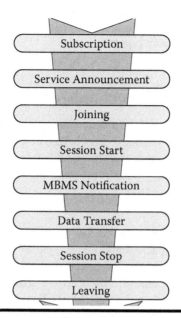

Figure 10.3 Phases of MBMS multicast service provision.

fading. During a fading period, the bit error rate of the wireless link may become very high, but normally after that period the wireless link is expected to recover.

The traditional congestion control mechanisms translate the packet loss as buffer overflow in the network nodes, i.e., as network congestion. Consequently, the action taken to resolve this situation is the reduction of the sender's transmission rate. Nevertheless, if the packet loss is caused by fading, the reduction of the transmission rate will not affect the packet loss. This is due to the fact that the packet loss does not depend on the arrival rate of the packets but on the wireless channel degradation. Finally, the packet loss will be resolved after the end of the fading period, without a transmission rate regulation.

Obviously, the wireless channel degradation may affect the performance of the TFMCC mechanism. If we suppose that a UE suffers from fading, then the packet loss probability for this UE will temporarily increase. This increment of packet loss may cause the selection of this UE as the CLR. The next step is that the transmission rate of the multicast server will be reduced according to the packet loss of the examined UE. The problem is that this reduction is unnecessary because it is wireless caused. After recovering from the bad wireless quality phase, the target throughput of the CLR will be improved. If a lot of UEs participate in the multicast group, there is a high probability that another UE suffers from fading. Soon, another UE suffering from channel degradation will be selected as CLR and will regulate the transmission rate. Eventually, the wireless channel degradation will cause a significant and steady degradation of the performance of the TFMCC mechanism and of the multicast service. During this analysis we shall refer to this problem as the CLR selection problem.

10.4 The Proposed Mechanism

As we have already mentioned, the proposed mechanism follows a design very similar to that of the TFMCC scheme. Nevertheless, new functionality has been added to the existing mechanism to deal with the CLR selection problem.

The basic principles that govern the proposed mechanism are the following:

1. Each UE measures its packet loss rate using the packet loss history scheme of TFMCC.
2. Each node B measures its packet loss rate. This information is written to the heading of the data packets and is then read by the UEs. This is a new functionality that combats the CLR selection problem in UMTS networks. This functionality does not exist in the TFMCC scheme and is explained below.
3. Each UE measures or estimates the RTT to the multicast server. This is achieved through an approach inherited from TFMCC. In more detail, time-stamped feedback is sent to the multicast server. The server then echoes the time-stamp and the corresponding UE_id in the header of a data packet. This approach causes minor traffic overhead in the network.
4. Each UE uses a control equation to calculate an acceptable sending rate from the sender back to it. The input parameters for the control equation are the loss rate and the RTT measured by the UE.
5. The feedback scheme of TFMCC is adopted. This scheme has devised a way for the feedback from the receiver calculating the slowest transmission rate to always reach the sender. In addition, the feedback is filtered using randomized timers to avoid a feedback implosion.

In the proposed mechanism, the nodes located at the border between wireless and wired network (i.e., nodes B) have an additional responsibility. This responsibility is to provide the receivers (i.e., the UEs) with information about their measured packet loss. This means that each UE is informed by its serving node B of the packet loss that node B measures. This information is piggybacked in the data packets of a multicast session.

This additional functionality of nodes B permits each UE to identify the reason of a packet loss. The UE compares the packet loss received from node B with its measured packet loss. In general, the following cases are distinguished:

- When the two values differ, the UE can conclude that the reason for the difference is losses at the wireless link caused by wireless channel degradation. This kind of packet loss is not related to the network congestion, and consequently, the reduction of the transmission rate will not affect this packet loss.

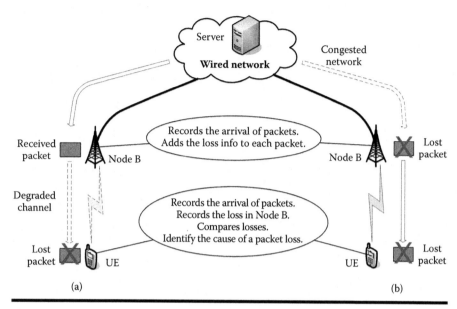

Figure 10.4 **Packet losses at the UE: (a) Packet loss due to wireless channel degradation and (b) Packet loss due to network congestion.**

In this case, the packet loss is not accounted for at the CLR selection. Figure 10.4(a) visualizes this functionality of the UE.

■ On the other hand, when both node B and the UE encounter a packet loss, this packet loss is considered to be caused due to network congestion. Consequently, this kind of packet loss is taken into consideration during the CLR selection. This scenario is depicted in Figure 10.4(b).

At this point we will present another aspect of the proposed mechanism. Consider the case that, under certain conditions, a permanent degradation of the wireless channel affects a specific UE. When a permanent degradation occurs on the wireless link, the buffer of node B will overflow and some packets will be rejected. Normally, this UE should be a CLR candidate. In the proposed mechanism, during permanent channel degradation, node B counts the rejected packets as general packet losses that happened due to network congestion. These packets are taken into account by the UE during the CLR selection. This scenario is illustrated in Figure 10.5.

As we mentioned above, this permanent degradation is not hidden from the UE. In fact, the UE is informed of the packet losses caused by buffer overflow. This functionality makes our proposed mechanism suitable not only with the CLR selection problem, but also with the permanent degradation of the wireless channel.

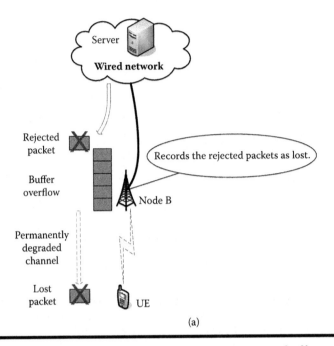

(a)

Figure 10.5   Permanent wireless channel degradation causes buffer overflow in node B.

10.5 Experiments

The above-described mechanism was implemented and subjected to extensive evaluation through simulation. The evaluation was conducted toward two directions. The first was to examine that the proposed mechanism preserves the benefits of TFMCC as they are presented in [4]. The other was to evaluate the behavior of the mechanism against the CLR selection problem.

10.5.1 Simulation Environment

For the verification of the proposed mechanism, the ns-2 network simulator [19] along with its EURANE extension were used. The Enhanced UMTS Radio Access Network Extensions (EURANE) for ns-2 [20] comprises extensions for the support of UMTS network functionality. The three UTRAN nodes that ns-2 does not support—RNC, node B, and UE—are implemented in EURANE. Moreover, EURANE supports the functionality of the following transport channels: FACH, Random Access Channel (RACH), DCH, and High Speed Downlink Shared Channel (HS-DSCH).

Given the fact that the ns-2 simulator does not support the multicast transmission in UMTS, we implemented the multicast packet forwarding mechanism described [21]. To use this multicast scheme, we had to introduce the routing lists in each node of the UMTS network except for the UEs.

The next step was the installation of the TFMCC scheme. The codes used to implement and evaluate the TFMCC by the authors of [4] are provided at [22]. We adopted this TFMCC implementation to evaluate our proposed variation of TFMCC.

Finally, the generic TFMCC was modified and extended to support the UMTS environment. In more detail, the implementation of the TFMCC was enhanced to support the functionality of node B and the UE as described in Section 10.4. The node B implementation was modified to provide the UEs with information about their measured packet loss. This means that each UE is informed by its serving node B of the packet loss that node B measures. This information is piggybacked in the data packets of a multicast session. One bit in the header of the data packet is enough for the provision of this information. On the other hand, the UE implementation was modified to read this bit and to take the decision whether a packet loss should be accounted for at the calculation of its acceptable sending rate.

10.5.2 Fairness

The first aspect that we examined was the TCP friendliness of TFMCC. In more detail, we considered the fairness of TFMCC toward the competing TCP flows when they share wired or wireless links. We tested the TFMCC fairness in various conditions. Below, we present the TFMCC behavior in a noncongested and in a congested UMTS network.

First, fairness toward competing TCP flows was analyzed using a noncongested UMTS network (Figure 10.6). We monitored the throughput over a wireless link connecting the UEx with node By. We supposed that UEx belongs to a multicast group and receives TFMCC traffic. At the same time, this UE receives TCP traffic from an external node.

Figure 10.7 illustrates the throughput of TFMCC flow against that of the TCP flow.

Due to our initial assumption that no congestion exists, the capacity of the wireless link poses a threshold of 384 Kbps for the throughput of the flows toward the examined UE. As it was expected, the available bandwidth is fairly shared between the flows. Figure 10.7 confirms that the average throughput of TFMCC flow closely matches the average TCP throughput. Actually, both the average throughputs match half of the available capacity of the UTRAN wireless link.

The next step of our experiment was to examine the fairness of the mechanism in a congested UMTS network. We considered the single-bottleneck topology

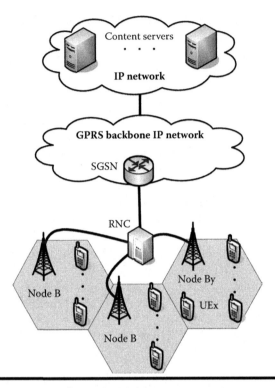

Figure 10.6   Noncongested UMTS topology.

depicted in Figure 10.8. The bottleneck was applied over a link that connects an SGSN with an RNC node (Iu interface).

As shown in Figure 10.8, a number of sending content servers are connected to a number of receiving UEs through a common bottleneck. In more detail, 15 servers send TCP traffic to as many UEs, whereas 5 multicast servers send TFMCC traffic to as many multicast groups.

Figure 10.9 shows the throughput of a TFMCC flow against two sample TCP flows (out of 15). The average throughput of TFMCC closely matches the average TCP throughput. Moreover, TFMCC achieves a smoother rate than the TCP.

Similar results can be obtained for many other scenarios, for example, if we suppose that congestion exists over a Gn interface (connects GGSN with SGSN nodes). In this case, the available throughput of the bottleneck link is evenly shared among the competing TFMCC and TCP flows. The TCP friendliness of the proposed scheme was therefore confirmed under all the congestion scenarios.

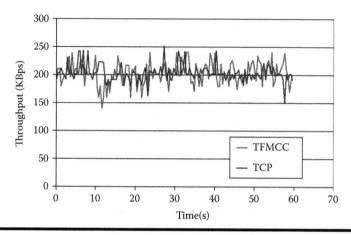

Figure 10.7 TFMCC flow versus TCP flow in a noncongested UMTS network.

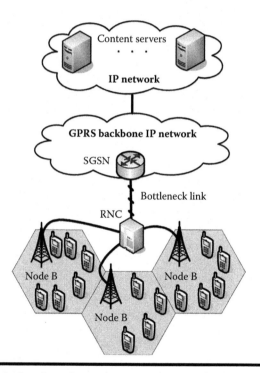

Figure 10.8 Single-bottleneck topology.

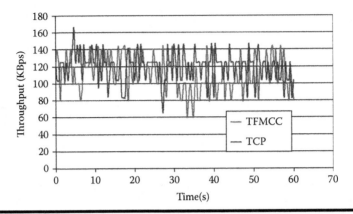

Figure 10.9 TFMCC flow versus TCP flow in a single-bottleneck UMTS network.

10.5.3 Responsiveness to Changes

An important concern in the design of congestion control protocols is their responsiveness to changes in the network conditions. This behavior was investigated using the single-bottleneck topology of Figure 10.8.

During the simulation we changed the applied loss rate of the bottleneck link. The simulation lasted 150 s. During this time interval three different loss rates were applied on the Iu interface. The TFMCC flow was monitored along with two TCP flows sharing the bottleneck link. The results of the simulation for the three competing flows are presented in Figure 10.10.

As shown in Figure 10.10, TFMCC matches closely the TCP throughput at all three loss levels. Moreover, the adaptation of the sending rate is fast enough. Actually, the simulator logs show that the UEs need 1500–2000 ms after the change of the loss rate to adapt to the new loss rate. These figures of response time are close enough to the corresponding time of TCP (about 1000–1500 ms).

A similar simulation setting was used to investigate the responsiveness to changes in the RTT. The results are similar to those above. The above experiment confirms the excellent reactivity of the TFMCC to changes in congestion level of the UMTS network. Moreover, it confirms that during the application of TFMCC over the UMTS, the properties and the benefits of this scheme are not affected.

10.5.4 Reaction to Wireless Channel Degradation

The next concern of our experiments was the evaluation of the proposed scheme when wireless-caused packet losses occur. We simulated a UMTS network and assumed a degradation of the wireless channels. In more detail, we simulated the wireless channel degradation by applying an error rate over the packets transmitted

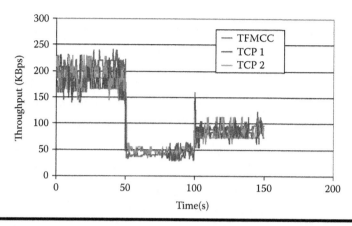

Figure 10.10   Responsiveness to changes in the loss rate.

via the wireless links. We examined the proposed scheme for different numbers of UEs belonging in the multicast group.

In Figure 10.11, our proposed scheme is referred to as modified TFMCC (mod_TFMCC). On the other hand, the typical TFMCC algorithm presented in [4] is referred to as TFMCC. The horizontal axis shows the number of UEs belonging in the examined multicast group. Both mechanisms were examined for up to 100 UEs participating in the multicast group. The vertical axis shows the average throughput, which is normalized to the corresponding TCP one. The results when 5 percent wireless-caused packet loss is applied are presented in Figure 10.10.

The results depicted in Figure 10.11 confirm the excellent behavior of our proposed scheme when wireless channel degradation occurs. The wireless-caused packet losses can be identified correctly at the UEs and be ignored at the calculation of the acceptable sending rate. This means that the CLR selection problem can be overcome and significant improvement is added on the TFMCC application over the UMTS.

10.5.5 Permanent Wireless Channel Degradation

The last concern of our experiments was the evaluation of the proposed scheme when wireless-caused packet losses occur in a permanent manner. We examined the behavior of the modified TFMCC when a wireless channel is permanently degraded so as to lead to buffer overflow and packet rejections in the corresponding node B. Figure 10.12 depicts the simulation setting. In the UMTS network of this setting we assumed a permanent degradation of the wireless link that connects the UEx with node By.

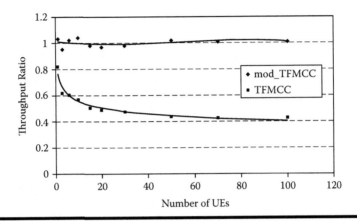

Figure 10.11 Throughput of our proposed scheme versus TFMCC when wireless-caused packet losses occur.

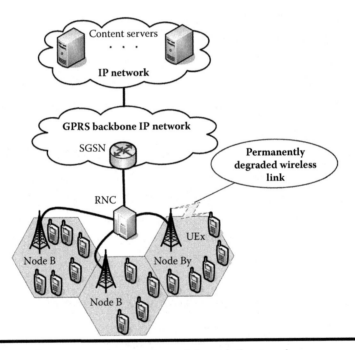

Figure 10.12 Permanent wireless channel degradation topology.

In more detail, we simulated the wireless channel degradation by applying an error rate of 50 percent over the packets transmitted via the corrupted wireless link. In the beginning of the simulation, no wireless channel degradation occurred.

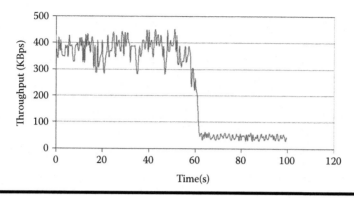

Figure 10.13   Throughput of our proposed scheme versus TFMCC when wireless-caused packet losses occur.

After 50 s of simulation, we applied the error rate over the wireless channel connecting the UEx with node By. We monitored the changes over the throughput of the corrupted wireless link for 100 s. The results of our experiment are presented in Figure 10.13.

The simulation results prove that our proposed scheme reacts to the permanent wireless channel degradation. In the beginning of the simulation, no congestion exists and the throughput matches the available bandwidth of the wireless link. After 50 s of simulation, the 50 percent packet error rate is applied. Obviously, the network does not immediately react to this degradation because it considers it as wireless caused.

During the period between 50 and 60 s of simulation, the buffer of node B overflowed and the network was able to distinguish the nature of the degradation. It took about 10 s to adapt to the new network conditions, but this time interval may differ according to the bit rate of the transmission and the size of the buffer in node B.

The results confirm the behavior of our proposed scheme when permanent wireless channel degradation occurs. The packet losses in node B are considered network congestion and are not ignored during the calculation of the acceptable sending rate in the UE. Eventually, this kind of packet loss causes reduction of the transmission rate.

In case that multiple wireless channels are degraded, the simulation results (not listed here) are similar. The CLR is a UE among the ones being connected with degraded wireless links and calculating the minimal acceptable sending rate.

10.6 Conclusions

We have described a congestion control scheme for multicast transmission over UMTS. The proposed scheme is based on the well-known TFMCC mechanism. The TFMCC is a TCP-friendly, single-rate multicast congestion control mechanism intended to scale to groups of several thousand receivers. Nevertheless, the legacy TFMCC algorithm was designed for fixed networks and had to be modified before being applied to wireless networks.

In wireless communication systems like UMTS, the CLR selection problem appears. This is because in this kind of network the packet loss may not mean network congestion but may be caused by wireless link degradation. The proposed scheme does not translate the wireless-caused packet loss as buffer overflow in the network. Consequently, no reduction of the sender's transmission rate is performed to resolve this situation.

We have evaluated the proposed scheme through simulation experiments. We concluded that it preserves the benefits of the TFMCC algorithm over the UMTS cellular network. The mechanism is fair toward competing TCP flows over congested and noncongested links. Moreover, the results of the experiments demonstrate the very good reactivity to changes in the congestion level for both loss rate and RTT.

Additionally, simulation experiments were performed to examine the proposed scheme against the CLR selection problem. The results confirm the excellent behavior of our proposed scheme when wireless channel degradation occurs. The wireless-caused packet losses can be identified correctly at the UEs and are ignored at the calculation of the acceptable sending rate. This means that the CLR selection problem can be overcome and significant improvement is added on the TFMCC application over the UMTS.

Finally, we examined the behavior of our proposed scheme against permanent wireless channel degradation. In this case, the packet losses due to buffer overflow in node B are considered network congestion and are not ignored during the calculation of the acceptable sending rate in the UE. Eventually, this kind of packet loss causes reduction of the transmission rate.

Minor modification in the UMTS architecture is needed. Actually, the impacts concern only two nodes of the UMTS network: nodes B and the UEs. The additional functionality in UE is to examine whether a packet loss is wireless caused. The reading of a single bit in each multicast packet header is sufficient for this purpose. On the other hand, each node B has to set this bit info over the packet header if it has encountered a packet loss. The proposed scheme therefore respects the limited computing power of the UEs, and no demanding operation is introduced in those network nodes.

10.7 Future Work

The step that follows this work is the evaluation of different congestion control schemes for UMTS networks. Other TCP-friendly multicast congestion control schemes like pgmcc [23] will be investigated and modified to meet the UMTS requirements. Additionally, the applicability over UMTS of different multicast congestion control approaches will be examined. Multicast architectures like multirate [9], layered [10], end-to-end [4, 10, 11], and active [12] services will be evaluated for their applicability. These emerging schemes will also be evaluated through comparison. This comparison will examine several aspects, like the efficiency and cost of implementation of each scheme.

Furthermore, we will try to formulate a multicast group control mechanism dedicated for the UMTS networks. In some cases, permanent wireless channel degradation may cause a large reduction to the transmission rate and eventually multicast service degradation. It will be specified under which circumstances wireless channel degradation will cause rejection of a corrupted UE from a multicast group.

Furthermore, we will try to take advantage of the broadcasting nature of the wireless channels. This broadcasting nature is a promising platform for enhancing the legacy multicast schemes and implementing efficient wireless multicast schemes.

References

[1] Holma, H., and Toskala, A. 2004. *WCDMA for UMTS: Radio access for third generation mobile communications.* 3rd ed. New York: Wiley.

[2] Varshney, U. 2002. Multicast over wireless networks. *Communications of the ACM* 45:31–37.

[3] 3GPP. *Multimedia broadcast/multicast service (MBMS): Stage 2.* Technical Report TR 23.846, version 6.1.0.

[4] Widmer, J., and Handley, M. 2001. Extending equation-based congestion control to multicast applications. *ACM SIGCOMM Computer Communication Review* 31:275–85.

[5] Fu, C. P., and Liew, S. C. 2003. TCP Veno: TCP enhancement for transmission over wireless access networks. *IEEE Journal on Selected Areas in Communications* 54:216–28.

[6] Pérez-Romero, J., Sallent, O., Agusti, R., and Diaz-Guerra, M. A. 2005. *Radio resource management strategies in UMTS.* New York: Wiley.

[7] Floyd, S., Handley, M., Padhye, J., and Widmer, J. 2000. Equation-based congestion control for unicast applications. In *Proceedings of the Conference on Applications, Technologies, Architectures, and Protocols for Computer Communication*, Stockholm, pp. 43–56.

[8] Liu, J., Li, B., and Zhang, Y.-Q. 2003. Adaptive video multicast over the Internet. *IEEE Multimedia* 10:22–33.

[9] Cheung, S., Ammar, M., and Li, X. 1996. On the use of destination set grouping to improve fairness in multicast video distribution. In *Proceedings of the INFOCOM '96 15th Annual Joint Conference of the IEEE Computer and Communications Societies*, 553–60.

[10] Vickers, B., Albuquerque, C., and Suda, T. 2000. Source-adaptive multilayered multicast algorithms for real-time video distribution. *IEEE/ACM Transactions on Networking* 8:720–32.

[11] Youssel, A., Abdel-Wahab, H., and Maly, K. 1998. A scalable and robust feedback mechanism for adaptive multimedia multicast systems. In *Proceedings of the 8th IFIP Conference on High Performance Networking (HPN '98)*, 113–37, Vienna.

[12] Amir, E., McCanne, S., and Katz, R. 1998. An active service framework and its application to real-time multimedia transcoding. In *Proceedings of the ACM SIGCOMM '98 Conference on Applications, Technologies, Architectures, and Protocols for Computer Communication*, Vancouver, 178–89, British Columbia.

[13] DeLucia, D., and Obraczka, K. 1997. Multicast feedback suppression using representatives. In *Proceedings of the INFOCOM '97 16th Annual Joint Conference of the IEEE Computer and Communications Societies*, 463–70, Kobe, Japan.

[14] Donahoo, M. J., and Ainapure, S. R. 2001. Scalable multicast representative member selection. In *Proceedings of the INFOCOM '01 Twentieth Annual Joint Conference of the IEEE Computer and Communications Societies*, 259–68, Anchorage, AK.

[15] Saito, T. and Yamamoto, M. 2005. Wireless-caused representative selection fluctuation problem in wireless multicast congestion control. *IECE Transactions on Communications* E88-B:2819–25.

[16] Korhonen, J. 2003. *Introduction to 3G mobile communications*. 2nd ed. Norwood, MA: Artech House.

[17] 3GPP. 2006. *Multimedia broadcast/multicast service (MBMS): Architecture and functional description*. Technical Specification TS 23.246, version 6.11.0.

[18] 3GPP. 2006. *Introduction of the multimedia broadcast/multicast service (MBMS) in the radio access network (RAN): Stage 2*. Technical Specification TS 25.346, version 7.1.0.

[19] *The Network Simulator—ns-2*. http://www.isi.edu/nsnam/ns (accessed March 2002).

[20] Ericsson Telecommunicatie B.V. 2005. *User manual for EURANE*. http://www.ti-wmc.nl/eurane/eurane_user_guide_1_6.pdf (accessed March 2007).

[21] Alexiou, A., Antonellis, D., Bouras, C., and Papazois, A. 2006. An efficient multicast packet delivery scheme for UMTS. In *Proceedings of the 9th IEEE/ACM International Symposium on Modeling, Analysis and Simulation of Wireless and Mobile Systems (MSWiM'06)*, 147–50, Torremolinos, Malaga, Spain.

[22] *TCP-friendly multicast congestion control (TFMCC)*. http://icapeople.epfl.ch/widmer/tfmcc/ (accessed March 2007).

[23] Rizzo, L. 2000. pgmcc: A TCP-friendly single-rate multicast congestion control scheme. In *Proceedings of the Conference on Applications, Technologies, Architectures, and Protocols for Computer Communication*, 17–28, Stockholm.

Chapter 11

QoS Service in Heterogeneous Wireless Networks

Torsha Banerjee, Bin Xie, and Dharma P. Agrawal

Contents

11.1 Introduction

The popularity of wireless communication systems can be envisaged almost everywhere in the form of wireless wide area networks (WWANs, e.g., 3G cellular network), wireless local area networks (WLANs, e.g., IEEE 802.11a/b/g/n), wireless metropolitan area networks (WMANs, e.g., IEEE 802.16e), and mobile ad hoc networks (MANETs). Many applications can benefit from a ubiquitous computing environment capable of accessing various heterogeneous wireless networks from single or different portable devices from any place and at any time. This has motivated researchers to integrate the Internet with various heterogeneous wireless platforms such as WWANs, WLANs, WMANs, and MANETs, and provide seamless service migration among them. Such an integrated network is called a heterogeneous wireless network (HWN). In a HWN, the mobile terminal (MT) may be equipped with multiple wireless network interfaces so that the integrated heterogeneous environment allows a user to access a particular network, depending on the specific application needs and types of available radio access networks [5].

Integration of different wireless technologies with distinct capabilities and functionalities is a multifaceted task and involves many issues at all layers of the Open System Interconnection (OSI) model [1]. It involves several critical issues to be addressed, particularly the ones related to quality of service (QoS) as they affect the overall network performance. Network performance is usually influenced by different factors at each of the physical, data link, network, and transport layers of the OSI protocol stack. In a standard real network scenario, and in particular in a heterogeneous setup, it is exceedingly difficult to define a general framework for performance evaluation of an experimental system and identifying the root of the results. In a heterogeneous network environment, there is a wide range of QoS parameters. As a rule of thumb, key parameters are latency, bandwidth, packet jitter, and packet loss. However, how to provide QoS for a service in the heterogeneous mobile network environment is not as easy as only one type of static network. Mobile communication today has HWNs providing varying levels of coverage and

QoS. Currently developed techniques concentrate on improving their accessibility and QoS guarantee. These methods enable MTs to communicate with each other by introducing changes in the network protocol stack. To improve flexibility, Internetworking, reliability, and robustness, they also support establishment and maintenance of connections between MTs using available links.

Based on the application, one or more of the QoS parameters could be more important than the others. For example, for multimedia applications, end-to-end delay is of increased concern compared to packet loss. One of the key challenges in providing the QoS in HWNs is to handle voice and video data effectively. It is really challenging to provide end-to-end QoS services in a HWN. The user may want to move from one network to another, trying to interact with different service providers, which might have drastically different network capacities, topologies, and policies. The interacting users may also be employing different wireless access technologies that differ in terms of bandwidth, loss, and delay of the channel, and thus have different underlying QoS capabilities. MTs might have different computing powers, display capabilities, and communication bandwidths. All of these factors can complicate any end-to-end QoS provisioning in such a heterogeneous scenario.

The objective of this chapter is to provide an overview of existing QoS challenges for a HWN and cover in detail state-of-the-art exploratory research being conducted. The rest of the chapter is organized as follows. Section 11.2 gives an overview of the HWN architecture and a simplified QoS model it can support. Section 11.3 discusses the QoS mechanisms for each layer of the OSI model, as well as the QoS-based mobility and connection management. Section 11.4 summarizes schemes used to achieve end-to-end QoS in HWNs and identifies various open research issues.

11.2 HWN Architecture and Its QoS Architecture

A HWN can be defined as an integrated system that provides QoS-guaranteed Internet connectivity across heterogeneous wireless platforms such as WWANs, WMANs, WLANs, and MANETs. According to the QoS requirements demand that is specified by applications, the HWN has the capability of providing connection to wireless networks through any radio interface of a multimode MT. A mobile user having a multimode MT can access the Internet services through cellular base stations (BSs), WMAN BSs, or WLAN access points (APs), and will also be able to support peer-to-peer communication with others using MANET connectivity. Figure 11.1 illustrates the HWN architecture with overlapping networks in the hotspot areas using different wireless technologies, and these different wireless networks are connected to the common Internet Protocol (IP) backbone. In a HWN, the basic components include MTs, BSs/APs and their access networks, and the core IP network (IP backbone) as follows.

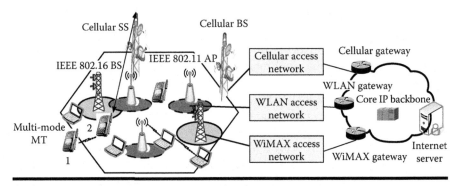

Figure 11.1   An example network of a HWN.

- **Single or multimode MTs:** A single-mode MT may have WWAN, WMAN, WLAN, or MANET capability, while a multimode MT may have some or all of these capabilities. In addition to a traditional cellular interface, a multimode MT can be equipped with WLAN or Bluetooth [18] access capabilities. At the same time, if a MT is outside the network coverage, MT can operate in a MANET mode, which allows it to connect to a network via a multi-hop connection or peer-to-peer communication.

- **SSs/BSs/APs and the access network:** Subscriber stations (SS), BSs (e.g., a cellular BS, an IEEE 802.16 BS), and APs (e.g., IEEE 802.11a/b/g/n AP) are fixed components that provide wireless access to MTs. In an access network, several SSs connect to the BS. The cellular BSs are connected to the cellular infrastructure networks, while an AP may be a part of a WLAN that is usually connected to the IP backbone through Internet gateway (IGWs). The IGWs serve as points for attachments of wireless networks to the Internet.

- **Core IP backbone:** As can be seen in Figure 11.1, HWNs are interconnected through an existing IP backbone. To support wireless communication, on the Internet, relevant network databases, registers (e.g., visiting location register [VLR] and home location register [HLR] for cellular networks), and other entities are needed to provide the functionalities, such as user profile management, pricing, billing, location coordination, authentication, and so on.

There are still many other components in the HWN. For example, IGWs provide the interface between the access networks (e.g., a cellular network) and the core IP backbone. The WLAN gateways allow interconnectivity between two WLANs and the Internet, and the cellular gateways offer Internet connectivity to the cellular networks. In each access network, there are various servers for the network management, such as QoS-based resource (network bandwidth, IP address) management. In the following simplified QoS model, for simplicity of explanation, these functionalities are assumed to be integrated with the IGW.

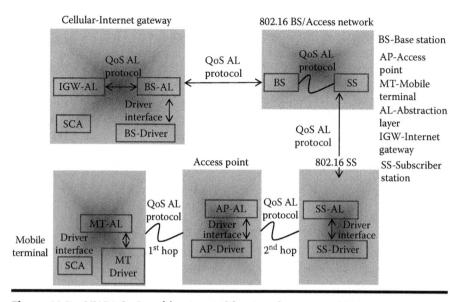

Figure 11.2   HWN QoS architecture with a two-hop scenario.

11.2.1 An Example QoS Architecture for Heterogeneous Network Access

The QoS architecture for the HWN, as well as its implementation, is an extremely complicated issue that involves different layers of the protocol stack and spans across heterogeneous network systems. In this section, we use a simplified QoS model [2] for the HWN to illustrate the basic QoS components. Figure 11.2 illustrates the QoS modules that reside in the MT, AP/SS, and BS, respectively. The IGW in the IEEE 802.16 access network is directly connected to the BS, as shown in Figure 11.2. Figure 11.2 shows a two-hop network access of MT-AP-IEEE 802.16 SS, and it connects the Internet by IEEE 802.16 BS-IGW. We first introduce these QoS modules, followed by their basic functionalities, in two-hop QoS architecture.

- **QoS modules in the MT:** In a single-mode MT, the QoS module consists of a MT driver module and MT–abstraction layer (MT-AL) module connected by a driver interface. In a multimode MT, an integrated MT driver module is required to handle multiple interfaces.
- **QoS modules in the AP/SS:** The MT wirelessly connects to the IEEE 802.11 AP followed by an IEEE 802.16 SS. The QoS module for an AP includes an AP driver module and AP–abstraction layer (AP-AL) module connected by a driver interface. In a SS, the QoS module has a SS driver module and SS–abstraction layer module (SS-AL) connected by a driver interface. A QoS

protocol runs between the AP–abstraction layer and the SS–abstraction layer for the purpose of QoS interoperation between them.

■ **QoS modules in the access network:** In this module, a QoS protocol runs between the SS and the BS. The BS is responsible for radio resource management and assignment for SS.

■ **QoS modules in the IGW:** Consists of a BS driver module and BS–abstraction layer (BS-AL) module connected by a driver interface. It also includes an IGW–abstraction layer (IGW-AL) that connects to the Internet. A QoS protocol runs between the IGW-AL and BS-AL.

In addition, some other modules and key interfaces between modules are as follows:

■ **Driver interface (DI):** This interface translates abstract QoS (e.g., 146 × 250 pixels for a mobile TV program, which may be specified by the application layer) to technology-specific QoS parameters (e.g., bandwidth and end-to-end delay to satisfy the QoS).

■ Apart from the QoS modules mentioned above, inter-QoS protocols run between MT and AP/SS, and AP/SS and the access network, respectively. The following inter-QoS protocol modules and QoS interfaces are defined:

■ **Abstraction layer (AL) module:** Performs the resource management functions in the QoS architecture. The application specifies the desired level of QoS at this layer and can be notified if this QoS demand cannot be provided by the driver module, e.g., the MT driver. The AL module gears the objective of resource management in the QoS architecture. The key functionalities of the AL module include resource reservation, querying for resources from the IGW to remotely located APs, QoS concatenation of multiple wireless technologies, handoff supports, and QoS notifications. For example, the QoS reservation can be done by creating a virtual channel between an MT and a BS with specific QoS guarantees. It is composed of the following variations of the same basic module:

    – **Source channel adaptation (SCA) module:** Located in both IGW and MT and used to improve application performance over the wireless medium by optimizing the usage of wireless resources and adapting the source channel accordingly.

    – **AL driver module:** Provides QoS support for a particular wireless technology by directly communicating with it and by translating general QoS parameters to network-specific QoS parameters. For different technologies it has different objectives. In IEEE 802.11e technology, it provides the mechanism to design an admission control and configuration guidelines for QoS guarantees. To implement these functionalities, the WLAN AL driver again consists of three modules:

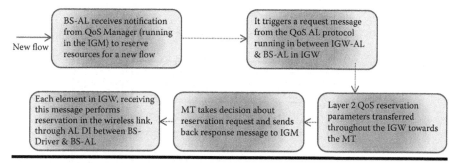

Figure 11.3   Steps of implementing an E2E QoS for a flow.

- Driver controller: Core module located at the AP and performs functions like executing admission control and reconfiguring algorithms.
- Monitoring module: Collects network performance information of all MTs and provides it to the driver controller.
- Configuration module: Sends the configurations computed by the driver controller to the configuration clients of the MTs. For IEEE 802.11e technology, once the configuration clients receive the configurations, the MTs then adjust the IEEE 802.11e driver with the given parameters.

Similar modules can be defined for other wireless technologies. For example, the IEEE 802.16 driver modules enforce the set of requested QoS parameters in the IEEE 802.16 wireless links. They perform the communication between the IEEE 802.16 SS-AL module and the IEEE 802.16 BS-AL.

Figure 11.3 illustrates an example implementation of QoS-based heterogeneous network access. If a new flow from the Internet is targeted for a MT in the HWN, the abstraction layer (AL) residing in the BS-AL module receives notification from the QoS manager to reserve resources for the new flow. This action triggers a request message from the QoS AL protocol running in between the MT and AP in Figure 11.2. Then, the QoS reservation parameters carried by the request are transferred from the SS to the BS. The BS determines whether the access network grants the QoS-demand connection, depending on the service type and system load. Once the SS receives a positive acknowledgment of the resource from the BS, the required resource (radio, buffer, priority) is reserved at the network access components (e.g., SS-AL, BS-AL). In the meantime, the AP coordinates with the SS to grant a connection. Each component in the entire path performs resource reservation in the wireless link, through the AL driver interfaces (e.g., between the SS-driver and BS-AL in Figure 11.2). The AL driver module in each component will perform the corresponding operation to guarantee the QoS requirement of the connection. Hereafter, a QoS-enabled path from the IGW to the destination MT is established for the incoming Internet flow. The proposed QoS architecture provides

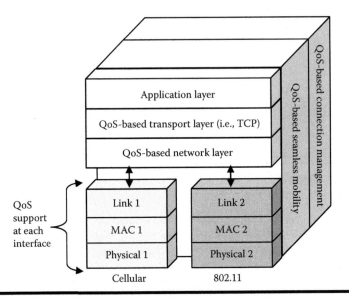

Figure 11.4   A dual-mode MT with QoS support.

scalability by merging any new access technology in the architecture by adding the corresponding AL-driver module. The two-hop architecture shown in Figure 11.2 allows concatenation of two diverse technologies and enables other functionalities by extending the access network.

11.3   QoS Mechanisms for HWNs

An overview of the existing QoS framework is given in this section, primarily focusing on the state-of-the-art research. As shown from the QoS architecture of Figure 11.2, the QoS mechanisms cannot be provided for each layer uniquely, as most of the functionalities are overlapping and thus require service spanning more than one layer. This gives rise to a cross-layer management that should be implemented across different interfaces. The QoS-based protocol stack for a dual-mode MT is given in Figure 11.4. As can be seen, the dual-mode MT has two physical, data link, and Medium Access Control (MAC) layers for the cellular and IEEE 802.11 radio interfaces, respectively.

The network layer selects a radio interface according to the availability of wireless access networks and the QoS requirements. The QoS solution for a dual-mode terminal should include a seamless mobility scheme that provides QoS support for mobile MT. An efficient QoS connection management functionality should be present as well. There are many challenges for supporting QoS in a HWN. Now, we discuss key points with regard to each layer of the OSI protocol stack.

11.3.1 QoS Mechanisms in the Physical Layer

The physical layers of all protocol stacks are responsible for modulation, packet transmission, and reception according to the wireless technologies in their own spectra. For example, the 802.11g orthogonal frequency division multiplexing (OFDM) uses the 2.4 GHz radio frequency band. To achieve higher throughput, some of the fundamental QoS questions in the physical layer include signal processing techniques (e.g., modulation, coding, and equalization), advanced antenna technologies (input multiple output (MIMO) e.g., use multiple antennas after transmission and reception to improve communication and smart antenna), advanced radio resource management techniques (e.g., QoS-based radio assignment algorithm), etc. For example, the efficient bandwidth utilization for a given packet error rate (say P_0) is achieved [4] with an adaptive modulation and coding (AMC) scheme that matches transmission parameters to the wireless channel settings adaptively.

- **Adaptive modulation and coding scheme:** An AMC has two components: AMC controller and AMC selector. The AMC controller follows the queue at the BS, which acts as the transmitter, and the AMC selector is implemented at the MT, which acts as the receiver. At the wireless link, multiple transmission modes exist, with each mode representing a particular modulation system and a forward error correction (FEC) code [4]. The goal of AMC is to maximize the data rate by adjusting transmission parameters according to variations in the channel, while maintaining the desired packet error rate, P_0. Based on the channel estimation at the receiver, the AMC selector decides on the particular modulation-coding pair to be used, which is sent back to the transmitter through a feedback channel, for the AMC controller to update the transmission mode. Coherent demodulation and maximum-likelihood (ML) decoding are implemented at the receiver. The decoded bit streams are mapped to the packets, which are forwarded upward to the data link layer. Different from nonadaptive modulations, AMC provides a dynamic, rather than deterministic, service process for the queue, which is capable of transmitting a variable number of packets per frame.
- **QoS metrics evaluated at physical layer:** The QoS metrics are evaluated in terms of the packet loss rate, ξ, the throughput, η, and the average delay, τ, over wireless links. These parameters can be evaluated as follows. Let P_d denote the packet dropping probability. It is derived as the ratio of the average number of dropping packets, $E\{D\}$, over the average number of arriving packets, $E\{A\}$, per frame [4]. A packet is said to be correctly received by the client (e.g., a MT) if it is not dropped from the queue with a probability $(1 - P_d)$ and is correctly received through the wireless channel with a probability $(1 - P_0)$.
- Therefore, the packet loss rate [4] is given by

$$\xi = 1 - (1 - P_d) \cdot (1 - P_0).$$

- The throughput is calculated as follows:

$$\eta = E\{A\}(1 - \xi) = \lambda\,(1 - \xi),$$

- where λ = the data arrival rate.
- Based on Little's theorem [6], the average delay per packet through the wireless link is

$$\tau = \frac{N_{wl}}{\lambda(1 - P_d)},$$

- where N_{wl} is the sum of the average number of packets in the queue and in transmission and is calculated with a stationary distribution.
- Based on the typical wireless link parameters, the QoS metrics can be calculated using these equations and provide a general QoS guideline for analysis, control, and design of wireless multimedia networks.

11.3.2 QoS Support in the Data Link and MAC Layers

When an interface is under the radio coverage of a cellular BS, the contention free MAC protocols such as Wideband code division multiple access (W-CDMA) may be used for coordinating the access from the MT's cellular interface to the BS/AP. To avoid any contention, radio resources are assigned via a central administrator (BS/AP). The contention-based MAC protocols, with a risk of collision, can be divided into two categories: random access (e.g., slotted ALOHA [i.e., "hello"], CSMA or Carrier Sense Multiple Access, etc.), and dynamical reservation/collision resolution (e.g., MACA or Multiple Access with Collision Avoidance, CSMA/CA, IEEE 802.11, etc.). Such contention-based MAC protocols can be used to support a MANET communication if the interface is outside the radio coverage of the BS/AP and operated in a MANET mode (e.g., between MT 1 and MT 2 in Figure 11.1).

11.3.2.1   QoS-Based MAC in the HWN

The need for a new generation of MAC layers stems from the fact that a large number of independent applications may utilize the same network, thereby requiring associated MAC layers to support a wide range of applications with diverse QoS requirements. The time division–code division multiple access (TD-CDMA) has been adopted by the Universal Mobile Telecommunications System (UMTS) for the third-generation (3G) wireless systems, while it has limited throughput capacity to adequately support broadband QoS-based services. A time division–space division CDMA (TSD-CDMA) is proposed [13] that is a QoS MAC protocol for fourth-generation (4G) integrated wireless networks. The three main channel accesses are differentiated ALOHA channel access, individualized polling channel access, and wireless tree channel access (WTCA).

- **Differentiated ALOHA channel access:** In a differentiated ALOHA channel access, some code-time minislots retain smaller collision rates by maintaining moderately small attempt rates for transmissions of requests, while others maintain higher utilization.
- **Individualized polling channel access:** For delay-sensitive and bursty traffic, a polling mechanism or wireless tree channel access is adopted. In TSD-CDMA, the time axis is partitioned into a contention interval and a number of transmission subintervals. Time minislots of the contention interval and time-slots of transmission subintervals can use orthogonal codes or random codes. Orthogonal codes and time-slots are allocated to MTs that have successfully requested resources on a per session/packet/burst basis. Once the requests are placed, the BS allocates space-code time-slots (SCTSs) efficiently, avoiding collisions with high probability. This ensures throughput optimization, and thus enhanced QoS provisioning is achieved. This can be further improved by defining a mutual interference metric (MIM) that gives a measurement of the interference caused by the transmission at a given power level from an active node to all other neighboring active nodes.
- **Wireless tree channel access (WTCA):** With WTCA, assignments of minislots for access requests are adjusted according to burstiness of the traffic, which leads to smaller overhead for channel access and reservation. It also offers higher channel utilization and small bounded channel access delay.

In the HWN, different MAC protocols have to reside in one device, such as the TSD-CDMA for cellular interface and IEEE 802.11 MAC for WLAN interface in the dual-mode MT, as shown in Figure 11.4. Except for the traditional QoS issues such as throughput, some new issues are introduced by integration of multiple MAC protocols. In a multimode MT, QoS-aware and power-efficient hardware and management strategies are necessary to avoid excessive power consumption by multiple interfaces. If the MT has no packet to send or receive, only one interface is needed to proactively detect the availability of its correspondent BS/AP, which enables the network layer to keep its location information updated and thereby maintain the Internet connectivity. Once there are packets for transmission or reception, the multiple MAC protocols can determine which interfaces are suitable for satisfying a particular QoS demand for an application. When data packets are simultaneously transmitted on multiple interfaces, a power management scheme on MAC layers should be used to efficiently support heterogeneous power levels and transmission ranges. When the multimode MT moves from a network to another heterogeneous network, such as from a cellular to an IEEE 802.16 network, the IEEE 802.16 MAC protocol should have a higher priority to schedule packet transmission for the MT.

11.3.2.2   Key QoS Functionalities of Data Link Layer

The key QoS-related functions provided by the data link layer consist of carrying reasonably sized packets, with realistic success rate, and performing error corrections at the receiver. The link layer model can be either point-to-point, where a link is established between a pair of nodes, or broadcast type, in which case multiple nodes can share a link. The data link layer provides a mechanism for signaling between the MT and the BS/AP, which sets up an association between them for resource reservation. The data link layer implements various channel modulation and error control schemes to prevent data loss. Again, reliability is improved by employing retransmission schemes like Automatic Repeat Request (ARQ). Delay is sometimes traded for bandwidth according to the nature of the channel and the application. This happens as ARQ schemes sometimes produce unpredictable delays. Thus, the upper layers may experience different delays or bandwidths at different times.

11.3.2.3   Scheduling a Flow

To achieve improved QoS, sometimes it becomes necessary to implement advanced functionality in the queuing mechanism of the data link layer. As shown in Figure 11.5, once the upper network layer has received the QoS capabilities of the link layer, it requests for QoS context establishment for a new IP flow. Then the underlying data link layer provides a certain number of QoS contexts (i.e., the QoS parameters to be used) [7] along with the unique context identifier to the network layer. These contexts can be used by both Differentiated Services (DiffServ) and Integrated Services (IntServ) for supporting a new flow.

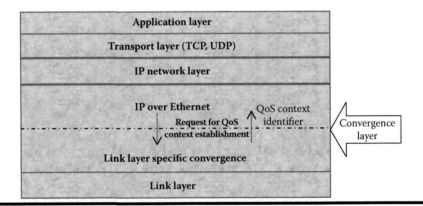

Figure 11.5   QoS context setup at link layer.

- **DiffServ:** Instead of strict bandwidth and delay requirements, the flow can be characterized by the packet loss sensitivity. A priority-based QoS can be specified with relative priority, delay bound, and loss rate. This kind of differentiated services with relative priority is called DiffServ. A unidirectional-link QoS context is first established, which is then attached to each data packet delivered to the convergence layer (shown in Figure 11.5) so that they are treated in a differentiated manner at the link layer queue. Though several flows may be encapsulated in the same service class, the packets themselves may be independent of each other. In DiffServ, if the IP network layer does not specify any QoS level, the QoS context identifier indicating a best-effort context (indicating best-effort service provided by IP by default) is attached to the packet sent upwards.

- **IntServ:** In an IntServ flow, a context is first advertised to the upper network layer, indicating changes in the QoS levels and QoS capabilities of the link layer. This enables the network layer to implement different scheduling mechanisms to handle different QoS-sensitive flows. The advertised context could involve creation of a new link layer QoS context with suitable QoS parameters so as to satisfy requests of the newly created network layer flow.

The network layer at first requests the data link layer to advertise its QoS capabilities during the link interface initialization. Further QS contexts ought to be established to determine whether the upper-level network flow is IntServ or DiffServ. Providing resource reservation by the data link layer ensures a desired level of bandwidth and a finite delay bound.

11.3.2.4 Flow Signaling via Resource Reservation Protocol

The Resource Reservation Protocol (RSVP) is employed to communicate desired IP flow (indicating the end-to-end resource reservation) requirements to the network [7]. Upon a successful reservation, a QoS context identifier is passed on to the network layer as a response. If the link layer cannot satisfy the demands of the network layer, it returns an error packet to the network layer.

Changes in QoS level may also occur due to handoff. In this case, the link layer informs the network layer of the QoS violation. It should also be able to indicate when new resources become available that might be useful for a MT intending to increase its reserved resources. This can be achieved by sending asynchronous packets to the network layer. A link-controlled approach is needed as packet flows have different reliabilities and delay requirements. Head-of-the-line blocking at the queue may cause undesirable delays in the link layer. Thus, the link layer needs to handle both flow classification and scheduling activities. The task of ensuring optimal performance by the data link layer scheduler should not adversely affect fairness of service provided by the network layer scheduler. Both schedulers should cooperate to hide the undesired effects of the underlying wireless link.

11.3.3 QoS-Based Network Selection on the Network Layer

At an interface, the network layer defines an Internetworking protocol for maintaining its corresponding connection from its interface to the destination. To provide continuous Internet connectivity, QoS-based network selection is an essential functionality so that the MT can be always connected to the best wireless network for services. In the HWN, QoS-based network selection is uniquely required due to the availability of multiple networks in a multimode MT. Given a set of available networks, the network selection determines the set of connected networks by the MT, as shown in Figure 11.6(a). For this purpose, always best connected (ABC) [14], multiconstraint dynamic access selection (MCDAS), and delay-sensitivity-based network selection (DNS) are proposed for achieving QoS-aware network selection in the HWN.

11.3.3.1   Always Best Connected Service

The notion of always best connected (ABC) [14] enables a multimode MT to seamlessly connect in a way to available networks that best suits the application needs. A network selection module is situated in the network layer and is a core component that fully explores available networks and determines the best set of interfaces (i.e.,

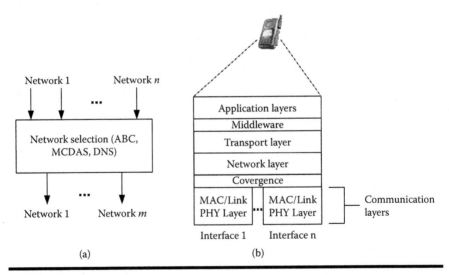

Figure 11.6   Network selection and always best connected.

networks) for service by a multimode MT. The definition of *best* depends on many QoS factors, such as service type, monetary cost, user mobility profile, user preferences, network coverage, bandwidth, delay, power consumption, operator, security level, and others [14–16]. These parameters are collected from each interface layer and are reported to the network layer by a cross-layer design. To provide the best communication to a MT, the network selection algorithm aims to evaluate these QoS factors from the user and network perspectives. From the user side, it has to consider what the required QoS of an ongoing application is and what quality levels can be provided on the MT's interfaces. At the same time, the network should evaluate the network performance (e.g., network utilization) and its stability on the interface, as per the network coverage and user mobility. In addition to the traditional layers in the OSI protocol stack, the ABC-enabled MT has three layers, which are shown in Figure 11.6(b): communication layers, convergence layer, and middleware layer. These layers define new functionalities for adaptive and intelligent ABC.

- **Communication layers:** As shown in Figure 11.6(b), this set of layers includes different access technologies like WLANs (e.g., IEEE 802.11a/b/g/n), 2.5G (i.e., general packet radio service [GPRS]), and 3G systems (e.g., Universal Mobile Telecommunications System [UMTS]). Similar to the dual-mode MT of Figure 11.4, each communication layer contains PHY, MAC, and link layers so as to equip the MT with multiple network capabilities.
- **Convergence layer:** As shown in Figure 11.6(b), the convergence layer is located between the link layer and network layer. The essential functionality of this layer is to provide the network layer the interface capabilities in terms of the offered QoS. Thus, there is a need for cooperation between the data link layer and the network layer, giving rise to the introduction of a convergence layer addressing such a cooperative framework. Radio resources being typically scarce and packet loss being so extensive, a need for cooperation becomes necessary.
- **Middleware layer:** This layer acts as an interface between the application layer and the access selection process. The key role of this layer is to pass application QoS requirements to the lower layers and notify the applications about the network state of affairs of the lower layers. Existing middleware technologies [14], such as transaction-oriented, message-oriented, or object-oriented middleware, have been formed that attempt to hide the differences as much as possible, and thus help in visualizing the system as a single integrated facility. The interaction primitives, such as distributed transactions, object requests, or remote procedure calls, assume a steady and continuous connection between components.

11.3.3.2   Multiconstraint Dynamic
Access Selection (MCDAS)

The multiconstraint dynamic access selection (MCDAS) approach [15] assigns multiple application traffic to a set of wireless interfaces so to minimize the power consumption at the MT while satisfying the required QoS and user preferences. For example, in Figure 11.1, the multimode MT can simultaneously employ the cellular and WLAN interfaces for packet transmission and reception. In MCDAS, the ABC scenario is formulated as a variant of the *bin packing* issue, which involves packing items (i.e., bandwidth requirements of applications) into a set of bins (i.e., total available bandwidth in accessing the networks) without violating the capacity limitations, while utilizing the minimum number of bins. If a flow cannot fit into any of the bins, it is discarded, and this type of flow is termed a rejected flow. In MCDAS, an application flow can be described as a 4-tuple, including access preference, partitionability, bandwidth, and delay requirement. On the other hand, each access network is defined with a 5-tuple, including total bandwidth capability of the network, maximum communication delay, and three power consumption parameters (background consumption and power consumption in transmission and reception, respectively).

11.3.3.3   MCDAS Algorithm

Upon receiving the data packets from applications, the MCDAS algorithm (first fit decreasing [FFD]) sorts the traffic in decreasing order of preference and bandwidth. Therefore, traffic with higher preference is assigned network access with a higher priority. The FFD algorithm is further optimized to get near-optimal solutions by doing substitution, partitioning, load awareness, and reallocation. Substitution is done by relocating some of the ongoing flows with the ones having stronger preferences in terms of power efficiency. A load-aware threshold t_h is defined to limit the loads on the access to a certain degree and achieve load balancing. Again, when the bandwidth and delay of a flow degrade due to mobility of the devices, instead of blindly dropping the flows, a reallocation method is devised to control the number of flows being dumped due to a sudden change in the network parameters.

The main metrics defined to evaluate the proposed heuristics of MCDAS are average power consumption cost and average preference dissatisfaction. A partitioned flow's dissatisfaction (where the dissatisfaction value for each traffic flow assignment is defined as the extent to which the assignment does not match the flow's access preference) is calculated by averaging dissatisfaction values for the two subflows. Similarly, the dissatisfaction value for a substituted flow is calculated by averaging the dissatisfactions of the assignments for that flow. Rejected flows are not considered while calculating the average power consumption. The percentage of rejected incoming flow is measured separately by defining the metric rejection rate. It is observed from simulation that when the service intensity is greater than 1, traffic flows arrive faster than they can be served, making access networks more and more crowded. This makes the average power cost, average dissatisfaction, and rejection rates increase with the rise in service intensity. This is due to the fact that when a network is very busy, some of the flows have to be assigned to high-cost accesses or nonpreferred accesses, and some are even rejected. However, when service intensity is less than one, this propensity is not observed, simply because no access network is busy, and therefore none of the aforementioned mechanisms need to be applied.

11.3.3.4   Delay-Sensitivity-Based Network Selection (DNS)

Delay-sensitivity-based network selection (DNS) [16] is a scheme for network selection for non-real-time data based on the minimum delay. In the HWN, a multimode MT may have multiple network options to access, depending on its location and interface configuration. Each network in the system employs a fixed price per byte, or charges different prices in terms of packet pattern. Therefore, each user would like his or her data to be delivered in a timely manner at the same time at the lowest price. A utility-based algorithm that accounts for user time constraints is proposed that estimates total file delivery time (for each of the offered access networks) and then chooses the most promising network based on the consumer surplus. This algorithm is implemented using a File Transfer Protocol (FTP) with a bit rate of 212 Kbps over two overlapping WLANs. Consumer surplus (CS) in microeconomic terms is defined as the difference between the actual price for communication and the value of the data to the user. For each transmission two decision strategies are supported:

- Always cheapest network selection (AC)
- Consumer surplus (CS) network selection

It is observed from simulation [16] that for users using the CS network selection strategy, the average completion time was considerably lower than for those users employing the AC selection strategy. Again, the expected average transfer completion time increases with an increase in file size, but the rate of increase is steeper for the AC network selection strategy. It is also observed that the proposed CS strategy is highly efficient in meeting the delay deadline compared to the AC network selection strategy. The results prove the effectiveness of the proposed CS algorithm in choosing the best network, compared to blindly employing the AC strategy without any consideration to real-time constraints.

11.3.4 Heterogeneous QoS Support in the Transport Layer

The Transport Control Protocol (TCP) is the most prevalent transport layer protocol. Earlier research showed that the TCP over cellular wireless systems suffers poor performance due to packet loss and corruption caused by wireless-induced errors. Some TCP mechanisms [17] are proposed to improve the TCP performance over cellular wireless systems. On the other hand, the MT in a HWN may establish a wireless TCP connection by a multi-hop path to the BS/AP, which involves MANET TCP connection. The MANET TCP has to address problems such as high bit error rates and frequent link failures due to mobility. To improve the TCP performance over MANET, many research efforts have been conducted in the last few years, and several TCP versions [8] have been proposed to overcome the performance degradation due to multi-hop connection. In these approaches, the explicit notification techniques allow the TCP sender to control its packet transmissions in a manner that counters the negative affects of the link failures.

Furthermore, the TCP performance has to be studied in terms of the mobility between heterogeneous components of a HWN. When a MT performs an inter-system handoff, the TCP connection suffers significant performance degradation if the MT uses the traditional TCP. To solve this problem, a TCP retransmission mechanism [8] is proposed to avoid packet dropping in the process of handoff. In this approach, a sender should temporarily halt its data transmission when a handoff is taking place. TCP employs a retransmission timer to trigger data retransmission when the feedback messages from the receiver are dropped. The duration of this timer is known as retransmission timeout (RTO) [8] and should be appropriately set. It is observed [8] that the TCP throughput is inversely proportional to RTT (i.e., round-trip time, which is defined as the time elapsed for a packet to traverse from a sender to the receiver and back), RTO, and error rate. The TCP throughput decreases with an increase in the value of RTO, and this degree of decrease in throughput increases as the error rate grows. The latter phenomenon can be explained by the fact that, with an increase in error rate, there is higher probability that the TCP sender will wait for the RTO to expire, and this produces a corresponding reduction in throughput, as no useful packet transmission occurs during this period.

TCP should also modify its operation depending on the type of handoff involved. When a handoff occurs between homogenous networks, the TCP sender needs to resume its data transfer at the same rate as before, once the handoff is completed. On the contrary, when a handoff occurs between two heterogenous networks, TCP needs to adjust all its parameters, such as RTT, before resuming data transmission. For example, when a MT is moving from a cellular network to the WLAN, the TCP sender re-estimates the available bandwidth of the WLAN immediately after completion of the handoff, quickly increasing its data rate and stabilizing its transmission rate. To facilitate this process, cross-layer design may be desirable. The physical layer of the MT reports its velocity and availability of a new connection to the network layer, which is then conveyed to the TCP layer. When the MT is working as a TCP sender, the TCP temporarily stops its transmission, until notification of the handoff is completed. When working as a TCP receiver, the

MT conveys the completion of handoff to the corresponding TCP sender, which stops the transmission until the MT finishes the handoff.

11.3.4.1   QoS Priority-Based TCP Management

To support QoS-differentiated services, the TCP QoS management in the HWN has to consider the QoS priorities for different applications by users. QoS priority-based TCP management [8] is referred to as a technique for the BS/AP to dynamically manage its resources according to QoS-differentiated services, thus decreasing the call dropping rate for higher-priority services. A rate adaptation TCP scheme with a queuing policy is followed based on the predefined QoS priority. Each user maintains a service level agreement (SLA) matrix for each of the traffic types when the user subscribes to that service. The SLA matrix includes QoS parameters, like maximum and minimum data rates, delay, call dropping rate, etc. Users who have better performance levels on the SLA are considered the higher-priority users. Based on this, three service levels are defined: SL 1, SL 2, and SL 3. SL 1 has the highest priority and SL 3 the lowest priority. The low-priority users need to voluntarily give up their resources to the higher-priority users within a specified degree, which is again dependent on the network load.

When a handoff call request arrives from either a homogenous network or a heterogeneous network, it is to be accepted only if resources are available. If not, the SL of the requested call is checked. If the call is for SL 1 traffic, the ongoing lower SL traffic is downgraded by decreasing the data rate of that traffic. An attempt is also made to downgrade the SL 3 traffic to reduce its data rates. If sufficient resources can be obtained after this effort, then the call is accepted. If none of the attempts succeed, the method is tried on the ongoing SL 2. If even this process does not work, the last attempt is to downgrade the service quality of the handoff call itself. As a result, the call dropping probability decreases as the degree of downgrade increases. Again, a user with real-time multimedia traffic experiences low-quality service by downgrading. It is also observed that the response time of Web traffic is mostly dependent on the page size, as well as the transmission bit rate. The effect of downgrading the data rate is not as serious to Web service users in terms of average response time as it is to real-time service users. This QoS priority-based TCP [8] guarantees a high QoS level to high-priority users with various differentiated services by reducing the call blocking rate.

11.3.5 QoS Support in the Application Layer

To provide end-to-end QoS services at the application layer, multiple system parameters need to be taken into account. The time-varying network conditions in the HWN (e.g., a handoff) can dramatically affect application performance. Thus, from the perspective of the application layer, it is important to negotiate with the network system about end-to-end QoS parameters. The coordination among the users and the network system includes the determination of local and peer resources, reservation of network resources, adaptation of multimedia streams, etc. For this purpose, an End-to-End Negotiation Protocol (E2ENP) [9] has been proposed as a mechanism for negotiating and coordinating QoS on an end-to-end basis at both the application and network layers. To maintain a QoS-demand service for a mobile user, the negotiation is done when a user moves to a heterogeneous network (e.g., handoff) or when spontaneous network reconfigurations occur. The E2ENP enables the efficient negotiation of system capabilities and allows the service provider to effectively influence the negotiation process. Furthermore, the E2ENP optimizes the efficiency of multimedia call setup in a HWN and reduces the time used for QoS renegotiations.

In E2ENP, application QoS parameters are used to describe E2E application performance with the corresponding software and hardware resources of end systems. These parameters are negotiated between the peers for coordinating E2E QoS in the form of QoS contracts (shown in Figure 11.7) at the application level [9], which can be regarded as the negotiation results, specifying the QoS parameters, video frame size, frame rate, and visual quality.

The QoS specification is represented as a hierarchical tree structure, as shown in Figure 11.7. A QoS specification related with a branch node of the tree is termed QoS context [9] and represents a high-level application QoS contract. It can be a session (video, audio, or data) or a stream (association). Each leaf of the tree represents a specification and is termed an application QoS contract [9]. At each level, siblings signify alternative QoS specifications that the application can select when provisioning QoS. For example, audio stream 1 and audio stream 2 in Figure 11.7 are siblings. Any given subtree originating from a particular branch node is linked with an adaptation rule predicate. By resolving this predicate, a child node is chosen (e.g., "if video configuration V11 is no longer enforceable, switch to video configuration V12" [4] in Figure 11.7), and this instructs the system to impose the QoS contract or QoS context connected with that child. Each of the contracts and contexts connected with the nodes of the QoS hierarchy is labeled with a specific identifier, which is used by peers to tackle subsequent contracts during the QoS negotiation/renegotiation process. This process of exchange only minimizes the QoS renegotiation traffic overhead.

The application uses the E2ENP description model to formally state proper system configurations and performance constraints. The SIP [10] framework used for QoS coordination identifies several roles, like end system, proxy, etc., for defining provider management. The three different negotiation modes used by E2ENP are:

■ **Prenegotiation:** Negotiates configuration information valid for more than one multimedia session. During this phase, the control parameters (like service configuration, QoS contracts, etc.) are exchanged between terminals for speeding up the overall negotiation process.

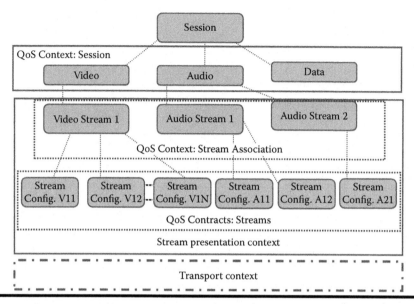

Figure 11.7 **Hierarchical QoS specification.**

- **Negotiation:** Information exchange occurs for establishing a specific multimedia session. The information exchange is a service information and specific session configuration. Due to low memory, if the terminals cannot perform prenegotiations (due to system policies), essential system configurations can be exchanged during this phase.
- **Renegotiation:** This phase enforces particular QoS contexts and contracts and the indication of adaptation conditions. If a different QoS context needs to be enforced due to changes in the resource availability, this phase is invoked.

From simulation, the time necessary for a prenegotiation, negotiation, or renegotiation is found to depend on the applied SIP transaction [9].

11.3.6 QoS-Based Seamless Mobility Support

To provide reliable and sustained QoS in the HWN, QoS-based mobility support is necessary to efficiently manage the wireless resources, adaptively cope with both temporal and spatial resource dynamics, and effectively address the collaboration between heterogeneous networks. Therefore, as shown in Figure 11.4, the solution for QoS-based mobility support requires an integrated design effort that spans heterogeneous wireless interfaces and every layer in the network protocol stack.

The mobile IP [19] is the key technology in supporting continuous Internet mobility, as shown in Figure 11.8, and its enhancements for wireless networks provide horizontal and vertical handoff for mobile users. The mobile IP protocol defines a home agent (HA) as the server on the MT's home network that maintains the information about the MT's current location, identified as care of address (CoA), billing, account, and security credentials. On the other hand, a foreign agent (FA) is the server on the visiting network providing the CoA and local mobility administration on the visiting network. As shown in Figure 11.8, on the core IP network, the mobile IP protocol supports Internet mobility where the mobile user changes its Internet attachments, e.g., from one FA to another FA. The mobile MT is associated with two distinct addresses. The home address is the permanent address of the MT and is assigned to the MT when it joins the home network for the first time. The CoA is the temporary address of the MT and is assigned by the FA when the MT roams in the foreign network. As shown in Figure 11.8, the data packets from the Internet host will first be delivered to the HA by using the MT's home address, and HA then forwards the packet to the FA that the MT is currently visiting. Upon receiving the packets from the HA, the FA again delivers the data packets to the destination MT via the local wireless network (e.g., the cellular network as shown in Figure 11.8). For continuous service, the HA must always have the location information of the MT's visiting network (i.e., CoA). The procedure by which the MT updates its location information (i.e., CoA) is called handoff. The delay for a handoff procedure [19] includes three parts:

■ **IP layer handover detection time:** Defined as the time taken by the MT to detect the availability of the new BS/AP and determine the handoff by evaluating the QoS of the corresponding access network. The availability can

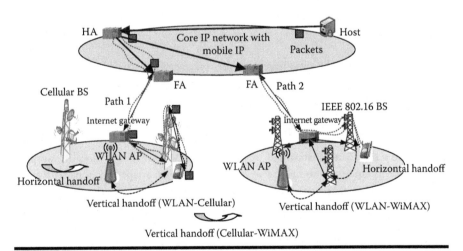

Figure 11.8   Handoff capability in the HWN.

be detected by proactively listening to the beacon from the new BS/AP. The QoS parameters include the access delay, available bandwidth per second, network coverage, service cost, user preference, and power consumption per bit. For example, the ABC scheme as illustrated in Section 11.3.3 can be used for the QoS evaluation.

■ **Access admission time:** Defined as the time taken by a MT to negotiate admission with a new BS/AP. It is the time interval between the sending of a request from the MT to the BS/AP and the reception of a CoA from the BS/AP, which indicates the handoff acceptance from the BS/AP.

■ **Mobility binding update time:** Defined as the round-trip delay between the sending of a handoff request by a MT and the reception of a positive acknowledgment. The request will be sent to the MT's HA depending on the handoff patterns. If it is a macromobility (interdomain mobility), the handoff request will carry the CoA and send to the FA, and the FA forwards the request to the HA of the MT. Upon receiving the handoff request, the HA updates the address mapping by using the new CoA, which means that the HA has notification of the new location of the moving MT. The HA then sends an acknowledgment message to the MT by using the reverse path. Path 1 in Figure 11.8 illustrates this. On the contrary, if the new BS/AP is located at the same domain as the previous one, a micromobility handoff (handoff within the same domain) is conducted, in which the MT updates its address binding at the domain without going through the MT's HA. For example, if the IEEE 802.11 BSs in Figure 11.8 are all located in the same domain by connecting to the same Internet gateway, path 2 in Figure 11.8 denotes the path for mobility address update. It can been seen that the micromobility improves the handoff QoS in terms of delay because it saves Internet traveling time for the handoff request.

The handoff of an MT can be further divided into two types in terms of network heterogeneity: horizontal and vertical. In the next section, we discuss the QoS issues based on these two types of handoff.

11.3.6.1   Seamless Vertical and Horizontal Handoff

In a HWN, the seamless handoff is referred to as the ability to automatically switch between homogenous or heterogeneous networks without packet loss between the MT and its peer communication host. The seamless handoff is also referred to as the ability for the MT to quickly discover and change its network connections to the best QoS network available. The handoff for an MT includes two types: (1) horizontal handoff and (2) vertical handoff. The horizontal handoff is defined as the transfer between two homogenous networks, i.e., from a cellular BS to a neighboring cellular BS, while the vertical handoff allows continuous and seamless migration between heterogeneous networks, i.e., from a cellular network to a WLAN, as shown in Figure 11.8. Most of the horizontal handoff schemes are performed by comparing received signal strength, signal-to-noise ratio (SNR), to set thresholds for making the handoff decision. In contrast to the horizontal handoff, the vertical handoff should consider not only the strength of signal power but also QoS provisioning, such as network bandwidth, service types, etc. Figure 11.9 illustrates the horizontal and vertical handoffs by which the MT can seamlessly hand off its connection between the WLAN and the cellular network. It is shown that the MT involves a QoS evaluation module before execution of a handoff. If the MT detects a new network that has a higher QoS connection, a handoff happens. The basic procedure [19] of a vertical handoff has three steps, which correspond to QoS evaluation, mobility updating, and data packet redirection, respectively, in Figure 11.9.

■ **Improved QoS network detection:** The QoS evaluation module in Figure 11.9 has the responsibility to detect the higher QoS network. For example, the dual-model MT, as shown in Figure 11.4, communicates with a BS/AP at an air interface (address: IP1) and detects the availability of another BS/AP with a new air interface (address: IP2), which provides a higher performance, such as a higher data rate, a lower cost, and lower power consumption. For example, MT in the WLAN evaluates the WLAN SNR. If the SNR is above a given threshold, the MT stays at the WLAN. If the SNR falls below the given threshold, the MT begins to detect the availability of the cellular network. If the SNR continuously decreases, the MT will switch its connection to the cellular network. On the other hand, if the initially connected network of the MT is the cellular network, the MT periodically scans the available

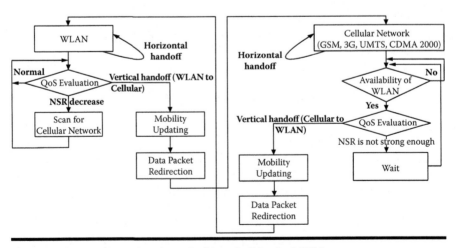

Figure 11.9 Horizontal and vertical handoff in the HWN.

WLAN. If the SNR of a WLAN is higher than the given threshold, the MT will switch its connection to the WLAN by the following process.

■ **Address and route binding:** The MT issues a vertical handoff request to the new BS/AP (from interface IP1 and to interface IP2). If the previous connection is lost before the new connection is established, to avoid packet loss, the packet sender (the MT or its communication peer host) halts the packet delivery at this stage until the completion of the handoff. If the new BS/AP accepts the handoff request, the new BS/AP forwards the handoff request to the FA so that the FA can create a mapping between IP1 and IP2. The request may be forwarded to the HA so that the HA can update the Internet route to the new BS/AP for the MT if it is a macromobility process. In the micromobility process, the new BS/AP and its previous BS/AP are attached to the same FA so that the location updating at the FA is sufficient and the HA maintains the same FA route for the MT. After creating an address binding between IP1 and IP2, the FA replies with a positive acknowledgment message to the MT through the new BS/AP. The above process is performed in the mobility updating module in Figure 11.9.

■ **Flow redirection:** After the process of address and route binding, all packets from the Internet with IP1 will be forwarded to the MT through the new interface (IP2) and the new BS/AP. The flow redirection modules (WLAN☒ cellular or cellular☒WLAN) in Figure 11.9 correspond to this process.

11.3.7 QoS-Based Connection Management

As shown in Figure 11.4, besides the QoS-based seamless mobility in the OSI protocol stack, connection management is the other management plane with the responsibilities of bandwidth reservation, connection admission control, packet scheduling, and buffer management. In a cellular network, for example, the BS may reserve fractional bandwidth of the BS/SS, and this reserved bandwidth is exclusively used for handoff to ensure the QoS of the serving connections, not new connection requests. On the other hand, the connection admission control aims to limit the number of connections admitted into the wireless network in a way that each individual connection can obtain its desired QoS. Packet scheduling is the technology for BS/AP to control the transmission sequence over the link according to the QoS demand of each connection. Furthermore, the buffer management scheme regulates the occupancy of the finite buffer in the system and decides whether to admit an incoming packet into the buffer or drop it.

The HWN has a new functionality in the connection management plane, termed packet multiplexing over multiple interfaces (i.e., networks). The main idea of packet multiplexing is to distribute the traffic over multiple interfaces such that the QoS requirements of the particular application are met. This scheme aggregates the bandwidth offered on multiple interfaces to improve the QoS-demanding applications that require higher bandwidth. For example, the video stream for a mobile TV application can be transmitted over multiple networks by exploring the multiple interfaces of the MT. The Earliest Delivery Path First (EDPF) algorithm is such a multiplexing scheme with an ability to reduce the overall delay due to packet reordering. The EDPF algorithm estimates the delivery time of the packets on each interface and schedules each packet on the path with minimum delay. By tracking the queues at the BSs and taking that into consideration while scheduling packets, the EDPF ensures that it utilizes all of the available path bandwidths, at the same time achieving minimal packet reordering. When packets are of fixed size, with EDPF algorithm, they will arrive in order at the client. Let A and B be two packets such that the size of A is larger than B. Packet A may arrive before B only if they were scheduled on different links. If the packet sizes of A and B are equal and the link on which A was transmitted delivers packets the earliest, EDPF, when scheduling A, would have picked that link for its transmission. This ensures that packets always arrive in order. For variable-sized packets, the EDPF algorithm distributes the bits across the links in proper order.

11.4 Limitations of Existing QoS Frameworks

This chapter makes an attempt to discuss the QoS issues involved in a HWN scenario. It discusses the layer-by-layer changes needed to support QoS capabilities in a HWN, along with cross-layer support. QoS support for homogeneous networks involves less complication as communication occurs in networks with similar characteristics. However, supporting QoS for HWNs still involves many open challenges:

- **Heterogeneous QoS collaboration and support:** In a heterogeneous environment, different types of networks with varying network capabilities and different values of QoS parameters are integrated. Therefore, the QoS parameters after integration are different from their homogeneous counterparts. This implies that the existing mechanism for supporting QoS for a homogeneous network will not hold well in the current scenario, and a new collaborative framework needs to be devised.

- **End-to-end and cross-layer QoS support:** QoS applications cannot be supported on an end-to-end basis across a HWN because there is no end-to-end QoS solution available, especially in the access network. A network infrastructure upgrade is needed to accommodate resource allocation in the access network. According to the type of service wanted, a basic service fee may be charged, or a higher fee may be charged for value-added services [3].

- **Network resource utilization:** Considering the limitation of bandwidth in wireless systems, and therefore HWNs, the most important target is to increase network resource utilization. It is well known that RTP, UDP, IP, and TCP have the problem of large header overhead on bandwidth-limited links. Header compression has been confirmed to be efficient for using those protocols. But unfortunately, existing header compression schemes [11] do not seem to work well on noisy links, particularly the one with high bit error rate (BER) and long round-trip time (RTT).

Furthermore, future mobile networks are expected to provide different types of multimedia services, and each service has different QoS requirements. Therefore, the resource management scheme should be able to adapt to each such requirement.

11.5 Future Directions

Despite several attempts by the industry and research groups, there are still some open issues in designing QoS schemes for HWNs. They are as follows:

- **Efficient heterogeneous QoS collaboration and support:** A QoS framework is needed that matches the current matching IP principles. Network services (QoS) should not be designed for a particular application and instead should be flexible enough to be used in any new application.
- **Multifaceted QoS support:** The user group that uses the QoS services in a HWN the most needs to be identified. Bandwidth guarantee is not so difficult compared to reducing the E2E delay, and it is crucial to see if the user group is willing to compromise on this issue. A user needing infrastructure upgrade to support basic E2E heterogeneous QoS needs to be determined, and also needs to be charged as per the value-added services that he or she needs.
- **Network resource utilization:** Because the amount of usable spectrum is finite, as more services are added, a saturation point will come when no spectrum is available for allocation. As the number, size, and complexity of operations are growing, the time taken for deployment is becoming extremely long. On the other hand, the rising data traffic, such as the IP traffic and high mobility of the user, is causing some variation in the spatial-temporal characteristics. Fixed-spectrum allocation is not suitable to be used in these types of changes, as it has shortcomings in terms of low spectrum efficiency in licensed bands and poor performance of radio devices in packed unlicensed bands. All these problems prompt toward a dynamic spectrum management framework.

There are many other open issues, such as designing an efficient traffic model. Classification of the dynamic traffic is also a challenging task and needs to be addressed effectively. A framework needs to be designed for scalable, dynamic traffic classification based on a statistical application signature, which should not be rigidly bounded by any particular application protocol [12].

References

[1] Zimmerman, H. 1980. OSI reference model: The ISO model of architecture for Open Systems Interconnection. *IEEE Transactions on Communication* 28(4).

[2] Carneiro, G., et al. 2005. The DAIDALOS architecture for QoS over heterogeneous wireless networks. In *IST Mobile & Wireless Communications Summit* (June 2005) 32, Dresden, Germany.

[3] Dugeon, O., et al. 2005. End to end quality of service over heterogeneous networks (EuQoS). In *Proceedings of the Network Control and Engineering for QoS, Security and Mobility, IFIP TC6 Conference (NetCon'05)*, 87–101. Lannion, France.

[4] Liu, Q., Zhou, S., and Giannakis, G. B. 2006. Cross-layer modeling of adaptive wireless links for QoS support in heterogeneous wired-wireless networks. *ACM/Kluwer Journal on Wireless Networks* (July 2006) 12:427–437.

[5] UMTS. http://www.umtsworld.com/technology/overview.htm.

[6] Little's theorem. http://cnx.org/content/m10747/latest/ (accessed November 20, 2006).

[7] Manner, J., et al. 2001. Exploitation of wireless link QoS mechanisms in IP QoS architectures. In *Proceedings of SPIE, Quality of Service over Next Generation Data Networks*, vol. 4524:273–83.

[8] Kim, S.-E. 2006. Efficient and QoS guaranteed data transport in heterogeneous wireless mobile networks. Doctoral thesis, Department of Electrical and Computer Engineering, Georgia Technological University.

[9] Guenkova-Luy, T., Kassler, A., and Mandato, D. 2004. End-to-end quality of service coordination for mobile multimedia applications. *IEEE Journal on Selected Areas in Communications* 22(5):889–903.

[10] Rosenberg, J., et al. 2002. *SIP: Session initiation protocol*. RFC 3261. IETF.

[11] Degermark, M., Nordgren, B., and Pink, S. 1999. *IP header compression*. RFC 2507. IETF.

[12] Pentikousis, K., and Huusko, M. 2005. Quality of service in heterogeneous networks, tutorial. In *Proceedings of the 8th Asia-Pacific Network Operations and Management Symposium (APNOMS 2005)*.

[13] Yeh, C.-H. 2003. TSD-CDMA: A QoS MAC protocol for 4G integrated mobile wireless systems. In *Proceedings of the IEEE International Symposium on Computer Communications*, 930–35.

[14] Passas, N. I., et al. 2006. Enabling technologies for the "always best connected" concept. *Wireless Communications and Mobile Computing* (March 2005) 5(2):523–40.

[15] Xing, B., and Venkatasubramania, N. 2005. Multi-constraint dynamic access selection in always best connected networks. In *Proceedings of IEEE MobiQuitous*, 56–64.

[16] Ormond, O., Muntean, G.-M., and Murphy, J. 2005. Network selection strategy in heterogeneous wireless networks. In *Proceedings of the IT&T Conference*, 175–84. Cork, Ireland.

[17] Balakrishnan, H., et al. 1996. A comparison of mechanisms for improving TCP performance over wireless links. In *IEEE/ACM Transaction on Networking* 5(6):756–765, December 1995.

[18] Bluetooth. http://www.bluetooth.org/.

[19] Xie, B., Kumar, A., Cavalcanti, D., et al. 2006. Multi-hop cellular IP: A new approach to heterogeneous wireless networks. *International Journal of Pervasive Computing and Communications.*

Index

343